Classics in Mathematics

Hans Grauert • Reinhold Remmert

Theory of Stein Spaces

Springer
Berlin
Heidelberg
New York
Hong Kong
London
Milan
Paris
Tokyo

Hans Grauert • Reinhold Remmert

Theory of Stein Spaces

Reprint of the 1979 Edition

Springer

Authors

Hans Grauert
Universität Münster
Mathematisches Institut
Einsteinstr. 62
48149 Münster, Germany

Reinhold Remmert
Universität Münster
Mathematisches Institut
Einsteinstr. 62
48149 Münster, Germany

Translator

Alan Huckleberry
University of Notre Dame
Department of Mathematics
Notre Dame, Indiana 46556, U.S.A.

Originally published as Vol. 236 of the
Grundlehren der mathematischen Wissenschaften

Mathematics Subject Classification (2000):
30A46, 32E10, 32J99, 32-02, 32A10, 32A20, 32C15

Inside front-cover photography courtesy of Klaus Peters.
From *The Mathematical Intelligencer Volume 2, Number 2, 1980.*

Library of Congress Cataloging-in-Publication Data

Grauert, Hans, 1930-
[Theorie der Steinschen Räume. English]
Theory of Stein spaces / H. Grauert, R. remmert.
p. cm. -- (Classics in mathematics, ISSN 1431-0821)
"Reprint of the edition 1979."
"Originally published as vol. 236 in the series: Grundlehren der mathematischen Wissenshcaften."
Includes bibliographical references and index.
ISBN 3-540-00373-8 (pbk. : acid-free paper)
1. Stein spaces. I. Remmert, Reinhold. II. Title. III. Series.

QA331.G68313 2003
515'.94--dc21

2003050517

ISSN 1431-0821
ISBN 3-540-00373-8 Springer-Verlag Berlin Heidelberg New York

Springer-Verlag is a part of Springer Science+Business Media
springeronline.com

Printed in Germany

Printed on acid-free paper 41/3142ck-5 4 3 2 1 0

H. Grauert R. Remmert

Theory of Stein Spaces

Translated by Alan Huckleberry

Springer-Verlag
Berlin Heidelberg New York

Hans Grauert
Mathematisches Institut
der Universität Göttingen
D-3400 Göttingen
Federal Republic of Germany

Reinhold Remmert
Mathematisches Institut der
Westfälischen Wilhelms-Universität
D-4400 Münster
Federal Republic of Germany

Translator:
Alan Huckleberry
Department of Mathematics
University of Notre Dame
Notre Dame, Indiana 46556
USA

AMS Subject Classifications: 30A46, 32E10, 32J99, 32-02, 32A10, 32A20, 32C15

Title of the German Original Edition: *Theorie der Steinschen Räume*,
Springer-Verlag Berlin Heidelberg 1977.

With 5 Figures

Library of Congress Cataloging in Publication Data
Grauert, Hans, 1930–
Theory of Stein spaces.

(Grundlehren der mathematischen Wissenschaften;
236)
Translation of Theorie der Steinschen Räume.
Includes index.
1. Stein spaces. I. Remmert, Reinhold, joint
author. II. Title. III. Series: Grundlehren der
mathematischen Wissenschaften in Einzeldarstellungen;
236.
QA331.G68313 515'.73 79-1430

Printed in the United States of America.

9 8 7 6 5 4 3 2 1

ISBN 3-540-90388-7 Berlin Heidelberg New York
ISBN 0-387-90388-7 New York Heidelberg Berlin

Dedicated to Karl Stein

Contents

Introduction . . . XV

Chapter A. Sheaf Theory

§ 0. Sheaves and Presheaves . . . 1
1. Sheaves and Sheaf Mappings . . . 1
2. Sums of Sheaves, Subsheaves, and Restrictions . . . 1
3. Sections . . . 2
4. Presheaves and the Section Functor Γ . . . 2
5. Going from Presheaves to Sheaves. The Functor Γ̌ . . . 3
6. The Sheaf Conditions $\mathscr{S}1$ and $\mathscr{S}2$ . . . 3
7. Direct Products . . . 4
8. Image Sheaves . . . 4
9. Gluing Sheaves . . . 5

§ 1. Sheaves with Algebraic Structure . . . 5
1. Sheaves of Groups, Rings, and $\mathscr{R}$-Modules . . . 5
2. Sheaf Homomorphisms and Subsheaves . . . 6
3. Quotient Sheaves . . . 7
4. Sheaves of Local k-Algebras . . . 8
5. Algebraic Reduction . . . 8
6. Presheaves with Algebraic Structure . . . 9
7. On the Exactness of Γ̌ and Γ . . . 9

§ 2. Coherent Sheaves and Coherent Functors . . . 10
1. Finite Sheaves . . . 10
2. Finite Relation Sheaves . . . 11
3. Coherent Sheaves . . . 11
4. Coherence of Trivial Extensions . . . 12
5. The Functors $\bigoplus^p$ and $\bigwedge^p$ . . . 12
6. The Functor $\mathscr{H}om$ and Annihilator Sheaves . . . 13
7. Sheaves of Quotients . . . 14

§ 3. Complex Spaces . . . 14
1. k-Algebraized Spaces . . . 15
2. Differentiable and Complex Manifolds . . . 15
3. Complex Spaces and Holomorphic Maps . . . 16
4. Topological Properties of Complex Spaces . . . 18
5. Analytic Sets . . . 18
6. Dimension Theory . . . 19
7. Reduction of Complex Spaces . . . 20
8. Normal Complex Spaces . . . 21

§ 4. Soft and Flabby Sheaves . . . 22
1. Soft Sheaves . . . 22
2. Softness of the Structure Sheaves of Differentiable Manifolds . . . 23
3. Flabby Sheaves . . . 25
4. Exactness of the Functor Γ for Flabby and Soft Sheaves . . . 25

Chapter B. Cohomology Theory

§ 1. Flabby Cohomology Theory . . . 28
1. Cohomology of Complexes . . . 28
2. Flabby Cohomology Theory . . . 30
3. The Formal de Rham Lemma . . . 32

§ 2. Čech Cohomology . . . 33
1. Čech Complexes . . . 34
2. Alternating Čech Complexes . . . 35
3. Refinements and the Čech Cohomology Modules $\check{H}^q(X, S)$ . . . 35
4. The Alternating Čech Cohomology Modules $\check{H}^q_a(X, S)$ . . . 37
5. The Vanishing Theorem for Compact Blocks . . . 37
6. The Long Exact Cohomology Sequence . . . 38

§ 3. The Leray Theorem and the Isomorphism Theorems . . . 40
1. The Canonical Resolution of a Sheaf Relative to a Cover . . . 40
2. Acyclic Covers . . . 42
3. The Leray Theorem . . . 43
4. The Isomorphism Theorem $\check{H}^q_a(X, \mathscr{S}) \cong \check{H}^q(X, \mathscr{S}) \cong H^q(X, \mathscr{S})$ . . . 43

Chapter I. Coherence Theory for Finite Holomorphic Maps

§ 1.′ Finite Maps and Image Sheaves . . . 45
1. Closed and Finite Maps . . . 45
2. The Bijection $f_*(\mathscr{S})_y \to \prod_{j=1}^{t} \mathscr{S}_{x_j}$ . . . 46
3. The Exactness of the Functor f_* . . . 46
4. The Isomorphisms $H^q(X, \mathscr{S}) = H^q(Y, f_*(\mathscr{S}))$ . . . 47
5. The $\mathscr{O}_y$-Module Isomorphism $\tilde{f}: f_*(\mathscr{S})_y \to \prod_1^t \mathscr{S}_{x_i}$ . . . 48

§ 2. The General Weierstrass Division Theorem and the Weierstrass Isomorphism . . . 48
1. Continuity of Roots . . . 48
2. The General Weierstrass Division Theorem . . . 49
3. The Weierstrass Homomorphism $\mathscr{O}_B^b \rightrightarrows \pi_*(\mathscr{O}_A)$ . . . 50
4. The Coherence of the Direct Image Functor π_* . . . 51

§ 3. The Coherence Theorem for Finite Holomorphic Maps . . . 52
1. The Projection Theorem . . . 52
2. Finite Holomorphic Maps (Local Case) . . . 53
3. Finite Holomorphic Maps and Coherence . . . 54

Chapter II. Differential Forms and Dolbeault Theory

§ 1. Complex Valued Differential Forms on Differentiable Manifolds . . . 56
1. Tangent Vectors . . . 56
2. Vector Fields . . . 58
3. Complex r-vectors . . . 59
4. Lifting r-vectors . . . 60
5. Complex Valued Differential Forms . . . 60
6. Exterior Derivative . . . 62
7. Lifting Differential Forms . . . 62
8. The de Rham Cohomology Groups . . . 63

§ 2. Differential Forms on Complex Manifolds . . . 64
1. The Sheaves $\mathscr{A}^{1,0}$, $\mathscr{A}^{0,1}$ and Ω^1 . . . 64
2. The Sheaves $\mathscr{A}^{p,q}$ and Ω^p . . . 66
3. The Derivatives ∂ and $\bar{\partial}$ . . . 67
4. Holomorphic Liftings of (p, q)-forms . . . 70

§ 3. The Lemma of Grothendieck . . . 71
1. Area Integrals and the Operator T . . . 72
2. The Commutivity of T with Partial Differentiation . . . 73
3. The Cauchy Integral Formula and the Equation $(\partial/\partial\bar{z})(Tf) = f$ . . . 74
4. A Lemma of Grothendieck . . . 75

§ 4. Dolbeault Cohomology Theory . . . 77
1. The Solution of the $\bar{\partial}$-problem for Compact Product Sets . . . 77
2. The Dolbeault Cohomology Groups . . . 79
3. The Analytic de Rham Theory . . . 80

Supplement to Section 4.1. A Theorem of Hartogs . . . 81

Chapter III. Theorems A and B for Compact Blocks $\mathbb{C}^m$

§ 1. The Attaching Lemmas of Cousin and Cartan . . . 83
1. The Lemma of Cousin . . . 83
2. Bounded Holomorphic Matrices . . . 85
3. The Lemma of Cartan . . . 87

§ 2. Attaching Sheaf Epimorphisms . . . 89
1. An Approximation Theorem of Runge . . . 90
2. The Attaching Lemma for Epimorphisms of Sheaves . . . 92

§ 3. Theorems A and B . . . 95
1. Coherent Analytic Sheaves on Compact Blocks . . . 96
2. The Formulations of Theorems A and B and the Reduction of Theorem B to Theorem A 96
3. The Proof of Theorem A for Compact Blocks . . . 98

Chapter IV. Stein Spaces

§ 1. The Vanishing Theorem $H^q(X, \mathscr{S}) = 0$ . . . 100
1. Stein Sets and Consequences of Theorem B . . . 100
2. Construction of Compact Stein Sets Using the Coherence Theorem for Finite Maps . . . 101

3. Exhaustions of Complex Spaces by Compact Stein Sets . . . 102
4. The Equations $H^q(H, \mathscr{S}) = 0$ for $q \geq 2$ . . . 103
5. Stein Exhaustions and the Equation $H^1(X, \mathscr{S}) = 0$ . . . 104

§ 2. Weak Holomorphic Convexity and Stones . . . 108
1. The Holomorphically Convex Hull . . . 108
2. Holomorphically Convex Spaces . . . 109
3. Stones . . . 111
4. Exhaustions by Stones and Weakly Holomorphically Convex Spaces . . . 112
5. Holomorphic Convexity and Unbounded Holomorphic Functions . . . 113

§ 3. Holomorphically Complete Spaces . . . 116
1. Analytic Blocks . . . 116
2. Holomorphically Spreadable Spaces . . . 117
3. Holomorphically Convex Spaces . . . 117

§ 4. Exhaustions by Analytic Blocks are Stein Exhaustions . . . 118
1. Good Semi-norms . . . 118
2. The Compatibility Theorem . . . 119
3. The Convergence Theorem . . . 120
4. The Approximation Theorem . . . 121
5. Exhaustions by Analytic Blocks are Stein Exhaustions . . . 123

Chapter V. Applications of Theorems A and B

§ 1. Examples of Stein Spaces . . . 125
1. Standard Constructions . . . 125
2. Stein Coverings . . . 127
3. Differences of Complex Spaces . . . 128
4. The Spaces $\mathbb{C}^2\backslash\{0\}$ and $\mathbb{C}^3\backslash\{0\}$ . . . 130
5. Classical Examples . . . 134
6. Stein Groups . . . 136

§ 2. The Cousin Problems and the Poincaré Problem . . . 136
1. The Cousin I Problem . . . 136
2. The Cousin II Problem . . . 138
3. Poincaré Problem . . . 139
4. The Exact Exponential Sequence $0 \to \mathbb{Z} \to \mathcal{O} \to \mathcal{O}^* \to 1$ . . . 142
5. Oka's Principle . . . 144

§ 3. Divisor Classes and Locally Free Analytic Sheaves of Rank 1 . . . 146
1. Divisors and Locally-Free Sheaves of Rank 1 . . . 146
2. The Isomorphism $H^1(X, \mathcal{O}^*) \to LF(X)$ . . . 147
3. The Group of Divisor Classes on a Stein Space . . . 148

§ 4. Sheaf Theoretical Characterization of Stein Spaces . . . 150
1. Cycles and Global Holomorphic Functions . . . 150
2. Equivalent Criteria for a Stein Space . . . 152
3. The Reduction Theorem . . . 152
4. Differential Forms on Stein Manifolds . . . 154
5. Topological Properties of Stein Spaces . . . 156

§ 5. A Sheaf Theoretical Characterization of Stein Domains in $\mathbb{C}^m$ . . . 157
1. An Induction Principle . . . 157
2. The Equations $H^1(B, \mathcal{O}_B) = \cdots = H^{m-1}(B, \mathcal{O}_B) = 0$ . . . 159
3. Representation of 1. . . . 161
4. The Character Theorem . . . 162

§ 6. The Topology on the Module of Sections of a Coherent Sheaf . . . 163
0. Fréchet Spaces . . . 163
1. The Topology of Compact Convergence . . . 164
2. The Uniqueness Theorem . . . 165
3. The Existence Theorem . . . 166
4. Properties of the Canonical Topology . . . 168
5. The topologies for $C^q(\mathfrak{U}, \mathcal{S})$ and $Z^q(\mathfrak{U}, \mathcal{S})$ . . . 170
6. Reduced Complex Spaces and Compact Convergence . . . 170
7. Convergent Series . . . 171

§ 7. Character Theory for Stein Algebras . . . 176
1. Characters and Character Ideals . . . 176
2. Finiteness Lemma for Character Ideals . . . 177
3. The Homeomorphism $\Xi: X \to \mathfrak{X}(T)$ . . . 180
4. Complex Analytic Structure on $\mathfrak{X}(T)$ . . . 181

Chapter VI. The Finiteness Theorem

§ 1. Square-integrable Holomorphic Functions . . . 187
1. The Space $\mathcal{O}_h(B)$ . . . 187
2. The Bergman Inequality . . . 188
3. The Hilbert Space $\mathcal{O}_h^k(B)$ . . . 189
4. Saturated Sets and the Minimum Principle . . . 190
5. The Schwarz Lemma . . . 190

§ 2. Monotone Orthogonal Bases . . . 191
1. Monotonicity . . . 191
2. The Subdegree . . . 192
3. Construction of Monotone Orthogonal Bases by Means of Minimal Functions . 193

§ 3. Resolution Atlases . . . 194
1. Existence . . . 194
2. The Hilbert Space $C_h^q(\mathfrak{U}, \mathcal{S})$ . . . 196
3. The Hilbert Space $Z_h^q(\mathfrak{U}, \mathcal{S})$ . . . 197
4. Refinements . . . 198

§ 4. The Proof of the Finiteness Theorem . . . 200
1. The Smoothing Lemma . . . 200
2. Finiteness Lemma . . . 201
3. Proof of the Finiteness Theorem . . . 202

Chapter VII. Compact Riemann Surfaces

§ 1. Divisors and Locally Free Sheaves . . . 204
0. Divisors . . . 205
1. Divisors of Meromorphic Sections . . . 205

2. The Sheaves $\mathcal{F}(D)$ . . . 206
3. The Sheaves $\mathcal{O}(D)$ . . . 207

§ 2. The Existence of Global Meromorphic Sections . . . 208
1. The Sequence $0 \to \mathcal{F}(D) \to \mathcal{F}(D') \to \mathcal{F} \to 0$ . . . 208
2. The Characteristic Theorem and the Existence Theorem . . . 209
3. The Vanishing Theorem . . . 210
4. The Degree Equation . . . 210

§ 3. The Riemann–Roch Theorem (Preliminary Version) . . . 211
1. The Genus of Riemann–Roch . . . 211
2. Applications . . . 212

§ 4. The Structure of Locally Free Sheaves . . . 213
1. Locally Free Subsheaves . . . 213
2. The Existence of Locally Free Subsheaves . . . 214
3. The Canonical Divisors . . . 214

Supplement to Section 4. The Riemann–Roch Theorem for Locally Free Sheaves . . . 215
1. The Chern Function . . . 215
2. Properties of the Chern Function . . . 216
3. The Riemann–Roch Theorem . . . 216

§ 5. The Equation $H^1(X, \mathcal{M})$ . . . 217
1. The $\mathbb{C}$-homomorphism $\mathcal{O}(np)(X) \to \operatorname{Hom}(H^1(X, \mathcal{O}(D)), H^1(X, \mathcal{O}(D+np)))$ . . . 217
2. The Equation $H^1(X, \mathcal{O}(D+np)) = 0$ . . . 218
3. The Equation $H^1(X, \mathcal{M}) = 0$ . . . 218

§ 6. The Duality Theorem of Serre . . . 219
1. The Principal Part Distributions with Respect to a Divisor . . . 219
2. The Equation $H^1(X, \mathcal{O}(D)) = I(D)$ . . . 220
3. Linear Forms . . . 221
4. The Inequality $\operatorname{Dim}_{\mathcal{M}(X)} J \leq 1$ . . . 221
5. The Residue Calculus . . . 222
6. The Duality Theorem . . . 223

§ 7. The Riemann–Roch Theorem (Final Version) . . . 225
1. The Equation $i(D) = l(K - D)$ . . . 225
2. The Formula of Riemann–Roch . . . 226
3. Theorem B for Sheaves $\mathcal{O}(D)$ . . . 227
4. Theorem A for Sheaves $\mathcal{O}(D)$ . . . 227
5. The Existence of Meromorphic Differential Forms . . . 228
6. The Gap Theorem . . . 229
7. Theorems A and B for Locally Free Sheaves . . . 229
8. The Hodge Decomposition of $H^1(X, \mathbb{C})$ . . . 231

§ 8. The Splitting of Locally Free Sheaves . . . 232
1. The Number $\mu(\mathcal{F})$ . . . 232
2. Maximal Subsheaves . . . 233
3. The Inequality $\mu(\mathcal{G}) = \mu(\mathcal{F}) + 2g$ . . . 234
4. The Splitting Criterion . . . 235
5. Grothendieck's Theorem . . . 237

6. Existence of the Splitting 237
7. Uniqueness of the Splitting 238

Bibliography 240

Subject Index 243

Table of Symbols 248

Addendum 251

Errors and Misprints 255

Introduction

1. The classical theorem of Mittag-Leffler was generalized to the case of several complex variables by Cousin in 1895. In its one variable version this says that, if one prescribes the principal parts of a meromorphic function on a domain in the complex plane $\mathbb{C}$, then there exists a meromorphic function defined on that domain having exactly those principal parts. Cousin and subsequent authors could only prove the analogous theorem in several variables for certain types of domains (e.g. product domains where each factor is a domain in the complex plane). In fact it turned out that this problem can not be solved on an arbitrary domain in $\mathbb{C}^m$, $m \geq 2$. The best known example for this is a "notched" bicylinder in $\mathbb{C}^2$. This is obtained by removing the set $\{(z_1, z_2) \in \mathbb{C}^2 \mid |z_1| \geq \frac{1}{2}, |z_2| \leq \frac{1}{2}\}$, from the unit bicylinder, $\Delta := \{(z_1, z_2) \in \mathbb{C}^2 \mid |z_1| < 1, |z_2| < 1\}$. This domain D has the property that every function holomorphic on D continues to a function holomorphic on the entire bicylinder. Such a phenomenon never occurs in the theory of one complex variable. In fact, given a domain $G \subset \mathbb{C}$, there exist functions holomorphic on G which are singular at *every* boundary point of G. In several complex variables one calls such domains (i.e. domains on which there exist holomorphic functions which are singular at every boundary point) *domains of holomorphy*. H. Cartan observed in 1934 that every domain in $\mathbb{C}^2$ where the above "Cousin problem" is always solvable is necessarily a domain of holomorphy. A proof of this was communicated by Behnke and Stein in 1937. Meanwhile it was conjectured that Cousin's theorem should hold on any domain of holomorphy. This was in fact proved by Oka in 1937: For every prescription of principal parts on a domain of holomorphy $D \subset \mathbb{C}^m$, there exists a meromorphic function on D having exactly those principal parts. In the same year, via the example of $\mathbb{C}^3 \setminus \{0\}$, H. Cartan showed that it is possible for the Cousin theorem to be valid on domains which are not domains of holomorphy.

As the theory of functions of several complex variables developed, it was often the case that, in order to have a chance of carrying over important one variable results, it was necessary to restrict to domains of holomorphy. This was particularly true with respect to the analog of the Weierstrass product theorem. Formulated as a question, it is as follows: Given a domain D in $\mathbb{C}^m$, can one prescribe the zeros (counting multiplicity) of a holomorphic function on D? It was soon realized that in some cases it is impossible to find even a continuous function which does the job. Conditions for the existence of a continuous solution of this problem, the so-called "second Cousin problem," were discussed by K. Stein in 1941. In fact he gave a sufficient condition which could actually be checked in particular examples. Nowadays this is stated in terms of the vanishing of the Chern class of the prescribed zero set. Stein, however, stated this in a dual and more intuitively geometric way. His condition is as follows: The "intersection number" of the zero surface (counting multiplicity) with any 2-cycle in D should always be zero.

It was similarly necessary to restrict to domains of holomorphy in order to prove the appropriate generalizations of the facts that, on a domain in $\mathbb{C}$, every meromorphic function is the ratio of (globally defined) analytic functions and, if the domain is simply connected, holomorphic functions can be uniformly approximated by polynomials (i.e. the Runge approximation theorem). Poincaré first posed the question about meromorphic functions of several variables being quotients of globally defined relatively prime holomorphic functions. He in fact answered this positively in certain interesting cases (e.g. for $\mathbb{C}^m$ itself).

2. It is not at all straightforward to generalize the notion of a Mittag-Leffler distribution (i.e. prescriptions of principal parts) to the several variable case. The main difficulty is that the set on which the desired function is to have poles is no longer discrete. In fact, in the case of domains in $\mathbb{C}^m$, $m \geq 2$, this set is a $(2m-2)$-dimensional real (possibly singular) surface. Thus one can no longer just prescribe points and pieces of Laurent series. This can be circumvented as follows: If G is a domain in $\mathbb{C}^m$ and $\mathfrak{U} = \{U_i\}$, $i \in I$, is an open covering of G, then the family $\{U_i, h_i\}$ is called an *additive Cousin distribution* on G, whenever each h_i is a meromorphic function on U_i, and on $U_{i_0 i_1} := U_{i_0} \cap U_{i_1}$ the difference $h_{i_0} - h_{i_1}$ is holomorphic for all choices of i_0 and i_1. In the case of $m = 1$, this means that h_{i_0} and h_{i_1} have the same principal parts. Thus one obtains a Mittag-Leffler distribution from the Cousin distribution. A meromorphic function h is said to have the Cousin distribution for its principal parts if $h - h_i$ is holomorphic on U_i for all i.

Different Cousin distributions can, on the same covering, define the same distribution of principal parts. This difficulty is overcome by introducing an equivalence relation. For this let $x \in G$. Let U be an open neighborhood of x in G and suppose that h is meromorphic on U. Then the pair (U, h) is called a *locally meromorphic function* at x. Two such pairs (U_1, h_1) and (U_2, h_2) are called equivalent if there exists a neighborhood V of x with $V \subset U_1 \cap U_2$ and $h_1 - h_2$ holomorphic on V. Each equivalence class is called a *germ of a principal part*. The set of all germs of principal parts at x is denoted by $\mathscr{H}_x$. We define $\mathscr{H} := \bigcup_{x \in X} \mathscr{H}_x$ and denote by $\pi: \mathscr{H} \to G$ the map which associates to every germ its base point $x \in G$. If $U \subset G$ is open and h is meromorphic on U then, for every $x \in U$, one has the associated principal part of h at x, $\bar{h}_x \in \mathscr{H}_x$. Consequently there exists a map s_h: $U \to \mathscr{H}$, $x \mapsto \bar{h}_x$, such that $\pi \cdot s_h = id$. It is easy to check that sets of the form $s_h(U)$, where U is any open set in G and h is any meromorphic function on U, form a basis for a topology on $\mathscr{H}$. Further, in this topology, π: $\mathscr{H} \to G$ is seen to be continuous and a local homeomorphism. In such a situation one calls $\mathscr{H}$ a sheaf over G. The fibers of π should be thought of as stalks with the open sets looking like transversal surfaces given by the maps s_h. The map $s_h\colon U \to \mathscr{H}$ is called a local section over U. Every Cousin distribution $\{U_i, h_i\}$ defines a *global* continuous map (section) $s\colon G \to \mathscr{H}$ with $\pi \cdot s = id$. This is locally defined by $s \mid U_i := s_{h_i}$. The condition that, for all i and j, $h_i - h_j$ is holomorphic on $U_i \cap U_j$ is equivalent to the fact that s is well-defined. Two Cousin distributions have the same principal parts if and only if they correspond to the same section in D over G. A meromorphic function h is a "solution" of the Cousin distribution s (i.e. has exactly the same principal parts as were prescribed) exactly when $s_h = s$.

It is clear from the above that the sheaf theoretic language is the ideal medium for the statement of the generalization of the Mittag-Leffler problem to the several variable situation. Of course for domains in $\mathbb{C}^n$ Oka had solved this without explicit use of sheaves. But even in this case the language of sheaves isolated the real problems and made the seemingly complicated techniques of Oka more transparent. This was also true in the case of the second Cousin problem, the Poincaré problem, etc. Furthermore this language was ideal for formulating new problems and for paving the road toward possible obstructions to their solutions. Theorems about sheaves themselves later gave rise to numerous interesting applications.

3. The germs of holomorphic functions form a sheaf which is usually denoted by $\mathcal{O}$. It has already been pointed out that the zero sets of analytic functions are important even in the study of the Cousin problems. Thus it should be expected that analytic sets, which are just sets of simultaneous zeros of finitely many holomorphic functions on domains in the various $\mathbb{C}^m$, would play an important role in the early development of the theory. In fact the totality of germs of holomorphic functions which vanish on a particular analytic set form a subsheaf of $\mathcal{O}$ which frequently comes into play in present day complex analysis. In 1950 Oka himself used the idea of distributions of ideals in rings of local holomorphic functions (idéaux de domaines indéterminés). This notion, which at the time of its conception seemed difficult and mysterious, just corresponds to the simple idea of a sheaf of ideals.

The use of germs and the idea of sheaves go back to the work of J. Leray. Sheaves have been systematically applied in the theory of functions of several complex variables ever since 1950/51. The idea of *coherence* is very important for many considerations in several complex variables. Roughly speaking, a sheaf of $\mathcal{O}$-modules is coherent if it is locally free except possibly on some small set where it is still finitely generated with the ring of relations again being finitely generated. Even in the early going it was necessary to prove the coherence of many sheaves. This was often quite difficult, because there were really no techniques around and most work had to be done from scratch. The most important coherence theorems originated with H. Cartan and K. Oka. After the foundations had been laid, coherent sheaves quickly enriched the theory of domains of holomorphy with new important results. In the meantime, in his memorable work "Analytische Functionen mehrerer komplexer Veränderlichen zu vorgegebenen Periodizitätsmoduln und das zweite Cousinsche Problem," Math. Ann. **123**(1951), 201–222, K. Stein had discovered complex manifolds which have basic (elementary) properties similar to domains of holomorphy. A domain $G \subset \mathbb{C}$ is indeed a domain of holomorphy if and only if it is a *Stein manifold*. The main point is that many theorems about coherent sheaves on domains of holomorphy can as well be proved for Stein manifolds. Cartan and Serre recognized that the language of sheaf cohomology, which had been developed only shortly before, is particularly suitable for the formulation of the main results: For every coherent sheaf $\mathcal{S}$ over a Stein manifold X, the following two theorems hold:

Theorem A. The $\mathcal{O}(X)$-module of global sections $\mathcal{S}(X)$ generates every stalk $\mathcal{S}_x$ as an $\mathcal{O}_x$-module for all $x \in X$.

Theorem B. $H^q(X, \mathcal{S}) = 0$ for all $q \geq 1$.

These famous theorems, which were first proved in the Seminaire Cartan 1951/52, contain, among many others, the results pertaining to the Cousin problems.

4. Following the original definition, a *paracompact* complex manifold is called a *Stein manifold* if the following three axioms are satisfied:

Separation Axiom: *Given two distinct points $x_1, x_2 \in X$, there exists a function f holomorphic on X such that $f(x_1) \neq f(x_2)$.*

Local Coordinates Axiom: *If $x_0 \in X$ then there exists a neighborhood U of x_0 and functions $f_1, \ldots, f_m$ which are holomorphic on X such that the restrictions $z_i := f_i \mid U$, $i = 1, \ldots, m$, give local coordinates on U.*

Holomorphic Convexity Axiom: *If $\{x_i\}$ is a sequence which "goes to ∞ in X" (i.e. the set $\{x_i\}$ is discrete) then there exists a function f holomorphic on X which is unbounded on $\{x_i\}$:* $\sup |f(x_i)| = \infty$.

It is clear that a domain in $\mathbb{C}^m$ is a Stein manifold if and only if it is holomorphically convex. However if one wants to study non-schlicht domains over $\mathbb{C}^m$ (i.e. ramified covers of domains in $\mathbb{C}^m$), then it is not apriori clear that two points lying over the same base point can be separated by global holomorphic functions. Likewise it is not obvious that neighborhoods of ramification points have local coordinates which are restrictions of global holomorphic functions. If one allows points which are not locally uniformizable (i.e. points where there is a genuine singularity and the "domain" is not even a manifold, as is the case at the point $(0, 0, 0) \in V := \{(x, y, z) \in \mathbb{C}^3 \mid x^2 = yz\}$, which is spread over the (y, z)-plane by projection) then the above definition is meaningless, because we assumed that X is a manifold. However, even in the non-locally uniformizable situation above, the following significant weakening of the separation and local coordinate axioms still holds:

Weak Separation Axiom: *For every point $x_0 \in X$ there exist functions $f_1, \ldots, f_n \in \mathcal{O}(X)$ so that x_0 is an isolated point in $\{x \in X \mid f_1(x) = \cdots = f_n(x) = 0\}$.*

Among other things, this allows the consideration of spaces with singularities. Due to the maximum principle, this weak separation implies that *every compact analytic subspace of X is finite.*

It turns out that, without losing the main results, the convexity axiom can also be somewhat weakened:

Weak Convexity Axiom: *Let K be a compact set in X and W an open neighborhood of K in X. Then $\hat{K} \cap W$ is compact, where $\hat{K}$ denotes the holomorphic hull of K in X:*

$$\hat{K} := \{x \in X \mid |f(x)| \leq \sup_{y \in K} |f(y)|, \quad \text{for all} \quad f \in \mathcal{O}(X)\}.$$

One way of strengthening the axiom immediately above is to require that $\hat{K}$ be compact in X. If one does this and further considers only the case where X is a manifold, then, without the use of deep techniques, one can show that the strengthened axiom is equivalent to the holomorphic convexity axiom (see Theorems IV.2.4 and IV.2.12).

For the purposes of this book, a Stein space is a *paracompact* (*not necessarily reduced*) *complex space for which Theorem B is valid.* It is proved that this condition is equivalent to the validity of Theorem A, and is also equivalent to the above weakened axioms. In particular it follows that if X is a manifold, the weakened axioms imply Stein's original axioms.

We will always assume that a complex space has *countable topology* and is thus *paracompact.* With a bit of work one can show that any irreducible complex space which satisfies the weak separation axiom is eo ipso paracompact (see 16, 24).

5. We conclude our introductory remarks with a short description of the contents of this book. We begin with two brief preliminary chapters (Chapters A and B) where we assemble the important information from sheaf theory and the related cohomology theories. The idea of coherence is explained in these chapters. A reader who is really interested in coherence proofs, can find such in our book, "Coherent Analytic Sheaves," which is presently in preparation. Complex spaces are introduced as special $\mathbb{C}$-algebraized spaces. Further we develop cohomology from the point of view of alternating (Čech) cochains as well as via flabby resolutions. Proofs which are easily accessible in the literature (e.g. [SCV], [TF], or [TAG]) are in general not carried out.

In Chapter I a short direct proof of the coherence theorem for finite holomorphic maps is given. It is based primarily on the Weierstrass division theorem and Hensel's lemma for convergent power series.

The Dolbeault cohomology theory is presented in Chapter II. As a consequence we obtain Theorem B for the structure sheaf $\mathcal{O}$ over a compact euclidean block (i.e. an m-fold product of rectangles), K, in $\mathbb{C}^m$. In other words, for $q \geq 1$, $H^q(K, \mathcal{O}) = 0$. It should be noted that, although we want to introduce Dolbeault cohomology in any case, this result follows directly and with less difficulty via the Čech cohomology.

Chapter III contains the proofs for Theorems A and B for coherent sheaves over euclidean blocks $K \subset \mathbb{C}^m$. One of the key ingredients for the proofs is the fact that, for every coherent sheaf $\mathcal{S}$, the cohomology groups, $H^q(K, \mathcal{S})$, vanish for all q large enough. The deciding factor in proving Theorem A is the "Heftungslemma" of Cartan. This is proved quite easily if while solving the Cousin problem, one simultaneously estimates the attaching functions.

In Chapter IV Theorems A and B are proved for an arbitrary Stein space, X. A summary of the proof is the following: First it is shown that X is exhausted by analytic blocks. (An analytic block is a compact set in X which can be mapped by a finite, proper, holomorphic map into an euclidean block in some $\mathbb{C}^m$.) The coherence theorem for finite maps along with the results in Chapter III yield the desired theorem free of charge. In order to obtain such theorems in the limit (i.e. for spaces exhausted by analytic blocks), an approximation technique, which is a generalization of the usual Runge idea, is needed.

Applications and illustrations of the main theorems, as well as examples of Stein manifolds, are given in Chapter V. The canonical Fréchet topology on the space of global sections $\mathcal{S}(X)$ of a coherent analytic sheaf is described in Section 4. By means of the normalization theorem, which we do not prove in this book, we give a simple proof for the fact that, for a reduced complex space X, the canonical Fréchet topology on $H^0(X, \mathcal{O})$ is the topology of compact convergence.

Chapter VI is devoted to proving that, for a coherent analytic sheaf $\mathcal{S}$ on a compact complex space X, $H^q(X, \mathcal{S})$, $q \geq 0$, are finite dimensional $\mathbb{C}$-vector space (Théorème de finitude of Cartan and Serre). In this proof we work with the Hilbert space of square-integrable holomorphic functions and make use of the orthonormal basis which was introduced by S. Bergman. The classical Schwarz lemma plays an important role, replacing the lemma of L. Schwartz on linear compact maps between Fréchet spaces.

In Chapter VII we attempt to entertain the reader with a presentation of the theory of compact Riemann surfaces which results from, among other considerations, the finiteness theorem of Chapter V. The celebrated Riemann-Roch and Serre duality theorems are proved. The flow of the proof is more or less like that in Serre [35], except that, in the analytic case, a real argument for $H^1(X, \mathcal{M}) = 0$ is needed. This is done in a simple way using an idea of R. Kiehl. The book closes with a proof of the Grothendieck theorem on the splitting of vector bundles over $\mathbb{C}P_1$.

The reader should be advised that, while the English version is not a word for word translation of Theorie der Steinschen Räume, there are no significant changes in the mathematics. There are a number of strategies for reading this book, depending on the experience and viewpoint of the reader. Those who are not currently working the field might first browse through the chapter on applications (Chapter V).

It gives us great pleasure to be able to dedicate this book to Karl Stein, who initiated the theory as well as collaborated in its development. Various preliminary versions of our texts were already in existence in the middle 60's. We would like to thank W. Barth for his help at that time.

It is our pleasure to express sincere thanks to Professor Dr. Alan Huckleberry from the University of Notre Dame, South Bend, Indiana, for translating this book into English.

Göttingen, Münster/Westf. H. Grauert R. Remmert

Chapter A. Sheaf Theory

In this chapter we develop sheaf theory only as far as is necessary for later function theoretic applications.

We mention [SCV], [TF], [TAG], and [FAC] as well as [CAS] as standard literature related to the material in this chapter.

The symbols X, Y will always denote topological spaces and U, V are open sets. It is frequently the case that $V \subset U$. Sheaves are denoted by $\mathscr{S}, \mathscr{S}_1, \mathscr{T}, \ldots$ and for the most part we use $S, S_1, T, \ldots$ for presheaves.

§ 0. Sheaves and Presheaves

1. Sheaves and Sheaf Mappings. A triple $(\mathscr{S}, \pi, X)$, consisting of *topological spaces* $\mathscr{S}$ and X and a *local homeomorphism* $\pi: \mathscr{S} \to X$ from $\mathscr{S}$ onto X is called a *sheaf on* X. Instead of $(\mathscr{S}, \pi, X)$ we often write $(\mathscr{S}, \pi)$, $\mathscr{S}_X$ or just $\mathscr{S}$. It follows that the *projection* π is open and every *stalk* $\mathscr{S}_x := \pi^{-1}(x)$, $x \in X$, is a *discrete* subset of $\mathscr{S}$.

If $(\mathscr{S}_1, \pi_1)$ and $(\mathscr{S}_2, \pi_2)$ are sheaves over X and $\varphi: \mathscr{S}_1 \to \mathscr{S}_2$ is a continuous map, then φ is said to be a *sheaf mapping* if it *respects the stalks* (i.e. if $\pi_2 \circ \varphi = \pi_1$). Since $\varphi(\mathscr{S}_{1x}) \subset \mathscr{S}_{2x}$, every mapping of sheaves $\varphi: \mathscr{S}_1 \to \mathscr{S}_2$ induces the stalk mappings $\varphi_x: \mathscr{S}_{1x} \to \mathscr{S}_{2x}$, $x \in X$. Since π_1 and π_2 are local homeomorphisms, it follows that a sheaf map $\varphi: \mathscr{S}_1 \to \mathscr{S}_2$ is always a local homeomorphism and is in particular an *open* map.

Let $(\mathscr{S}_3, \pi_3)$ be another sheaf over X and suppose that $\psi: \mathscr{S}_2 \to \mathscr{S}_3$ and $\varphi: \mathscr{S}_1 \to \mathscr{S}_2$ are sheaf mappings. Then $\psi \circ \varphi: \mathscr{S}_1 \to \mathscr{S}_3$ is likewise a sheaf mapping. Since $\mathrm{id}: \mathscr{S} \to \mathscr{S}$ is a sheaf mapping, this shows that *the set of sheaves over X, with sheaf maps as morphisms, is a category.*

2. Sums of Sheaves, Subsheaves, and Restrictions. Let $(\mathscr{S}_1, \pi_1)$ and $(\mathscr{S}_2, \pi_2)$ be sheaves over X. We equip

$$\mathscr{S}_1 \oplus \mathscr{S}_2 := \{(p_1, p_2) \in \mathscr{S}_1 \times \mathscr{S}_2 : \pi_1(p_1) = \pi_2(p_2)\} = \bigcup_{x \in X} (\mathscr{S}_{1x} \times \mathscr{S}_{2x})$$

with the relative topology in $\mathcal{S}_1 \times \mathcal{S}_2$. Defining $\pi\colon \mathcal{S}_1 \oplus \mathcal{S}_2 \to X$ by $\pi(p_1, p_2) := \pi_1(p_1)$, it follows that $(\mathcal{S}_1 \oplus \mathcal{S}_2, \pi)$ is a sheaf over X. It is called the *direct* or *Whitney sum* of $\mathcal{S}_1$ and $\mathcal{S}_2$.

A subset $\mathcal{S}'$ of a sheaf $\mathcal{S}$, equipped with the relative topology is called a *subsheaf* of S whenever $(\mathcal{S}', \pi \mid \mathcal{S}')$ is a sheaf over X. Thus $\mathcal{S}'$ is a subsheaf of $\mathcal{S}$ if and only if it is an open subset of $\mathcal{S}$ and $\pi \mid \mathcal{S}'$ is surjective.

Again let $\mathcal{S}$ be a sheaf over X and take Y to be a topological subspace of X. Then, with the relative topology on $\mathcal{S} \mid Y := \pi^{-1}(Y) \subset \mathcal{S}$, the triple $(\mathcal{S} \mid Y, \pi \mid (\mathcal{S} \mid Y), Y)$ is a sheaf over Y. It is called the restriction of $\mathcal{S}$ to Y and is denoted by $\mathcal{S} \mid Y$ or $\mathcal{S}_Y$.

3. Sections. Let $\mathcal{S}$ be a sheaf on X and $Y \subset X$ be a subspace. A continuous map $s\colon Y \to \mathcal{S}$ is called a *section* over Y if $\pi \circ s = \mathrm{id}_Y$. For $x \in Y$, we denote the "value" of s at x by s_x (in the literature the symbol $s(x)$ is also used for this purpose). Certainly $s_x \in \mathcal{S}_x$ for all $x \in Y$. The set of all sections over Y in the sheaf $\mathcal{S}$ is denoted by $\Gamma(Y, \mathcal{S})$. Quite often we use the shorter symbol $\mathcal{S}(Y)$. A section, $s \in \mathcal{S}(U)$, over an open set $U \subset X$ is a local homeomorphism. The collection $\{s(U) = \bigcup_{x \in U} s_x \mid U \subset X \text{ open}, s \in \mathcal{S}(U)\}$ forms a basis for the topology of $\mathcal{S}$.

If $\varphi\colon \mathcal{S}_1 \to \mathcal{S}_2$ is a sheaf mapping then, for every $s \in \mathcal{S}_1(Y)$, $\varphi \circ s \in \mathcal{S}_2(Y)$. Hence φ induces a mapping $\varphi_Y\colon \mathcal{S}_1(Y) \to \mathcal{S}_2(Y)$, $s \mapsto \varphi \circ s$. On the other hand, one can easily show the following:

A map $\varphi\colon \mathcal{S}_1 \to \mathcal{S}_2$ is a sheaf map if, for every $p \in \mathcal{S}_1$, there exists an open set $U \subset X$ and a section $s \in \mathcal{S}_1(U)$ with $p \in s(U)$ so that the map $\varphi \circ s\colon U \to \mathcal{S}_2$ is a section in $\mathcal{S}_2$ (i.e. $\varphi \circ s \in \mathcal{S}_2(U)$).

4. Presheaves and the Section Functor Γ. Suppose that for every open set U in X there is associated some set $S(U)$. Further suppose that for every pair of open sets U, $V \subset X$ with $\emptyset \neq V \subset U$ we have a *restriction map* $r_V^U\colon S(U) \to S(V)$ satisfying

$$r_U^U = \mathrm{id} \quad \text{and} \quad r_W^V \cdot r_V^U = r_W^U,$$

whenever $W \subset V \subset U$. Then $S := \{S(U), r_V^U\}$ is called a *presheaf* over X. We note that a presheaf on X is just a contravariant functor from the category of open subsets of X to the category of sets.

A *map of presheaves* $\Phi\colon S_1 \to S_2$, where $S_i = \{S_i(U), r_{iV}^U\}$, $i = 1, 2$, is a set of maps $\Phi = \{\phi_U\}$, $\phi_U\colon S_1(U) \to S_2(U)$, such that, for all pairs of open sets U, V with $V \subset U$, $\phi_V \cdot r_{1V}^U = r_{2V}^U \circ \phi_U$. *Thus the presheaves on X form a category.*

For every sheaf $\mathcal{S}$ over X we have the *canonical presheaf* $\Gamma(\mathcal{S}) := \{\mathcal{S}(U), r_V^U\}$, where $r_V^U(s) := s \mid V$. Every sheaf map $\varphi\colon \mathcal{S}_1 \to \mathcal{S}_2$ determines a map of presheaves $\Gamma(\varphi)\colon \Gamma(\mathcal{S}_1) \to \Gamma(\mathcal{S}_2)$ where $\Gamma(\varphi) := \{\varphi_U\}$. The following is immediate:

Γ is a covariant functor from the category of sheaves into the category of presheaves.

5. Going from Presheaves to Sheaves. The Functor Γ. Every presheaf $S = \{S(U), r_V^U\}$ over X determines in a natural way a sheaf $\mathscr{S}$ *which is defined as follows*: *For every* $x \in X$ *the* subsystem $\{S(U), r_V^U, x \in X\}$ is *directed* with respect to inclusion of open neighborhoods of x. Thus the *direct limit* $\mathscr{S}_x := \lim\limits_{x \in U} S(U)$ and the maps r_x^U: $S(U) \to \mathscr{S}_x$ are defined. We let $\mathscr{S} := \bigcup\limits_{x \in X} \mathscr{S}_x$ and define π: $\mathscr{S} \to X$ by $\pi(p) = x$ when $p \in \mathscr{S}_x$. Every element $s \in S(U)$ determines the set $s_U := \bigcup\limits_{x \in U} r_x^U(s) \subset \mathscr{S}$. The system of subsets of $\mathscr{S}$, $\{s_U | U$ open in $X, s \in S(U)\}$, is a basis for a topology on $\mathscr{S}$. We equip $\mathscr{S}$ with this topology. Then it is easy to verify that $(\mathscr{S}, \pi)$ is a sheaf over X. We call $\check{\Gamma}(S) := \mathscr{S}$ the sheaf associated to the presheaf S.

Let ϕ: $S_1 \to S_2$ be a map of presheaves, where $S_i = \{S_i(U), r_{iV}^U\}$, $i = 1, 2$, and $\phi = \{\phi_U\}$. Then ϕ determines a sheaf map $\check{\Gamma}(\phi)$: $\check{\Gamma}(S_1) \to \check{\Gamma}(S_2)$ in the following way: For $p \in \mathscr{S}_{1x}$ one chooses $s \in S_1(U)$ with $r_{1x}^U s = p$ and sets $\check{\Gamma}(\phi)(p) := r_{2x}^U \phi_U(s)$. It is easy to show that this definition is independent of the choice of s and that $\check{\Gamma}(\phi)$ is in fact a sheaf map. Thus

$\check{\Gamma}$ *is a covariant functor from the category of presheaves into the category of sheaves.*.

For every sheaf $\mathscr{S}$ we have the associated sheaf $\check{\Gamma}(\Gamma(\mathscr{S}))$. One obtains a natural map φ: $\mathscr{S} \to \check{\Gamma}(\Gamma(\mathscr{S}))$ as follows: Let $p \in \mathscr{S}_x$. Take U to be an open neighborhood of p so that there exists $s \in \mathscr{S}(U)$ with $p = s_x$. Now define $\varphi(p) := r_x^U(s)$. Then φ is independent of the choice of U and s, and it is clear that φ is a sheaf map. It is quite easy to check that

φ: $\mathscr{S} \to \check{\Gamma}(\Gamma(\mathscr{S}))$ *is a sheaf isomorphism and the functors* $\check{\Gamma}\Gamma$ *and* id *are naturally isomorphic.*

6. The Sheaf Conditions $\mathscr{S}1$ and $\mathscr{S}2$. For every presheaf $S = \{S(U), r_V^U\}$ we have the associated presheaf $\Gamma(\check{\Gamma}(S))$. There is an explicit map between these presheaves: For every $s \in S(U)$ the map $x \mapsto r_x^U(s) \in \mathscr{S}_x$, $x \in U$, is a section over U in $\mathscr{S} := \check{\Gamma}(S)$. This defines a natural map ϕ_U: $S(U) \to S(U)$. One has no trouble verifying that

$\phi := \{\phi_U\}$ *is a presheaf map* ϕ: $S \to \Gamma(\check{\Gamma}(S))$ *which induces the identity* $\check{\Gamma}(\phi)$: $\check{\Gamma}(S) \to \check{\Gamma}(S)$.

A presheaf map $\{\phi_U\}$ is called a mono-, epi-, or isomorphism whenever *all* of the maps ϕ_U are respectively injective, surjective, or bijective. The map ϕ above is in general not an isomorphism. It is easy to see that ϕ_U: $S(U) \to S(U)$ is *injective if and only if the following condition is satisfied*:

$\mathscr{S}1$. *If* $s, t \in S(U)$ *are such that there exists an open cover* $\{U_\alpha\}$ *of* U *with* $r_{U_\alpha}^U s = r_{U_\alpha}^U t$ *for all* α, *then* $s = t$.

In order to guarantee the bijectivity of ϕ_U, we must require even more: Let ϕ_V: $S(V) \to S(V)$ *be injective for every open* $V \subset U$. *Then* ϕ_U *is surjective* (and thus bijective) *if and only if the following condition is satisfied:*

$\mathscr{S}2$. Given an open cover $\{U_\alpha\}$ of U and $s_\alpha \in S(U_\alpha)$ satisfying $r^{U_\alpha}_{U_\alpha \cap U_\beta} s_\alpha = r^{U_\beta}_{U_\alpha \cap U_\beta} s_\beta$ for all α and β, there exists $s \in S(U)$ with $r^U_{U_\alpha} s = s_\alpha$ for all α.

Thus a presheaf is isomorphic to the canonical presheaf associated to its sheaf if and only if the conditions $\mathscr{S}1$ and $\mathscr{S}2$ are satisfied for *all* open sets. In the literature, a sheaf is often defined as a presheaf satisfying $\mathscr{S}1$ and $\mathscr{S}2$ for all open sets in X (e.g. [TF], p. 109).

Remark: One can formulate $\mathscr{S}1$ and $\mathscr{S}2$ in an instructive way by requiring that the sequence

$$0 \longrightarrow S(U) \xrightarrow{u} \prod_\alpha S(U_\alpha) \overset{v}{\underset{w}{\rightrightarrows}} \prod_{\alpha,\beta} S(U_\alpha \cap U_\beta),$$

where u, v, w are obtained in the obvious way by restriction, is *exact*. This means that u maps $S(U)$ bijectively onto the set of $x \in \prod S(U_\alpha)$ satisfying $v(x) = w(x)$.

Example: For every open U contained in $\mathbb{R}$, let $S(U)$ be the set of real-valued continuous functions on $U \times U$. Using the natural restriction maps, S is a presheaf. It is easy to check that S satisfies neither of the above axioms.

7. Direct Products. The interplay between the functors Γ and $\check{\Gamma}$ is clarified by the way direct products are defined. If $(\mathscr{S}_i)$, $i \in I$, is a *family of sheaves* on X then one defines $S(U) := \prod_{i \in I} \mathscr{S}_i(U)$ as the direct product of sets of sections and r^U_V as the product of all of the restriction mappings $r^U_{iV}\colon S_i(U) \to S_i(V)$. Then $S := \{S(U), r^U_V\}$ is a *presheaf* over X. We set $\mathscr{S} := \check{\Gamma}(S)$ and call $\mathscr{S}$ the *direct product of the sheaves* $\mathscr{S}_i$. It is clear that $\mathscr{S}$ fulfills conditions $\mathscr{S}1$ and $\mathscr{S}2$ and thus it is the canonical presheaf of S. We write

$$\mathscr{S} = \prod_{i \in I} \mathscr{S}_i.$$

Warning: For every $x \in X$ one has a canonical injection $\mathscr{S}_x \to \prod_i \mathscr{S}_{ix}$. But, for infinite index sets, it is in general *not surjective*. The point is that germs (p_i), $i \in I$, of sections in S_{ix} are not necessarily simultaneously realized as the restriction of sections on some fixed open neighborhood of x.

Sheaves are frequently constructed using roughly the same procedure as we did for products: One begins with a sheaf, goes to the presheaf level via Γ, defines the new presheaf and then returns to the sheaf level by means of $\check{\Gamma}$. In the next section we use this principle to introduce image sheaves. Later on, tensor product sheaves (but not Hom sheaves) are obtained in this way as well.

8. Image Sheaves. Let $\mathscr{S}$ be a sheaf over X and $f\colon X \to Y$ a continuous mapping from X into a topological space Y. To every open set $V \subset Y$ we associate the set $\mathscr{S}(f^{-1}(V))$. If $V' \subset V$ then we let $\rho^V_{V'}\colon \mathscr{S}(f^{-1}(V)) \to \mathscr{S}(f^{-1}(V'))$ be the restriction mapping for sections. Then it is clear that *the family* $\{\mathscr{S}(f^{-1}(V)), \rho^V_{V'}\}$ *is a presheaf over Y which satisfies conditions $\mathscr{S}1$ and $\mathscr{S}2$.*

The associated sheaf $\check{\Gamma}(\mathscr{S}(f^{-1}(V)))$ is denoted by $f_*(\mathscr{S})$ and is called the *(0-th) image sheaf of* $\mathscr{S}$ with respect to f. Due to the natural bijection $\mathscr{S}(f^{-1}(V)) \to (f_*(\mathscr{S}))(V)$, we always identify $(f_*(\mathscr{S}))(V)$ with $\mathscr{S}(f^{-1}(V))$.

Every germ $\sigma \in f_*(\mathscr{S})_{f(x)}$ is represented in a neighborhood V of $f(x)$ by a section $s \in \mathscr{S}(f^{-1}(V))$. Since $f^{-1}(V)$ is a neighborhood of x in X, s determines a germ $s_x \in \mathscr{S}_x$ which is independent of the choice of the representation and uniquely determined by σ. Thus it is clear that,

For every point $x \in X$, there exists a natural map $\hat{f}_x$:

$$f_*(\mathscr{S})_{f(x)} \to \mathscr{S}_x, \ \sigma \mapsto s_x.$$

If $\varphi: \mathscr{S}_1 \to \mathscr{S}_2$ is a mapping of sheaves then, for every open set $V \subset Y$, one has the map $\varphi_f - 1_{(V)}: (f_*(\mathscr{S}_1))(V) \to (f_*(\mathscr{S}_2))(V)$. The family $\{\varphi_{f^{-1}(V)}\}$ is a map of presheaves. We denote the associated sheaf map by $f_*(\varphi)$. One sees that f_* *is a covariant functor from the category of sheaves over X into the category of sheaves over Y.*

If, along with f, another continuous map $g: Y \to Z$ of Y into a topological space Z is given, then one has the sheaves $(gf)_*(\mathscr{S})$ and $g_*(f_*(\mathscr{S}))$ over Z. For every open set $W \subset Z$,

$$\begin{aligned}(g_*(f_*(\mathscr{S})))(W) &= (f_*(\mathscr{S}))(g^{-1}(W)) = \mathscr{S}(f^{-1}(g^{-1}(W))) = \mathscr{S}((gf)^{-1}(W)) \\ &= ((gf)_*(\mathscr{S}))(W).\end{aligned}$$

Thus

$$g_*(f_*(\mathscr{S})) = (gf)_*(\mathscr{S}).$$

9. Gluing Sheaves. Let $\{U_i\}_{i \in I}$ be a covering of X by open subsets, and suppose that on each U_i a sheaf $\mathscr{S}_i$ is given. Defining $U_{ij} := U_i \cap U_j$, we further assume that for each (i, j) we have a sheaf isomorphism $\Theta_{ij}: \mathscr{S}_j | U_{ij} \to \mathscr{S}_i | U_{ij}$. The family $\{\mathscr{S}_i\}$ is said to be *glued together* by $\{\Theta_{ij}\}$ whenever the following "cocycle condition" is satisfied:

$$\Theta_{ij}\Theta_{jk} = \Theta_{ik} \quad \text{on} \quad U_i \cap U_j \cap U_k \quad \text{for all} \quad i, j, k \in I.$$

From such a family one canonically constructs a new sheaf (see [FAC], p. 201):

For every family of sheaves $\{\mathscr{S}_i\}$ on X which is glued together by $\{\Theta_{ij}\}$ there exists a sheaf $\mathscr{S}$ on X and a family $\{\theta_i\}_{i \in I}$ of sheaf isomorphisms $\theta_i: \mathscr{S} | U_i \to \mathscr{S}_i$ so that $\Theta_{ij} = \theta_i \cdot \theta_j^{-1}$ on U_{ij}. Up to an isomorphism the sheaf $\mathscr{S}$ and the family $\{\theta_i\}$ are uniquely determined by $\{\mathscr{S}_i\}$ and $\{\Theta_{ij}\}$.

§ 1. Sheaves with Algebraic Structure

In most applications the stalks of a sheaf carry additional algebraic structures. Sheaves of local $\mathbb{C}$-algebras are particularly important for us.

1. Sheaves of Groups, Rings, and $\mathscr{R}$-Modules. A sheaf $\mathscr{S}$ over X is called a *sheaf of abelian groups* if, for all $x \in X$, the stalk $\mathscr{S}_x$ is an (additively written) abelian group and "subtraction" $\mathscr{S} \oplus \mathscr{S} \to \mathscr{S}$, $(p, q) \to p - q$, is continuous. Note that if $(p, q) \in \mathscr{S} \oplus \mathscr{S}$, then $p, q \in \mathscr{S}_x$ with $x := \pi(p) = \pi(q)$. Thus $p - q$ is a well-defined element of $\mathscr{S}_x$.

If $\mathscr{S}$ is such a sheaf of abelian groups and 0_x is the identity element in $\mathscr{S}_x$, then the map $0: X \to \mathscr{S}$, $x \to 0_x$, is a section in $\mathscr{S}$ over X and is called the zero section. The set

$$\operatorname{supp} \mathscr{S} := \{x \in X: \mathscr{S}_x \neq \{0_x\}\}$$

is *called the support of* $\mathscr{S}$.

Using the additive structure in the stalks, $\mathscr{S}(U)$ is in a natural way an abelian group for all open $U \subset X$ (e.g. for all $s, t \in \mathscr{S}(U)$, $s - t \in \mathscr{S}(U)$ is given by $(s - t)_x := s_x - t_x$ for $x \in U$).

In reality, sheaves occur with even more algebraic structure. The further operations are defined analogously, the key point being that they are stalk-wise defined and are continuous.

A sheaf of abelian groups $\mathscr{R}$ over X is called a *sheaf of commutative rings*, if, along with the additive structure, there is a further sheaf mapping $\mathscr{R} \oplus \mathscr{R} \to \mathscr{R}$, $(p, q) \to p \cdot q$ (*multiplication*), which makes every stalk $\mathscr{R}_x$ a commutative ring. If moreover every stalk $\mathscr{R}_x$ possesses a multiplicative identity 1_x, and if the mapping $x \to 1_x$ is a section in $\mathscr{R}$ (the identity section), then $\mathscr{R}$ is called a *sheaf of rings with identity*. In the following, $\mathscr{R}$ denotes a sheaf of rings with identity over X. Obviously $1_x \neq 0_x$ for all $x \in \operatorname{supp} \mathscr{R}$.

A sheaf $\mathscr{S}$ of abelian groups is called a *sheaf of modules over* $\mathscr{R}$, or simply an $\mathscr{R}$*-sheaf* or an $\mathscr{R}$*-module*, if a sheaf mapping $R \oplus S \to S$ is defined in such a way to induce the structure of an $\mathscr{R}_x$-module on $\mathscr{S}_x$ for all $x \in X$. Obviously $\mathscr{R}$ is itself an $\mathscr{R}$-module.

As in the case of sheaves of groups, the algebraic structure of a sheaf induces the same structure on the set of sections via point-wise definitions. Thus $\mathscr{R}(U)$ is likewise a ring and, if $\mathscr{S}$ is a sheaf of $\mathscr{R}$-modules, the set $\mathscr{S}(U)$ is an $\mathscr{R}(U)$-module.

If $\mathscr{S}_1, \ldots, \mathscr{S}_p$ are sheaves of $\mathscr{R}$-modules then the Whitney sum $\mathscr{S}_1 \oplus \cdots \oplus \mathscr{S}_p$ is an $\mathscr{R}$-module with the operations being component-wise defined. In particular, for every natural number p, $\mathscr{R}^p := \mathscr{R} \oplus \cdots \oplus \mathscr{R}$ is an $\mathscr{R}$-module.

2. Sheaf Homomorphisms and Subsheaves. We introduce here the relevant notions for sheaves of $\mathscr{R}$-modules. The analogous ideas for sheaves with other algebraic structures go more-or-less along the same lines and will not be discussed. In these considerations $\mathscr{S}$, $\mathscr{S}_1$, and $\mathscr{S}_2$ are always $\mathscr{R}$-sheaves.

A sheaf mapping $\varphi: \mathscr{S}_1 \to \mathscr{S}_2$ is called a *sheaf homomorphism* or an $\mathscr{R}$-homomorphism if, for every $x \in X$, the induced mapping $\varphi_x: \mathscr{S}_{1x} \to \mathscr{S}_{2x}$ is an $\mathscr{R}_x$-module homomorphism.

The sheaves of $\mathscr{R}$*-modules over* X *along with the* $\mathscr{R}$*-homomorphism form a category.*

In this category $\mathscr{S}_1$ and $\mathscr{S}_2$ are isomorphic if and only if there is a sheaf mapping $\varphi: \mathscr{S}_1 \to \mathscr{S}_2$ so that $\varphi_x: \mathscr{S}_{1x} \to \mathscr{S}_{2x}$ is an $\mathscr{R}_x$-isomorphism for all $x \in X$.

A subset $\mathscr{S}'$ of $\mathscr{S}$ is called an *$\mathscr{R}$-submodule of $\mathscr{S}$*, if $\mathscr{S}'$ is a subsheaf of $\mathscr{S}$ and every stalk $\mathscr{S}'_x$ is an $\mathscr{R}_x$-submodule of $\mathscr{S}_x$.

If $\mathscr{S}'_x$ is an $\mathscr{R}_x$-submodule of $\mathscr{S}_x$ for all $x \in X$ then $\mathscr{S}' := \bigcup_{x \in X} S'_x$ is an $\mathscr{R}$-submodule of $\mathscr{S}$ if and only if $\mathscr{S}'$ is open in $\mathscr{S}$.

It follows immediately that if $\mathscr{S}'$ and $\mathscr{S}''$ are $\mathscr{R}$-submodules of $\mathscr{S}$, then their *sum* $\mathscr{S}' + \mathscr{S}'' := \bigcup_{x \in X} (\mathscr{S}'_x + \mathscr{S}''_x)$ and their intersection $\mathscr{S}' \cap \mathscr{S}'' := \bigcup_{x \in X} (\mathscr{S}'_x \cap \mathscr{S}''_x)$ are likewise $\mathscr{R}$-submodules of $\mathscr{S}$.

A sheaf of ideals $\mathscr{I}$, or for short an ideal, is an $\mathscr{R}$-submodule of the $\mathscr{R}$-module $\mathscr{R}$. For every ideal $\mathscr{I} \subset \mathscr{R}$ one defines the *product*

$$\mathscr{I} \cdot \mathscr{S} := \bigcup_{x \in X} \mathscr{I}_x \cdot \mathscr{S}_x \subset \mathscr{S},$$

where $\mathscr{I}_x \cdot \mathscr{S}_x$ consists of linear combinations $\sum_{1}^{<\infty} a_{\nu x} \cdot s_{\nu x}$, $a_{\nu x} \in \mathscr{I}_x$, $s_{\nu x} \in \mathscr{S}_x$. Thus $\mathscr{I}_x \cdot \mathscr{S}_x$ is an $\mathscr{R}_x$-submodule of $\mathscr{S}_x$ and, since $\mathscr{I} \cdot \mathscr{S}$ is open in $\mathscr{S}$, $\mathscr{I} \cdot \mathscr{S}$ is an $\mathscr{R}$-submodule of $\mathscr{S}$.

If $\varphi\colon \mathscr{S}_1 \to \mathscr{S}_2$ is an $\mathscr{R}$-homomorphism, then the sets $\mathscr{K}er\, \varphi := \bigcup_{x \in X} \ker \varphi_x$ and $\mathscr{I}m\, \varphi := \bigcup_{x \in X} \operatorname{im} \varphi_x$ are $\mathscr{R}$-submodules of $\mathscr{S}_1$ and $\mathscr{S}_2$ respectively. If $\rho\colon \mathscr{R}_1 \to \mathscr{R}_2$ is a sheaf homomorphism between sheaves of rings (i.e. each mapping $\rho_x\colon \mathscr{R}_{1x} \to \mathscr{R}_{2x}$ is a ring homomorphism with $\rho_x(1_x) = 1_x$), then $\mathscr{K}er\, \rho$ is an ideal in $\mathscr{R}_1$.

A system of $\mathscr{R}$-sheaves and $\mathscr{R}$-homomorphisms

$$\cdots \longrightarrow \mathscr{S}_{i-1} \xrightarrow{\varphi_{i-1}} \mathscr{S}_i \xrightarrow{\varphi_i} \mathscr{S}_{i+1} \longrightarrow \cdots, \qquad i \in \mathbb{Z},$$

is called an *$\mathscr{R}$-sequence*. An $\mathscr{R}$-sequence is *called exact* at $\mathscr{S}_i$ if $\mathscr{I}m\, \varphi_{i-1} = \mathscr{K}er\, \varphi_i$. It is said to be *exact* if it is exact at every $\mathscr{S}_i$.

3. Quotient Sheaves. Let $\mathscr{S}$ be an $\mathscr{R}$-module and $\mathscr{S}' \subset \mathscr{S}$ an $\mathscr{R}$-submodule of $\mathscr{S}$. We set

$$\mathscr{S}/\mathscr{S}' := \bigcup_{x \in X} \mathscr{S}_x/\mathscr{S}'_x$$

and define $q\colon \mathscr{S} \to \mathscr{S}/\mathscr{S}'$ stalkwise via the canonical quotient homomorphism $q_x\colon \mathscr{S}_x \to \mathscr{S}_x/\mathscr{S}'_x$. We use the finest topology on $\mathscr{S}/\mathscr{S}'$ for which q is continuous. Thus a set $W \subset \mathscr{S}/\mathscr{S}'$ is open if and only if $q^{-1}(W)$ is open. Since $q(\mathscr{S}_x) = \mathscr{S}_x/\mathscr{S}'_x$, we have the natural projection $\bar{\pi}\colon \mathscr{S}/\mathscr{S}' \to X$ so that $\bar{\pi} \circ q = \pi$. Thus we have the following:

The triple $(\mathscr{S}/\mathscr{S}', \bar{\pi}, X)$ is a sheaf of $\mathscr{R}$-modules and $q\colon \mathscr{S} \to \mathscr{S}/\mathscr{S}'$ is an $\mathscr{R}$-epimorphism with $\mathscr{K}er\, q = \mathscr{S}'$.

We call $\mathscr{S}/\mathscr{S}'$ the *quotient sheaf of $\mathscr{S}$ by $\mathscr{S}'$*.

Every $\mathscr{R}$-homomorphism $\varphi\colon \mathscr{S}_1 \to \mathscr{S}_2$ determines the *exact* $\mathscr{R}$-sequences

$$0 \to \mathscr{K}er\,\varphi \to \mathscr{S}_1 \to \mathscr{I}m\,\varphi \to 0, \quad \text{and} \quad 0 \to \mathscr{I}m\,\varphi \to \mathscr{S}_2 \to \mathscr{S}_2/\mathscr{I}m\,\varphi \to 0,$$

where 0 denotes the zero sheaf.

4. Sheaves of Local k-algebras. Let k be a commutative field and $\mathscr{K} := X \times k$ the *constant* sheaf of fields over X (i.e. $\pi\colon \mathscr{K} \to X$, $(x, a) \to x$ is the projection). A sheaf of rings $\mathscr{R}$ is called a *sheaf of k-algebras* if $\mathscr{R}$ is a $\mathscr{K}$-sheaf with supp $\mathscr{R} = X$ such that $c(r_1 r_2) = (cr_1)r_2$ for all $c \in \mathscr{K}_x$ and $r_1, r_2 \in \mathscr{R}_x$. In particular, the identity section $1 \in \mathscr{R}(X)$ is nowhere zero and $\imath\colon \mathscr{K} \to \mathscr{R}$, $(x, a) \to a \cdot 1_x$ is a sheaf *monomorphism* (of rings). We identify $\mathscr{K}$ with $\imath(K) \subset \mathscr{R}$ and k with $k1_x \subset \mathscr{R}_x$.

A sheaf $\mathscr{R}$ of k-algebras is called a *sheaf* of local k-algebras if every stalk $\mathscr{R}_x$ is a local ring with maximal ideal $\mathfrak{m}(\mathscr{R}_x)$ so that the quotient epimorphism $\mathscr{R}_x \to \mathscr{R}_x/\mathfrak{m}(\mathscr{R}_x)$ always maps k onto $\mathscr{R}_x|\mathfrak{m}(\mathscr{R}_x)$.

One identifies $\mathscr{R}_x/\mathfrak{m}(\mathscr{R}_x)$ with k and has a canonical decomposition $\mathscr{R}_x = k \oplus \mathfrak{m}(\mathscr{R}_x)$ as a k-vector space.

Example: Every topological space X carries the sheaf $\mathscr{C}$ of *germs of complex-valued continuous functions*: The $\mathbb{C}$-algebra $\mathscr{C}(U)$ of continuous functions $f\colon U \to \mathbb{C}$ is defined for all open sets $U \subset X$ and $r_V^U\colon \mathscr{C}(U) \to \mathscr{C}(V)$, $V \subset U$, is the natural restriction. The system $\{\mathscr{C}(U), r_V^U\}$ is a presheaf of $\mathbb{C}$-algebras which satisfies $\mathscr{S}1$ and $\mathscr{S}2$ and determines the sheaf $\mathscr{C}$. This is a sheaf of local $\mathbb{C}$-algebras such that maximal ideal $\mathfrak{m}(\mathscr{C}_x)$ consists of the germs $f_x \in \mathscr{C}_x$ which are represented in neighborhoods of x by continuous function f which vanish at x.

If $\mathscr{R}$ is a sheaf of local k-algebras and $s \in \mathscr{R}(Y)$ is a section over a subset $Y \subset X$, then s has a *value* $s(x)$ in k for all $x \in Y$, namely the equivalence class of the germ $s_x \in \mathscr{R}_x$ in k. Thus every section $s \in \mathscr{R}(Y)$ defines a k-valued function $[s]\colon Y \to k$. The homomorphism $s \to [s]$ is not in general injective. In other words, *a section s is more than the function* $[s]$.

A sheaf mapping $\varphi\colon \mathscr{R}_1 \to \mathscr{R}_2$ between sheaves of k-algebras is called a k-homomorphism if every induced map $\varphi_x\colon \mathscr{R}_{1x} \to \mathscr{R}_{2x}$ is a k-algebra homomorphism. It is clear that k-homomorphisms between sheaves of local rings over k are automatically stalk-wise local (i.e. $\varphi_x(\mathfrak{m}(\mathscr{R}_{1x})) \subset \mathfrak{m}(\mathscr{R}_{2x})$).

5. Algebraic Reduction. We let $\mathfrak{n}(\mathscr{R}_x)$ denote the *nilradical* (i.e. the ideal of nilpotent elements) of the stalk $\mathscr{R}_x$. Then

$$\mathfrak{n}(\mathscr{R}) := \bigcup_{x \in X} \mathfrak{n}(\mathscr{R}_x) \subset \mathscr{R}$$

is open in $\mathscr{R}$ and is consequently a sheaf of ideals. We call $\mathfrak{n}(\mathscr{R})$ the *nilradical of* $\mathscr{R}$.

The sheaf of rings $\mathscr{R}ed\,\mathscr{R} := \mathscr{R}/\mathfrak{n}(\mathscr{R})$ is called the (*algebraic*) *reduction of* $\mathscr{R}$. If $\mathscr{R}$ is a sheaf of local rings over k, then since $\mathfrak{n}(\mathscr{R}_x) \subset \mathfrak{m}(\mathscr{R}_x)$, $\mathscr{R}ed\,\mathscr{R}$ is likewise such a sheaf. We say that $\mathscr{R}$ is *reduced* whenever $\mathfrak{n}(\mathscr{R}) = 0$. For example, the sheaf $\mathscr{C}$ is reduced.

Remark: In the case where $\mathcal{R}$ is a sheaf of local rings the set $\bigcup_{x \in X} \mathfrak{m}(\mathcal{R}_x)$ is not necessarily open. Thus it is not in general a subsheaf of $\mathcal{R}$ and a construction analogous to the above, replacing $\mathfrak{n}(\mathcal{R}_x)$ with $\mathfrak{m}(\mathcal{R}_x)$, does not make sense.

6. Presheaves with Algebraic Structure. A presheaf $S = \{S(U), r_V^U\}$ over X is called a *presheaf of abelian groups* if $S(U)$ is always an abelian group and r_V^U is always a group homomorphism. A *presheaf of rings* $R = \{R(U), \bar{r}_V^U\}$ is defined analogously. In the following R denotes a fixed presheaf of rings.

A presheaf S is called a presheaf of R-modules (an R-presheaf) if every $S(U)$ is an $R(U)$-module and for all $a \in R(U)$, $s \in S(U)$ it follows that $r_V^U(as) = \bar{r}_V^U(a) r_V^U(s)$. If $\mathcal{S}$ is an $\mathcal{R}$-sheaf then $\Gamma(\mathcal{S})$ is an $\Gamma(\mathcal{R})$-presheaf. On the other hand, if S is an R-presheaf, then $\check{\Gamma}(S)$ is an $\check{\Gamma}(R)$-sheaf. One just carries over the algebraic structure via the direct limit map. The continuity of the operations is evident.

A *presheaf homomorphism* $\phi: S_1 \to S_2$, $\phi = (\phi_U)$ is a presheaf mapping with every ϕ_U being a homomorphism of the underlying algebraic structure. The mapping $\check{\Gamma}(\phi)$ is thus a sheaf homomorphism. Conversely, every sheaf homomorphism $\varphi: \mathcal{S}_1 \to \mathcal{S}_2$ determines a presheaf homomorphism $\Gamma(\varphi)$.

In the category of R-presheaves, just as in the case of $\mathcal{R}$-sheaves, we have subpresheaves and quotient presheaves. An R-presheaf $S' = \{S'(U), r_V'^U\}$ is called an R-subpresheaf of the R-presheaf S if every $S'(U)$ is an $R(U)$-submodule of $S(U)$ and $r_V'^U$ is always the restriction of r_V^U to $S'(U)$. If S' is an R-subpresheaf of S, then $\tilde{S}(U) := S(U)/S'(U)$ is always an $R(U)$-module and, for every open set $V \subset U$, the map $r_V^U: S(U) \to S(V)$ induces an $R(U)$-homomorphism $\tilde{r}_V^U: \tilde{S}(U) \to \tilde{S}(V)$. Obviously $\tilde{S} := \{\tilde{S}(U), \tilde{r}_V^U\}$ is an R-presheaf. It is called the R-quotient presheaf of S by S' and we write $\tilde{S} = S/S'$.

Every R-presheaf homomorphism $\phi: S_1 \to S_2$ determines the R-presheaves

$$\operatorname{Ker} \phi = \{\operatorname{Ker} \phi_V, \rho_V^U\}$$

and

$$\operatorname{Im} \phi = \{\operatorname{Im} \phi_U, \sigma_V^U\}$$

where ρ_V^U and σ_V^U are defined in the natural way.

We say that an R-sequence $S_1 \xrightarrow{\phi} S_2 \xrightarrow{\psi} S_3$ of R-presheaves is *exact* whenever $\operatorname{Im} \phi = \operatorname{Ker} \psi$. Thus every R-presheaf homomorphism $\phi: S_1 \to S_2$ determines two exact sequences of R-presheaves:

$$0 \to \operatorname{Ker} \phi \to S_1 \to \operatorname{Im} \phi \to 0 \quad \text{and} \quad 0 \to \operatorname{Im} \phi \to S_2 \to S_2/\operatorname{Im} \phi \to 0.$$

7. On the Exactness of $\check{\Gamma}$ and Γ. Since the direct limit of exact sequences is again exact, every exact R-presheaf sequence $S_1 \xrightarrow{\phi} S_2 \xrightarrow{\psi} S_3$ induces an exact $\check{\Gamma}(R)$-sheaf sequence $\check{\Gamma}(S_1) \xrightarrow{\Gamma(\phi)} \check{\Gamma}(S_2) \xrightarrow{\Gamma(\psi)} \check{\Gamma}(S_2)$. In other words $\check{\Gamma}$ is an *exact* functor from the category of R-presheaves to the category of sheaves of $\check{\Gamma}(R)$-modules.

In contrast to this, the section functor Γ is *only left-exact*. A short exact $\mathcal{R}$-

sequence $0 \to \mathcal{S}' \to \mathcal{S} \to \mathcal{S}'' \to 0$ does induce the exact $\Gamma(\mathcal{R})$-sequence

$$0 \to \Gamma(\mathcal{S}') \to \Gamma(\mathcal{S}) \to \Gamma(\mathcal{S}'')$$

of canonical presheaves. However the last homomorphism is not in general surjective. Its image is the quotient presheaf $S := \Gamma(\mathcal{S})/\Gamma(\mathcal{S}')$ and, since $0 \to \Gamma(\mathcal{S}') \to \Gamma(\mathcal{S}) \to S \to 0$ is exact and $\check{\Gamma}$ is an exact functor, it follows that $\check{\Gamma}(S) = \mathcal{S}''$. Nevertheless S is in general a proper subpresheaf of $\Gamma(\mathcal{S}'')$. The phenomenon of nonexactness of Γ is the beginning point of cohomology theory.

§ 2. Coherent Sheaves and Coherent Functors

The notion of coherence of a sheaf of modules is of fundamental importance in the theory of complex spaces. In this section we compile the general properties of coherent sheaves. The symbol $\mathcal{R}$ always stands for a sheaf of rings over a topological space X. We use $\mathcal{S}$, $\mathcal{S}'$, etc. to denote sheaves of $\mathcal{R}$-modules.

1. Finite Sheaves. Finitely many sections $s_1, \ldots, s_p \in \mathcal{S}(U)$ define an $\mathcal{R}_U$-sheaf homomorphism.

$$\sigma_U: \mathcal{R}^p_U \to \mathcal{S}_U, \quad (a_{1x}, \ldots, a_{px}) \to \sigma(a_{1x}, \ldots, a_{px}) := \sum_{i=1}^{p} a_{ix} s_{ix}, \quad x \in U.$$

We say that $\mathcal{S}_U$ is *generated* by the section $s_1, \ldots, s_p$ if σ is surjective. In this case every stalk $\mathcal{S}_x$, $x \in U$ is a finitely generated (by $s_{1x}, \ldots, s_{px}$) $\mathcal{R}_x$-module. An $\mathcal{R}$-sheaf $\mathcal{S}$ is said to be *finite* at $x \in X$ if there is an open neighborhood U of x and with finitely many sections $s_1, \ldots, s_p \in \mathcal{S}(U)$ which generate $\mathcal{S}_U$. This condition is equivalent to the existence of a neighborhood U of x, a natural number p, and an exact sequence $\mathcal{R}^p_U \xrightarrow{\sigma} \mathcal{S}_U \longrightarrow 0$. The germs of the σ-images of the basis $(\delta_{i1}, \ldots, \delta_{ip}) \in \mathcal{R}^p(U)$, $i = 1, \ldots, p$, generate $\mathcal{S}_U$ as a sheaf of $\mathcal{R}_U$-modules. If it is possible to choose U so that σ is an isomorphism, then one says that $\mathcal{S}$ is free at the point x. In this case, the exponent p such that $\mathcal{S}_U = \mathcal{R}^p_U$ is uniquely determined by $\mathcal{S}$ and is called the *rank* of $\mathcal{S}$ at x. An $\mathcal{R}$-sheaf $\mathcal{S}$ is called *finite* (resp. *locally free*) on X if it is finite (resp. free) at each $x \in X$. It is *free* if it is globally isomorphic to $\mathcal{R}^p$ for some $p \in \mathbb{N}$.

Quotient sheaves of finite sheaves are finite. On the other hand, subsheaves of finite sheaves are not necessarily finite (even in the case every stalk $\mathcal{S}_x$ is a Noetherian $\mathcal{R}_x$-module). Thus finiteness at x says more than the fact that the stalks are finitely generated in some neighborhood of x.

Among the important properties of finite sheaves is the following:

If $\mathcal{S}$ is finite at x and $s_1, \ldots, s_p \in \mathcal{S}(U)$ are such that $s_{1x}, \ldots, s_{px}$ generate $\mathcal{S}_x$ as an $\mathcal{R}_x$-module, then there is a neighborhood $V \subset U$ of x so that $s_1|V, \ldots, s_p|V$ generate the sheaf $\mathcal{S}_V$. In particular, the support supp $\mathcal{S}$ *of a finite sheaf is closed in X.*

2. Finite Relation Sheaves. If $\sigma\colon \mathcal{R}_U^p \to \mathcal{S}_U$ is an $\mathcal{R}_U$-homomorphism which is determined by sections $s_1, \ldots, s_p \in \mathcal{S}(U)$, then the sheaf of $\mathcal{R}_U$-submodules in $\mathcal{R}_U^p$,

$$\mathcal{R}el(s_1, \ldots, s_p) := \operatorname{Ker} \sigma = \bigcup_{x \in U} \left\{(a_{1x}, \ldots, a_{px}) \in \mathcal{R}_x^p \,\middle|\, \sum_1^p a_{ix} s_{ix} = 0\right\},$$

is called the *sheaf of relations* of $s_1, \ldots, s_p$. One says that $\mathcal{S}$ is a *finite relation sheaf at* $x \in X$ if, for every open neighborhood U of x and for arbitrary sections $s_1, \ldots, s_p \in \mathcal{S}(U)$, the sheaf of relations $\mathcal{R}el(s_1, \ldots, s_p)$ is finite at x. This is the case if and only if, for every sheaf homomorphism $\sigma\colon \mathcal{R}_U^p \to \mathcal{S}_U$, $\operatorname{Ker} \sigma$ is finite at x. A sheaf of $\mathcal{R}$-modules $\mathcal{S}$ is called a *finite relation sheaf* if it is a finite relation sheaf at all $x \in X$.

3. Coherent Sheaves. A sheaf of $\mathcal{R}$-modules $\mathcal{S}$ over X is called *coherent* if it is finite and a finite relation sheaf. Thus $\mathcal{S}$ is coherent if it is coherent at every $x \in X$ (i.e. whenever for every $x \in X$ there exists a neighborhood $U = U(x)$ so that $\mathcal{S}|_U$ is coherent). If $\mathcal{S}$ is coherent then every finite subsheaf of $\mathcal{R}$-submodules of $\mathcal{S}$ is likewise coherent.

A sheaf of rings $\mathcal{R}$ is called coherent if $\mathcal{R}$ is coherent as an $\mathcal{R}$-module. This is the case precisely when $\mathcal{R}$ is a finite relation sheaf. A sheaf of ideals $\mathcal{J}$ in $\mathcal{R}$ is said to be coherent if it is coherent as an $\mathcal{R}$-submodule of $\mathcal{R}$. If $\mathcal{R}$ is coherent, then the product $\mathcal{J}_1 \cdot \mathcal{J}_2$ of coherent ideal sheaves $\mathcal{J}_1, \mathcal{J}_2$ is likewise coherent ($\mathcal{J}_1 \cdot \mathcal{J}_2$ is finite!).

If $\mathcal{S}$ is a coherent $\mathcal{R}$-module then, for every $x \in X$, there exists an open neighborhood U of x with positive integers p and q such that

$$\mathcal{R}_U^p \to \mathcal{R}_U^q \to \mathcal{S}_U \to 0$$

is exact.

The following is basic for many relations between coherent sheaves:

Five Lemma: *Suppose that*

$$\mathcal{S}_1 \xrightarrow{\varphi_1} \mathcal{S}_2 \xrightarrow{\varphi_2} \mathcal{S}_3 \xrightarrow{\varphi_3} \mathcal{S}_4 \xrightarrow{\varphi_4} \mathcal{S}_5$$

is an exact sequence of sheaves of $\mathcal{R}$-modules such that $\mathcal{S}_1$, $\mathcal{S}_2$, $\mathcal{S}_4$ and $\mathcal{S}_5$ are coherent. Then $\mathcal{S}_3$ is likewise coherent.

This remark is equivalent to the following:

Three Lemma (Serre). *If, in the exact $\mathcal{R}$-sequence*

$$0 \to \mathcal{S}' \to \mathcal{S} \to \mathcal{S}' \to 0,$$

two sheaves are coherent, then the third is also coherent.

We note some important consequences of the Five Lemma:

a) *Let $\varphi\colon \mathscr{S} \to \mathscr{S}'$ be an $\mathscr{R}$-homomorphism between the coherent sheaves $\mathscr{S}, \mathscr{S}'$. Then $\mathscr{K}er\,\varphi$, $\mathscr{I}m\,\varphi$ and $\mathscr{C}oker\,\varphi = S'/\mathscr{I}m\,\varphi$ are coherent sheaves of $\mathscr{R}$-modules.*

b) *The Whitney Sum of finitely many coherent sheaves is coherent.*

c) *Let $\mathscr{S}'$ and $\mathscr{S}''$ be coherent $\mathscr{R}$-submodules of a coherent $\mathscr{R}$-module $\mathscr{S}$. Then the $\mathscr{R}$-sheaves $\mathscr{S}' + \mathscr{S}''$ and $\mathscr{S}' \cap \mathscr{S}''$ are coherent.*

d) *Let $\mathscr{R}$ be a coherent sheaf of rings. Then the sheaf of $\mathscr{R}$-modules $\mathscr{S}$ is coherent if and only if, for every $x \in X$, there exist a neighborhood U of x, with positive integers p and q and an exact sequence*

$$\mathscr{R}_U^q \to \mathscr{R}_U^p \to \mathscr{S}_U \to 0.$$

In particular every locally-free sheaf of $\mathscr{R}$-modules is coherent.

4. Coherence of Trivial Extensions. If $\mathscr{J}$ is an ideal in $\mathscr{R}$ and $\mathscr{R}/\mathscr{J}$ is the associated quotient sheaf of rings over X, then every sheaf of $\mathscr{R}/\mathscr{J}$-modules $\mathscr{S}$ is in a canonical way a sheaf of $\mathscr{R}$-modules. One can study the coherence of $\mathscr{S}$ as an $\mathscr{R}/\mathscr{J}$-module as well as an $\mathscr{R}$-module. In this situation we have the following remark:

Let $\mathscr{R}$ and $\mathscr{J}$ be coherent $\mathscr{R}$-modules and $\mathscr{S}$ a coherent $\mathscr{R}/\mathscr{J}$-sheaf. Then $\mathscr{S}$ is coherent as an $\mathscr{R}/\mathscr{J}$-sheaf if and only if it is $\mathscr{R}$-coherent. In particular $\mathscr{R}/\mathscr{J}$ is a coherent sheaf of rings.

This implies that,

if $\mathscr{R}$ and the nilradical $\mathfrak{n}(\mathscr{R})$ are coherent, then $\mathscr{R}ed\ \mathscr{R} = \mathscr{R}/\mathfrak{n}(\mathscr{R})$ is a coherent sheaf of rings.

The coherence of $\mathscr{R}/\mathscr{J}$ implies that $X' := \operatorname{supp}(\mathscr{R}/\mathscr{J})$ is a closed subspace of X and that $\mathscr{R}' := (\mathscr{R}/\mathscr{J})\,|\,X'$ is a coherent sheaf of rings on X'. Every $\mathscr{R}'$-module $\mathscr{T}'$ on X' has a *trivial extension* $\mathscr{T}$ *on* X (i.e. $\mathscr{T}_x = 0$ for $x \in X\backslash X'$). In a natural way $\mathscr{T}$ is an $\mathscr{R}$-sheaf. Denoting the embedding by $i\colon X' \to X$, one can identify $\mathscr{T}$ with the image sheaf $i_*(\mathscr{T}'_*)$. The following fact (which is a special case of the finiteness theorem, Chapter I.3) is particularly useful for applications in function theory:

Let $\mathscr{R}$ and $\mathscr{J}$ be coherent, $X' := \operatorname{supp}(\mathscr{R}/\mathscr{J})$, and $\mathscr{R}' := (\mathscr{R}/\mathscr{J})\,|\,X'$. Let $\mathscr{T}'$ be an $\mathscr{R}'$-sheaf. Then $\mathscr{T}'$ is coherent if and only if the trivial extension $\mathscr{T}$ is a coherent sheaf $\mathscr{R}$-modules on X.

5. The Functors $\otimes^p$ and $\bigwedge^p$. The system $T := \{\mathscr{S}(U) \otimes_{\mathscr{R}(U)} \mathscr{S}'(U), r_V^U \otimes r_V'^U\}$ is an $\Gamma(\mathscr{R})$-presheaf which satisfies $S1$ and $S2$ (for the definition of $\otimes$, see the standard literature). The associated $\mathscr{R}$-sheaf $\mathscr{S} \otimes_{\mathscr{R}} \mathscr{S}' := \check{\Gamma}(T)$ is called the *tensor product* of $\mathscr{S}$ and $\mathscr{S}'$ (over $\mathscr{R}$). It always follows that

$$(\mathscr{S} \otimes_{\mathscr{R}} \mathscr{S}')(U) = \mathscr{S}(U) \otimes_{\mathscr{R}(U)} \mathscr{S}'(U)$$

and

$$(\mathscr{S} \otimes_{\mathscr{R}} \mathscr{S}')_x = \mathscr{S}_x \otimes_{\mathscr{R}_x} \mathscr{S}'_x.$$

The tensor product functor is *covariant in both entries, additive, and right exact.* Moreover, *if $\mathscr{S}$ and $\mathscr{S}'$ are coherent* (resp. *locally-free*) *then $\mathscr{S} \otimes_{\mathscr{R}} \mathscr{S}'$ is coherent* (resp. *locally-free*).

One defines the *p*-fold tensor product, $\otimes^p$, $p = 0, 1, 2, \ldots$, inductively by

$$\otimes^p \mathscr{S} := (\otimes^{p-1} \mathscr{S}) \otimes_{\mathscr{R}} \mathscr{S}$$

with $\otimes^0 \mathscr{S} := \mathscr{R}$.

In $(\otimes^p \mathscr{S})_x$ we consider the $\mathscr{R}_x$-submodule $\mathscr{M}_x$ which is generated by $a_1 \otimes \cdots \otimes a_p$ where $a_\mu = a_\nu$ for some pair (μ, ν) with $\mu \neq \nu$.

Then $\mathscr{M} := \bigcup_{x \in X} \mathscr{M}_x$ is an $\mathscr{R}$-subsheaf of $\otimes^p \mathscr{S}$. The quotient sheaf

$$\wedge^p \mathscr{S} := (\otimes^p \mathscr{S})/\mathscr{M}, \quad p = 0, 1, 2, \ldots,$$

is called the *p-fold exterior product of $\mathscr{S}$*. Note that $\wedge^0 \mathscr{S} = \mathscr{R}$ and $\wedge^1 \mathscr{S} = \mathscr{S}$. If $\varepsilon\colon \otimes^p \mathscr{S} \to \wedge^p \mathscr{S}$ is the quotient homomorphism, then $\varepsilon(a_1 \otimes \cdots \otimes a_p) =: a_1 \wedge \cdots \wedge a_p$. It follows that

$\wedge^p$ is a covariant functor and, if $\mathscr{S}$ is coherent (resp. locally-free), *then $\wedge^p \mathscr{S}$ is likewise coherent* (resp. locally-free).

6. The Functor $\mathscr{H}om$ and Annihilator Sheaves. For $\mathscr{R}$-sheaves $\mathscr{S}$ and $\mathscr{S}'$, the set $H(U) := \mathscr{H}om_{\mathscr{R}|U}(\mathscr{S}|U, \mathscr{S}'|U)$ of all $\mathscr{R}|U$-homomorphism $\mathscr{S}|U \to \mathscr{S}'|U$ is always an $\mathscr{R}(U)$-module. The restrictions $r_V^U\colon H(U) \to H(V)$ are canonically at hand and the resulting system $H := \{H(U), r_V^U\}$ is a $\Gamma(\mathscr{R})$-presheaf which satisfies S1 and S2. The associated $\mathscr{R}$-sheaf $\mathscr{H}om_{\mathscr{R}}(\mathscr{S}, \mathscr{S}') := \check{\Gamma}(H)$ is called the *sheaf of germs of $\mathscr{R}$-homomorphisms from $\mathscr{S}$ to $\mathscr{S}'$*. It is always the case that

$$\mathscr{H}om_{\mathscr{R}}(\mathscr{S}, \mathscr{S}')(U) = \mathscr{H}om_{\mathscr{R}|U}(\mathscr{S}|U, \mathscr{S}'|U).$$

The functor $\mathscr{H}om_{\mathscr{R}}$ is contravariant in the first argument and covariant in the second argument. Additionally, $\mathscr{H}om_{\mathscr{R}}$ is left exact.

Remark: For all $x \in X$ there exists a natural $\mathscr{R}_x$-homomorphism $\rho_x\colon \mathscr{H}om_{\mathscr{R}}(\mathscr{S}, \mathscr{S}')_x \to \mathscr{H}om_{\mathscr{R}_x}(\mathscr{S}_x, \mathscr{S}'_x)$ which is in general certainly neither injective nor surjective. The reader should note that the $\mathscr{R}(U)$-module $\mathscr{H}om_{\mathscr{R}(U)}(\mathscr{S}(U), \mathscr{S}'(U))$ cannot be used for the definition because, among other things, the restrictions r_V^U do not exist.

As in the case of the tensor functors, the $\mathscr{H}om$-functor is also coherent:

If $\mathscr{S}$, $\mathscr{S}'$ are coherent (resp. locally-free) then $\mathscr{H}om_{\mathscr{R}}(\mathscr{S}, \mathscr{S}')$ is also coherent (resp. locally-free).

For an $\mathscr{R}$-sheaf $\mathscr{S}$, we define $\mathscr{A}n\,\mathscr{S}_x := \{r_x \in \mathscr{R}_x \mid r_x \cdot \mathscr{S}_x = 0\}$ and $\mathscr{A}n\,\mathscr{S} :=$

$\bigcup_{x \in X} \mathscr{A}n\, \mathscr{S}_x$. Then $\mathscr{A}n\, \mathscr{S}$ is open in $\mathscr{S}$ and is consequently a sheaf of $\mathscr{R}$-ideals. One calls $\mathscr{A}n\, \mathscr{S}$ the *annihilator of* $\mathscr{S}$.

If $\mathscr{R}$ is coherent then the annihilator of every coherent $\mathscr{R}$-sheaf is a coherent sheaf of ideals.

7. Sheaves of Quotients. A set $\mathscr{M} \subset \mathscr{R}$ is called *multiplicative* if $\mathscr{M}$ is an open subset of $\mathscr{R}$ and $\mathscr{M}_x$ is multiplicative in the ring $\mathscr{R}_x$ for all $x \in X$).[1] If $\mathscr{M}$ is such a set, then, for every non-empty open set U, $\mathscr{M}(U) := \{r \in \mathscr{R}(U) \mid r_x \in \mathscr{M}_x$ for all $x \in U\}$ is multiplicative in the ring $\mathscr{R}(U)$. The ring of quotients $\mathscr{R}(U)_{\mathscr{M}(U)}$ is an $\mathscr{R}(U)$-module and, since one has the natural restrictions $r_V^U\colon \mathscr{R}(U)_{\mathscr{M}(U)} \to \mathscr{R}(V)_{\mathscr{M}(V)}$, the system $\{\mathscr{R}(U)_{\mathscr{M}(U)}, r_V^U\}$ is a presheaf of rings. The associated sheaf of rings is denoted by $\mathscr{R}_{\mathscr{M}}$ and is called the *sheaf of quotients of $\mathscr{R}$ with respect to $\mathscr{M}$*. It is clear that $\mathscr{R}_{\mathscr{M}}$ is also an $\mathscr{R}$-module.

Since $\mathscr{M}$ is open in $\mathscr{R}$, every stalk $(\mathscr{R}_{\mathscr{M}})_x$, $x \in X$, is canonically isomorphic to the ring $(\mathscr{R}_x)_{\mathscr{M}_x}$. In particular, if $\mathscr{R}$ is a sheaf of local k-algebras, then $\mathscr{R}_{\mathscr{M}}$ is likewise a sheaf of local k-algebras. One frequently identifies $\mathscr{R}_{\mathscr{M}}$ with $\bigcup_{x \in X} (\mathscr{R}_x)_{\mathscr{M}_x}$, defining a basis for the topology as follows: For every open set $U \subset X$ and every pair $f \in \mathscr{R}(U)$, $g \in \mathscr{M}(U)$, let $[f, g, U]$ be the set of all germs $f_x/g_x \in (\mathscr{R}_x)_{\mathscr{M}_x}$, $x \in U$. The collection of all such $[f, g, U]$ is a basis of open sets.

For the theory of *meromorphic functions* (see Chapter V.3.1), the following set, no element of which is a zero-divisor, is an important multiplicative set:

$$\mathscr{N} := \bigcup_{x \in X} \mathscr{N}_x, \quad \mathscr{N}_x := \text{the elements of } \mathscr{R}_x \text{ which are not zero-divisors.}$$

Every $\mathscr{N}_x$ is multiplicative in $\mathscr{R}_x$. Furthermore,

if $\mathscr{R}$ is coherent, then $\mathscr{N}$ is open and consequently is multiplicative in $\mathscr{R}$.

Proof: Let $\mathfrak{n} \in \mathscr{R}(U)$ be such that $\mathfrak{n}_x \in \mathscr{N}_x$ for some $x \in U$. Define the $\mathscr{R}_U$-homomorphism $\rho\colon \mathscr{R}_U \to \mathscr{R}_U$ by $r_y \to \mathfrak{n}_y r_y$, for $y \in U$. Then $(\mathscr{K}er\, \rho)_x = 0$. Since $\mathscr{K}er\, \rho$ is coherent, there exists a neighborhood V of x in U with $\mathscr{K}er\, \rho_V = 0$. This means that $\mathfrak{n}_y \in \mathscr{N}_y$ for all $y \in V$ (i.e. $\mathfrak{n}_V \in \mathscr{N}$). □

§ 3. Complex Spaces

In this section, X, Y, Z always denote Hausdorff spaces equipped with sheaves $\mathscr{R}_X$, $\mathscr{S}_Y$, ... When there is no confusion we drop the indices which indicate the space. We reserve k for a field.

[1] A subset T of a commutative ring B with unit 1 is called *multiplicative* if $1 \in T$ and, whenever a, $b \in T$, $ab \in T$. Every multiplicative set $T \subset B$ determines the *ring of quotients* B_T whose elements are the equivalence classes of the following equivalence relation on $B \times T$: Two pairs $(b_i, t_i) \in B \times T$ are called equivalent if there exists a $t \in T$ so that $t(t_2 b_1 - t_1 b_2) = 0$. One writes each equivalence class as a "fraction" b/t and carries out the arithmetic operations in the usual way.

1. k-algebraized Spaces. A space X together with a sheaf $\mathscr{R}$ of *local* k-algebras is called a k-algebraized space. Thus, for example, the pair $(X, \mathscr{C})$, where $\mathscr{C}$ is the sheaf of germs of complex valued continuous functions on X, is a $\mathbb{C}$-algebraized space.

If $(X, \mathscr{R})$ is a k-algebraized space and $f: X \to Y$ is a continuous map, then the image sheaf $f_*(\mathscr{R})$ is a sheaf of k-algebras. We emphasize that this is not in general a sheaf of local k-algebras. For every $x \in X$ we have (see 0.8) a canonical map $\hat{f}_*: f_*(\mathscr{R})_{f(x)} \to \mathscr{R}_x$. It is always a k-algebra homomorphism.

A k-morphism $(f, \tilde{f}): (X, \mathscr{R}_X) \to (X, \mathscr{R}_Y)$ of k-algebraized spaces is a continuous map $f: X \to Y$ together with a k-algebra homomorphism $\tilde{f}: \mathscr{R}_Y \to f_*(\mathscr{R}_X)$ so that every map which arises from the composition

$$\mathscr{R}_{Y,y} \xrightarrow{\tilde{f}_y} f_*(\mathscr{R}_X)_y \xrightarrow{\hat{f}_x} \mathscr{R}_{X,x}$$

where x is arbitrary in X and $y := f(x)$, is *local*. In other words, $\mathfrak{m}(\mathscr{R}_{Y,y})$ is mapped into $\mathfrak{m}(\mathscr{R}_{X,x})$.

The above formulation of a k-morphism is due to Grothendieck. An equivalent definition which uses neither image nor preimage sheaves is possible. However it is less elegant.

Example: Every *continuous* map $f: X \to Y$ between topological spaces determines a $\mathbb{C}$-morphism $(X, \mathscr{C}_X) \to (Y, \mathscr{C}_Y)$. The map $\tilde{f}: \mathscr{C}_Y \to f_*(\mathscr{C}_X)$ is obtained by lifting continuous functions: $\tilde{f}_V: \mathscr{C}_Y(V) \to f_*(\mathscr{C}_X)(V) = \mathscr{C}_X(f^{-1}(V))$ by $g \to g \circ f$.

If $(f, \tilde{f}): (X, \mathscr{R}_X) \to (Y, \mathscr{R}_Y)$ is a k-morphism and $\mathscr{S}_X$ is an $\mathscr{R}_X$-module, then $f_*(\mathscr{S}_X)$ is an $f_*(\mathscr{R}_X)$-sheaf and consequently, via $\tilde{f}$, an $\mathscr{R}_Y$-module. Every $\mathscr{R}_X$-homomorphism $\varphi: \mathscr{S}_X \to \mathscr{S}'_X$ uniquely determines an $\mathscr{R}_Y$-homomorphism $f_*(\varphi): f_*(\mathscr{S}_X) \to f_*(\mathscr{S}'_X)$. Thus f_* *is a covariant functor from the category of* $\mathscr{R}_X$*-modules into the category of* $\mathscr{R}_Y$*-modules.*

The identity k-morphism $(X, \mathscr{R}_X) \to (X, \mathscr{R}_X)$ is given by the identity maps $\mathrm{id}: X \to X$ and $\mathrm{id}: \mathscr{R}_X \to \mathrm{id}_*(\mathscr{R}_X) = \mathscr{R}_X$. If $(f, \tilde{f}): (X, \mathscr{R}_X) \to (Y, \mathscr{R}_Y)$ and $(g, \tilde{g}): (Y, \mathscr{R}_Y) \to (Z, \mathscr{R}_Z)$ are k-morphisms, then $(h, \tilde{h}): (X, \mathscr{R}_X) \to (Z, \mathscr{R}_Z)$, where $h := g \circ f$ and $h := g_*(\tilde{f}) \circ \tilde{g}: R_Z \to g_*(f_*(R_X))$, is also a k-morphism. Thus

the k-algebraized spaces with k-morphisms form a category.

A k-morphism $(f, \tilde{f}): (X, \mathscr{R}_X) \to (Y, \mathscr{R}_Y)$ is an isomorphism if and only if f is a homeomorphism and $\tilde{f}$ is a sheaf isomorphism.

Remark: If $(f, \tilde{f})$ is a k-morphism, then the map $\tilde{f}$ is strongly bound to the underlying continuous map f. For example,

if the sheaves $\mathscr{R}_X$, $\mathscr{R}_Y$ are reduced, then for a given continuous map $f: X \to Y$, there exists at most one map $\tilde{f}: \mathscr{R}_Y \to f_*(\mathscr{R}_X)$ so that $(f, \tilde{f})$ is a k-morphism.

2. Differentiable and Complex Manifolds. Let U be a non-empty open subset in $\mathbb{R}^m$, the space of n-tuples of real numbers. Then one has the $\mathbb{R}$-algebra $\mathscr{E}^{\mathbb{R}}(U)$ and the $\mathbb{C}$-algebra $\mathscr{E}^{\mathbb{C}}(U)$ of $\mathbb{R}$-valued and $\mathbb{C}$-valued infinitely often differentiable functions. Obviously $\mathscr{E}^{\mathbb{C}}(U) = \mathscr{E}^{\mathbb{R}}(U) + i\mathscr{E}^{\mathbb{R}}(U)$. In both cases one has the natural restrictions r_V^U for $V \subset U$. The system $\{\mathscr{E}^k(U), r_V^U\}$, where k stands for either $\mathbb{R}$ or $\mathbb{C}$, is a presheaf of k-algebras which satisfies $S1$ and $S2$. The associated sheaves on $\mathbb{R}^m$, $\mathscr{E}^{\mathbb{R}}$ and $\mathscr{E}^{\mathbb{C}}$, are sheaves of local rings over $\mathbb{R}$ and $\mathbb{C}$ respectively. Thus $(\mathbb{R}^m, \mathscr{E}^{\mathbb{R}})$ is an $\mathbb{R}$-algebraized space and $(\mathbb{R}^m, \mathscr{E}^{\mathbb{C}})$ is a $\mathbb{C}$-algebraized space.

An $\mathbb{R}$-algebraized space $(X, \mathscr{E}_X^{\mathbb{R}})$ is called *differentiable manifold* if, for every $x \in X$, there exists a neighborhood U of x in X and a domain D in $\mathbb{R}^m$ so that the $\mathbb{R}$-algebraized spaces $(U, \mathscr{E}_U^{\mathbb{R}})$ and $(D, \mathscr{E}_D^{\mathbb{R}})$ are isomorphic. The sheaf $\mathscr{E}^{\mathbb{R}}$ is called the *sheaf of germs of real-valued differentiable functions on* X. It uniquely determines the sheaf $\mathscr{E}_X^{\mathbb{C}} = \mathscr{E}_X^{\mathbb{R}} + i\mathscr{E}_X^{\mathbb{R}}$ *of complex-valued differentiable* functions on X. Obviously $\mathscr{E}_X^{\mathbb{C}}$ is a subsheaf of local $\mathbb{C}$-algebras of the sheaf $\mathscr{C}_X$. Further one has a natural conjugation isomorphism (linear over $\mathbb{R}$), $^-: \mathscr{E}_X^{\mathbb{C}} \to \mathscr{E}_X^{\mathbb{C}}$, which fixes $\mathscr{E}_X^{\mathbb{R}}$ element-wise.

The morphisms between differentiable manifolds are the *differentiable maps* (i.e. arbitrary continuous maps which lift differentiable functions to differentiable functions).

In the space $\mathbb{C}^m$ of m-tuples $(z_1, \ldots, z_m)$ of complex numbers, the $\mathbb{C}$-algebra $\mathcal{O}(U)$ of holomorphic functions is defined for every non-empty open set U. Again one has the natural restrictions r_V^U so that the system $\{\mathcal{O}(U), r_V^U\}$ is a presheaf which satisfies $S1$ and $S2$. The associated sheaf of local $\mathbb{C}$-algebras is denoted by $\mathcal{O}$. Thus $(\mathbb{C}^m, \mathcal{O})$ is a $\mathbb{C}$-algebraized space. If one identifies $\mathbb{C}^m$ with the space $\mathbb{R}^{2m}$ of real $2m$-tuples $(x_1, y_1, \ldots, x_m, y_m)$, $x_\mu := \mathrm{Re}\, z_\mu$, $y_\mu := \mathrm{Im}\, z_\mu$, then $\mathcal{O}$ is a $\mathbb{C}$-subsheaf of $\mathscr{E}_{\mathbb{C}^m}^{\mathbb{C}} := \mathscr{E}_{\mathbb{R}^{2m}}^{\mathbb{C}}$.

A $\mathbb{C}$-algebraized space $(X, \mathcal{O}_X)$ is called a *complex manifold* if, for every $x \in X$ there exists a neighborhood U of x in X and a domain D in $\mathbb{C}^m$ so that the $\mathbb{C}$-algebraized spaces $(U, \mathcal{O}_U)$ and $(D, \mathcal{O}_D)$ are isomorphic. The sheaf $\mathcal{O}_X$ is called the *sheaf of germs of holomorphic functions on* X (the structure sheaf of the complex manifold). Since $\mathcal{O}_{\mathbb{C}^m} \subset \mathscr{E}_{\mathbb{C}^m}^{\mathbb{C}}$, it is clear that every complex manifold $(X, \mathcal{O})$ determines a differentiable manifold $(X, \mathscr{E}^{\mathbb{R}})$ with $\mathcal{O} \subset \mathscr{E}^{\mathbb{C}} \subset \mathscr{C}$.

A fundamental theorem of K. Oka [see CAS] states that,

for every complex manifold $(X, \mathcal{O})$, *the structure sheaf* $\mathcal{O}$ *is coherent.*

Morphisms between complex manifolds are called *holomorphic* maps (i.e. arbitrary continuous maps which lift holomorphic functions to holomorphic functions). In local coordinates holomorphic maps are represented by systems of holomorphic functions.

3. Complex Spaces and Holomorphic Maps. If $D \subset \mathbb{C}^m$ is a domain and $\mathscr{J} \subset \mathcal{O}_D$ is a coherent ideal, then the support A of the coherent $\mathcal{O}_D$-sheaf $\mathcal{O}_D/\mathscr{J}$ is a closed set in D. The sheaf $\mathcal{O}_A := \mathcal{O}_D/\mathscr{J} \mid A$ is a sheaf of local $\mathbb{C}$-algebras on A which, by the results in Section 2.4 is coherent. The $\mathbb{C}$-algebraized space $(A, \mathcal{O}_A)$ is called a *closed complex subspace* of D. The injection $\imath: A \to D$ determines a holomorphic embedding $(\imath, \tilde{\imath})$: $(A, \mathcal{O}_A) \to (D, \mathcal{O}_D)$ where $\tilde{\imath}: \mathcal{O}_D \to i_*(\mathcal{O}_A) \cong \mathcal{O}_D/\mathscr{J}$ is the quotient map.

A *complex space* $X = (X, \mathcal{O}_X)$ is a $\mathbb{C}$-algebraized space in which every point has a neighborhood U so that $(U, \mathcal{O}_U)$ is isomorphic to a complex subspace $(A, \mathcal{O}_A)$ of a domain in some $\mathbb{C}^n$. *The structure sheaf* $\mathcal{O}_X$ *of a complex space* X *is coherent.* Sheaves of $\mathcal{O}_X$-modules are called analytic sheaves.

Complex spaces with *holomorphic mappings* as morphisms form a category. We let $\mathrm{Hol}(X, Y)$ denote the set of all holomorphic maps $(f, \tilde{f})\colon (X, \mathcal{O}_X) \to (Y, \mathcal{O}_Y)$. We often write $f\colon X \to Y$ for short.

If X is a complex space and X' is a non-empty open subset of X, then $(X', \mathcal{O}_{X'})$ is likewise a complex space. One has a natural holomorphic map $(i, \tilde{i})\colon (X', \mathcal{O}_{X'}) \to (X, \mathcal{O}_X)$, where $i := \mathrm{id}\,|\,X'$ and $\tilde{i}$ is the canonical map from $\mathcal{O}_X$ to $i_*(\mathcal{O}_{X'})$, the latter being the trivial extension of $\mathcal{O}_{X'}$ to X. One calls X' together with i an *open complex subspace of* X.

Every section $s \in \mathcal{O}(X)$ in the structure sheaf is called a *holomorphic function* on X. The equivalence class of the germ $s_x \in \mathcal{O}_x$ in $\mathcal{O}_x/\mathfrak{m}_x = \mathbb{C}$ is called the (*complex*) *value* $s(x) \in \mathbb{C}$ of s at $x \in X$. Since $\mathcal{O} = \mathcal{O}_X$ can contain non-zero nilpotent elements, a holomorphic function s on X is in general more than this $\mathbb{C}$-valued function $x \to s(x)$.

If $(f, \tilde{f})\colon X \to \mathbb{C}$ is a holomorphic mapping and $z \in \mathcal{O}(\mathbb{C})$ is the identity function, then we obtain a section by lifting z as follows:

$$s := \tilde{f}(z) \in f_*(\mathcal{O}_X)(\mathbb{C}) = \mathcal{O}_X(X); \qquad s(x) = f(x).$$

It is routine to show that

the association $\mathrm{Hol}(X, \mathbb{C}) \to \mathcal{O}(X)$, $(f, \tilde{f}) \to s$, *is bijective.*

Therefore one identifies the holomorphic functions on X with the associated holomorphic maps from X to $\mathbb{C}$.

If $(X, \mathcal{O}_X)$ is a complex space and $\mathcal{I} \subset \mathcal{O}_X$ is a coherent ideal then $(Y, \mathcal{O}_Y)$, with $Y := \mathrm{supp}(\mathcal{O}_X/\mathcal{I})$ and $\mathcal{O}_Y := (\mathcal{O}_X/\mathcal{I})\,|\,Y$, is again a complex space. The injection $\iota;\ Y \to X$ together with the quotient homomorphisms $\tilde{\iota}\colon \mathcal{O}_X \to \iota_*(\mathcal{O}_Y) \cong \mathcal{O}_X/\mathcal{I}$ is a "holomorphic embedding" $(Y, \mathcal{O}_Y) \to (X, \mathcal{O}_X)$. One calls $(Y, \mathcal{O}_Y)$ *the closed complex subspace of* $(X, \mathcal{O}_X)$ *associated to* $\mathcal{I}$. □

There exists a *product* in the category of complex spaces:

For every two complex spaces X_1, X_2 there exists a complex space X and maps $\pi_i\colon X \to X_i$, $i = 1, 2$, so that if Z is any complex space, the map

$$\mathrm{Hol}(Z, X) \to \mathrm{Hol}(Z, X_1) \times \mathrm{Hol}(Z, X_2), \qquad f \to (\pi_1 f, \pi_2 \circ f),$$

is bijective. The space X and the maps π_i, $i = 1, 2$, are uniquely determined up to isomorphism.

Topologically speaking, the above product is just the Cartesian product $X_1 \times X_2$. Thus we use the suggestive notation $X = X_1 \times X_2$. □

Holomorphic maps factor through their graphs:

Let $f\colon X \to Y$ be a holomorphic map. Then there exists a canonical closed complex subspace $\mathrm{Gph}(f)$ *of $X \times Y$, the "graph" of f, and a biholomorphic map $\iota\colon X \to$* $\mathrm{Gph}(f)$ *so that $f = \pi \circ \iota$, where π is the restriction of the projection $X \times Y \to Y$ to* $\mathrm{Gph}(f)$.

4. Topological Properties of Complex Spaces. Every complex space is *locally compact*. From here on we will only consider complex spaces and differentiable manifolds which have *countable bases of open sets for their topologies*. Such space are *metrizable*.

At this point we remind the reader of some of the basic notions and results of general topology which are used, for example, in the cohomology theory. A covering $\mathfrak{V} = \{V_j\}_{j \in J}$ (the elements of a covering are always *open* sets) of a topological space X is called a refinement of a cover $\mathfrak{U} = \{U_i\}_{i \in I}$ of X (symbolized by $\mathfrak{V} < \mathfrak{U}$), if every V_j is contained in some U_i. Thus there exists a map $\tau\colon J \to I$ of the index sets so that $V_j \subset U_{\tau(j)}$ for all $j \in J$. We call τ a *refinement map*.

A Hausdorff space is called *paracompact* if for every cover $\mathfrak{U}$ of X there exists a refinement $\mathfrak{V}$ of $\mathfrak{U}$ which is *locally finite* (i.e. every point $x \in X$ has a neighborhood which intersects only finitely many of the sets $V \in \mathfrak{V}$).

Every paracompact space X is *normal*. Every closed subspace of a paracompact space is paracompact. Every *metrizable* space, in particular a complex space or a differentiable manifold, is paracompact. A space which is *countable at infinity* (i.e. a space which is a countable union of compact subsets) is paracompact whenever it is *locally compact*.

The following has numerous applications:

Shrinking Theorem. *For every locally finite covering $\{U_i\}_{i \in I}$ of a normal space X, there exists a covering $\{V_i\}_{i \in I}$ of X (with the same index set) so that $\bar{V}_i \subset U_i$ for all $i \in I$.*

5. Analytic Sets. If $(X, \mathcal{O}_X)$ is a complex space and $\mathcal{I} \subset \mathcal{O}_X$ is a coherent ideal, then $A := \operatorname{supp}(\mathcal{O}_X/\mathcal{I})$ is an *analytic set in X*. One also refers to A as the *zero set of the ideal $\mathcal{I}$*. Analytic sets are therefore just the supports of closed complex subspaces of $(X, \mathcal{O}_X)$. The support of every coherent $\mathcal{O}_X$-sheaf $\mathcal{S}$ is an analytic set, because $\operatorname{supp} \mathcal{S} = \operatorname{supp}(\mathcal{O}_X \mid A \cap S)$.

A given analytic set is the zero set of many different coherent ideals (e.g. $\operatorname{supp}(\mathcal{O}_X/\mathcal{I}) = \operatorname{supp}(\mathcal{O}_X/\mathcal{I}^n)$ for $n \geq 1$). However there exists a *largest* coherent ideal. This and more is the content of the following famous theorem of H. Cartan and K. Oka.

Coherence Theorem for Ideal Sheaves. *Let A be a closed subset of a complex space $(X, \mathcal{O}_X)$ such that for every point $a \in A$ there is a neighborhood V of a in X and holomorphic functions $f_1, \ldots, f_q \in \mathcal{O}_X(V)$ so that*

$$A \cap V = \{x \in V \mid f_1(x) = \cdots = f_q(x) = 0\}.$$

Then A is an analytic set in X. More precisely, the system $\{\mathcal{I}(U)\}$, where U is open in X and $\mathcal{I}(U) := \{f \in \mathcal{O}_X(U) \mid f(A \cap U) = 0\}$, forms a presheaf for the coherent ideal sheaf $\mathcal{I}$. It follows that $A = \operatorname{supp}(\mathcal{O}_X/\mathcal{I})$, and the pair $(A, \mathcal{O}_A)$ with $\mathcal{O}_A := (\mathcal{O}_X/\mathcal{I}) \mid A$ is a closed reduced complex subspace of $(X, \mathcal{O}_A)$.

The sheaf $\mathcal{I}$ is called the *nullstellen ideal* of A. For every $\mathcal{O}_X$-ideal $\mathcal{I}$, one defines the radical ideal rad $\mathscr{I}$ stalkwise by

$$(\mathrm{rad}\ \mathscr{I})_x := \{f_x \in \mathcal{O}_{X,x} \mid f_x^n \in \mathscr{I}_x \text{ for some } n \in \mathbb{N}\}.$$

It follows that rad $\mathscr{I}$ is likewise an $\mathcal{O}_X$-ideal. We now formulate the so-called Nullstellensatz of Hilbert and Rückert. The proof of this as well as that for the coherence theorem can be found in [CAS].

Nullstellensatz: *For every coherent ideal $\mathscr{I} \subset \mathcal{O}_X$, the radical ideal* rad $\mathscr{I}$ *is the nullstellen ideal of* $A := \mathrm{supp}(\mathcal{O}_X/\mathscr{I})$.

The coherence theorem for ideal sheaves therefore implies that *if $\mathscr{I}$ is a coherent $\mathcal{O}_X$-ideal then* rad $\mathscr{I}$ *is also coherent.*

It is well-known that every topological space can be decomposed into its connected components. A much stronger *decomposability lemma* can be proved for analytic sets. A non-empty analytic set A in a complex space X is said to be *reducible* in X if there exist non-empty analytic sets B, C in X with $A \neq B$ and $A \neq C$ so that $A = B \cup C$. If this cannot be done, then A is called *irreducible.*

Decomposition Lemma. *Every non-empty analytic set A in X has a decomposition $A = \bigcup_{i \in I} A_i$ with the following properties:*

0) *The index set I is at most countable.*
1) *For every $i \in I$ the set A_i is an irreducible analytic set in X, and the family $\{A_i\}_{i \in I}$ is locally finite in X.*
2) *For all $i, j \in I$ with $i \neq j$ the intersection $A_i \cap A_j$ is nowhere dense in A_i. The family $\{A_i\}_{i \in I}$ is uniquely determined by A up to a change of indexing. One calls the sets A_i the irreducible components (branches) of A in X.*

If U is a non-empty open subset of X and A is an analytic set in X, then $A \cap U$ is an analytic set in U. If A_i is an irreducible component of A in X, then $A_i \cap U$ is not in general irreducible in U. The branches of $A \cap U$ are the branches of the non-empty intersection $A_i \cap U$ for all irreducible components A_i of A in X.

Identity Theorem for Analytic Sets. *Let A, A' be analytic sets in X, and suppose that U is an open set in X so that $A \cap U$ and $A' \cap U$ have a common branch as analytic subsets of U. Then A and A' have a common branch in X.*

6. Dimension Theory. Every complex space X and more generally every analytic set A in X has at every point $x \in A$ a well-defined *topological dimension,* dim $\mathrm{top}_x\, A \in \mathbb{N}$. This number is always *even*, and one defines the *complex dimension* of A *at* x by

$$\dim_x A := \tfrac{1}{2} \dim \mathrm{top}_x A$$

The number

$$\mathrm{codim}_x A := \dim_x X - \dim_x A$$

is called the *complex codimension at x of A in X.*

Every stalk $\mathcal{O}_x$, $x \in X$, is a local $\mathbb{C}$-algebra and thus has an *algebraic dimension* $\dim \mathcal{O}_x$ (see [AS], Chapter II.4). A non-trivial theorem says that

$$\dim_x X = \dim \mathcal{O}_x \quad \text{for all } x \in X.$$

The following are standard (useful!) results from the dimension theory of analytic sets:

a) *The analytic set A is nowhere dense in X if and only if it is at least 1-codimensional.*
b) *A point $p \in A$ is an isolated point in A if and only if* $\dim_p A = 0$.
c) *If A is irreducible, then A is pure dimensional (i.e. the function* $\dim_x A$, $x \in A$, *is constant).*

The number $\dim A := \sup\{\dim_x A \mid x \in A\}$ is called the complex dimension of A. We note that the case $\dim A = \infty$ is possible.

d) *If $\{A_i\}_{i \in I}$ is the set of irreducible components of A, then* $\dim A = \sup\{\dim A_i \mid i \in I\}$.
If $\dim A < \infty$, *then there exists $j \in I$ with* $\dim A = \dim A_j$.
e) *If B is likewise analytic in X and $B \subset A$, then* $\dim B \le \dim A$. *If* $\dim B = \dim A < \infty$, *then A and B have a common branch.*

7. Reduction of Complex Spaces. The following is a fundamental theorem of H. Cartan and K. Oka [see CAS]:

The nilradical $\mathfrak{n}(\mathcal{O}_X)$ of a complex space $X = (X, \mathcal{O}_X)$ is always a coherent ideal.

Since $\operatorname{supp}(\mathcal{O}_X/\mathfrak{n}(\mathcal{O}_X) = X$, it follows that

$$\operatorname{red} X := (X, \mathcal{O}_{\operatorname{red} X}), \quad \text{with} \quad \mathcal{O}_{\operatorname{red} X} := \mathcal{O}_X/\mathfrak{n}(\mathcal{O}_X),$$

is a complex subspace of X. It is called the *reduction* of X. The associated holomorphic mapping $\operatorname{red} X \to X$ is likewise denoted by "red" and is called the *reduction mapping*.

The structure sheaf of red X is in a natural way a subsheaf of the sheaf $\mathscr{C}_X$ of germs of continuous, complex valued functions (i.e. $\mathcal{O}_{\operatorname{red} X} \subset \mathscr{C}_X$). For all $s \in \mathcal{O}(X)$, $s(x) = (\operatorname{red} s)(x)$ for all $x \in X$.

Every holomorphic map $f\colon X \to Y$ between complex spaces canonically determines a map $\operatorname{red} f\colon \operatorname{red} X \to \operatorname{red} Y$ of the reductions such that the following is commutative:

$$\begin{array}{ccc} \operatorname{red} X & \xrightarrow{\operatorname{red} f} & \operatorname{red} Y \\ \downarrow{\scriptstyle \operatorname{red}} & & \downarrow{\scriptstyle \operatorname{red}} \\ X & \xrightarrow{\ f\ } & Y \end{array}$$

One easily verifies that red is a covariant functor.

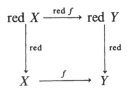

A complex space X is said to be reduced at the point $x \in X$ if the stalk $\mathcal{O}_x$ is reduced (i.e. $\mathfrak{n}(\mathcal{O}_x) = 0$). The set of non-reduced points of X is the support of $\mathfrak{n}(\mathcal{O}_x)$, and is therefore an *analytic* set in X. The space X is called reduced whenever it is reduced at all of its points (i.e. whenever $X = \text{red } X$). The space red X is reduced, and thus red (red X) = red X.

A point x in a complex space X is called *regular* or *non-singular* if the stalk $\mathcal{O}_x$ is regular (i.e. isomorphic to a $\mathbb{C}$-algebra of convergent power series). Every regular point is reduced. The non-regular or *singular* points form an analytic set in X, S, the so-called *singularity set* of X. The set $X \setminus S$ is a (possibly emp) complex manifold. If X is reduced, then S is *nowhere dense* in X and is in particular everywhere at least 1-codimensional.

8. Normal Complex Spaces. For every complex space $(X, \mathcal{O})$ the set $N \subset \mathcal{O}$ of elements which do not divide zero is *multiplicative* (see Section 2.7). Thus the sheaf of quotients

$$\mathcal{M} := \mathcal{O}_N \quad \text{with} \quad \mathcal{M}_x = (\mathcal{O}_x)_{N_x}, \qquad x \in X,$$

is well-defined and is an $\mathcal{O}$-module. One calls $\mathcal{M}$ the *sheaf of germs* of *meromorphic functions* on X. The sections in $\mathcal{M}$ (over X) are the *meromorphic functions* on X. The reader should note that $\mathcal{M}$ is *not* a coherent $\mathcal{O}$-sheaf.

If X is reduced at x, then a germ $f_x \in \mathcal{O}_x$ is in N_x if and only if there is a neighborhood U of x and a representative $f \in \mathcal{O}(U)$ of f_x so that the zero set of f is nowhere dense in U. In this case it follows in particular that $f^{-1} \in \mathcal{M}(U)$.

If X is *irreducible* at x (i.e. $\mathcal{O}_x$ is an integral domain), then $\mathcal{M}_x$ is the quotient field of $\mathcal{O}_x$.

A complex space X is said to be *normal at* $x \in X$ whenever X is reduced at x and the ring $\mathcal{O}_x$ is *integrally closed* in $\mathcal{M}_x$. Every regular point is normal. The following is a famous theorem of Oka (see [CAS]):

The set of non-normal points of a complex space X is an analytic set in X.

A complex space X is called *normal* if it is normal at each of its points. The singularity set S of a normal complex space is everywhere at least 2-codimensional.

We will use the following Chapter V:

Riemann Continuation Theorem. *Let X be a normal complex space and A an analytic set in X.* Then the following hold:

1) *If A is everywhere at least 1-codimensional, then every function which is continuous on X and holomorphic on $X \setminus A$ is holomorphic on X:*

$$\mathcal{O}(X) = \mathscr{C}(X) \cap \mathcal{O}(X \setminus A)$$

2) *If A is everywhere at least 2-codimensional, then the restriction homomorphism $\mathcal{O}(X) \to \mathcal{O}(X \setminus A)$ is bijective.*

Normalization Theorem. *For every reduced complex space X with singularity set S there is a normal complex space $\tilde{X}$ and a finite*[2] *surjective holomorphic mapping $\xi\colon \tilde{X} \to X$. The manifold $\tilde{X} \setminus \xi^{-1}(S)$ is biholomorphically mapped onto $X \setminus S$, and $\xi^{-1}(S)$ is nowhere dense in $\tilde{X}$.*

The pair $(\tilde{X}, \xi)$ is called a *normalization* of X. Normalizations are uniquely determined up to analytic isomorphisms.

If $(\tilde{X}, \xi)$ is a normalization of X, then the sheaf $\mathcal{O}_X$ is an analytic subsheaf of the coherent direct image $\mathcal{O}_X$-sheaf $\xi_*(\mathcal{O}_{\tilde{X}})$.

The reader can find proofs of these theorems in [CAS].

§ 4. Soft and Flabby Sheaves

Many important theorems in classical analysis have the following form: Sections of a certain sheaf which are only defined on certain subsets can be extended to the entire space. The material in this section is devoted to this and related matters.

1. Soft Sheaves. The section continuation problem is satisfactorily solved "in the small" by the following:

Theorem 1. *Let $\mathcal{S}$ be a sheaf (no algebraic structure implied) on a metrizable space X and let t be a section in $\mathcal{S}$ over a set $Y \subset X$. Then there exists a neighborhood W of Y in X and a section $s \in S(W)$ with $s \mid Y = t$.*

The proof uses methods of general topology (see, for example [TF], p. 150).

A sheaf $\mathcal{S}$ on X is called *soft* if, for every closed set $A \subset X$, the restriction map $\mathcal{S}(X) \to \mathcal{S}(A)$ is surjective (i.e. if every section over A is continuable to a section over the entire X). For sheaves of modules one has a handy softness criterion in the form of a separation condition:

Theorem 2. *Let X be metrizable and $\mathcal{R}$ a sheaf of rings (with identity) over X. Suppose that, for every closed subset A in X and every open neighborhood W of A in X, there exists a section $f \in \mathcal{R}(X)$ such that*

$$f \mid A = 1 \quad \text{and} \quad f \mid X \setminus W = 0.$$

Then every $\mathcal{R}$-module $\mathcal{S}$ is soft.

Proof: Let $A \subset X$ be closed and $t \in \mathcal{S}(A)$. By Theorem 1 there exists an open neighborhood W of A in X and a section $s' \in \mathcal{S}(W)$ with $s' \mid A = t$. By assumption

---

[2] Finite holomorphic maps are studied in detail in Chapter I.

there exists $f \in \mathscr{R}(X)$ with $f \mid A = 1$ and $f \mid X \backslash W = 0$. Define $s \in \mathscr{S}(X)$ by

$$s(x) := f(x)s'(x) \quad \text{for} \quad x \in W \quad \text{and} \quad s(x) := 0 \quad \text{for} \quad x \in X \backslash W.$$

Then s is the desired extension of t. □

2. Softness of the Structure Sheaves of Differential Manifolds. The sheaf $\mathscr{C}$ of real-valued, continuous function germs on a metrizable space satisfies the separation condition of Theorem 2 (such a space is normal!). Thus on such spaces every sheaf of $\mathscr{C}$-modules is soft. In the following it will be shown that the sheaf of real-valued infinitely often differentiable functions on a differentiable manifold also fulfills the separation condition. The proof is based on the following lemma:

Lemma 3. *In $\mathbb{R}^m$ with coordinates $x_1, \ldots, x_m$, let Q and Q' be two open "blocks",*

$$Q := \{x \in \mathbb{R}^m \mid a_\mu < x_\mu < b_\mu\}, \qquad Q' := \{x \in \mathbb{R}^m \mid a'_\mu < x_\mu < b'_\mu\},$$

such that $\bar{Q} \subset Q'$ (i.e. $a'_\mu < a_\mu$ and $b_\mu < b'_\mu$ for all μ). Then there exists a function $r \in \mathscr{E}^{\mathbb{R}}(\mathbb{R}^m)$ with the following properties:

a) $0 < r(x) < 1$ *for* $x \in \mathbb{R}^m$,
b) $r(x) = 1$ *for* $x \in Q$ *and* $r(x) = 0$ *for* $x \in \mathbb{R}^m \backslash Q'$

Proof: We first consider the case $m = 1$. For every two real numbers c, d with $c < d$ we associate the real valued function

$$q_{cd}(x) := \exp\left(\frac{1}{x-d} - \frac{1}{x-c}\right), \quad \text{for } c < x < d,$$

and

$$q_{cd}(x) := 0 \text{ otherwise.}$$

By a well-known theorem in calculus, q_{cd} is infinitely often differentiable on $\mathbb{R}^1$. By definition $q_{cd}(x) \geq 0$ for $x \in \mathbb{R}^1$ and $\int_c^d q_{cd}(x)\,dx > 0$. Thus

$$p_{cd}(x) := \left(\int_c^x q_{cd}(x)\,dx\right) \cdot \left(\int_c^d q_{cd}(x)\,dx\right)^{-1} \in \mathscr{E}^{\mathbb{R}}(\mathbb{R}^1).$$

Furthermore

$$0 \leq p_{cd} \leq 1 \quad \text{for} \quad x \in \mathbb{R},$$
$$p_{cd}(x) = 0 \quad \text{for} \quad x \leq c,$$

and

$$p_{cd}(x) = 1 \quad \text{for} \quad x \geq d.$$

Since $a_1' < a_1$ and $b_1 < b_1'$, the function $r(x) := p_{a_1 a_1'}(x)$ for $x < b_1$ and $r(x) := 1 - p_{b_1 b_1'}(x)$ for $x \geq b_1$ is well defined. Furthermore r has the desired properties.

Now let $m > 1$. Let r_μ be an infinitely often differentiable function of x_μ alone so that a) and b) are fulfilled for the intervals

$$\{x_\mu \in \mathbb{R}^1 \,|\, a_\mu < x_\mu < b_\mu\} \quad \text{and} \quad \{x_\mu \in \mathbb{R}^1 \,|\, a'_\mu < x'_\mu < b'_\mu\}.$$

Then $r(x_1, \ldots, x_m) := r_1(x_1) \cdot \cdots \cdot r_m(x_m)$ is the desired function. □

Theorem 4 (Separation theorem): *Let X be a differentiable manifold with (real) structure sheaf $\mathscr{E} = \mathscr{E}^{\mathbb{R}}$ (of course with countable topology). Let $A \subset X$ be closed and $W \subset X$ an open neighborhood of A. Then there exists a section $f \in \mathscr{E}(X)$ so that*

$$f \,|\, A = 1, \qquad f \,|\, X \backslash W = 0, \quad \text{and} \quad 0 \leq f(x) \leq 1 \quad \text{for all} \quad x \in X.$$

Proof: Let $p \in A$ and choose a neighborhood $W' \subset W$ of p which is diffeomorphic to a bounded domain in $\mathbb{R}^m$. In W' we choose neighborhoods U, U' of p with $U \Subset U' \Subset W'$ such that, with respect to the embedding $W' \hookrightarrow \mathbb{R}^m$, they are mapped onto open blocks Q and Q' respectively. Thus $Q \Subset Q'$ and, by lifting the function constructed in Lemma 3 and extending it trivially to all of X, we obtain a function $g \in \mathscr{E}(X)$ with

$$g \,|\, U = 1, \qquad g \,|\, X \backslash U' = 0, \quad \text{and} \quad 0 \leq g(x) \leq 1 \quad \text{for all} \quad x \in X.$$

Suppose that A is compact. Then there are finitely many points $p_1, \ldots, p_a \in A$ so that the neighborhoods that we just constructed, $U_1, \ldots, U_a$ $(U_1', \ldots, U_a')$, cover the set A. Let g_α be the function associated to U_α and let

$$f := 1 - \prod_{\alpha=1}^{a} (1 - g_\alpha) \in \mathscr{E}(X).$$

Then all the values of f lie between 0 and 1. Furthermore

$$f(x) = 1 \quad \text{for} \quad x \in \bigcup_1^a U_\alpha \quad \text{and} \quad f(x) = 0 \quad \text{for} \quad x \in X \backslash \bigcup_1^m U'_\alpha,$$

where $U_\alpha' \Supset U_\alpha$. Since $A \subset \bigcup_1^a U_\alpha$ and $\bigcup_1^a U_\alpha' \Subset W$, f has the desired properties.

Now let A be an arbitrary closed set in X. We begin with a locally-finite covering $\{U_i\}_{i \in I}$ of X by relatively compact open sets and (via the Shrinking Theorem) we choose a covering $\{V_i\}_{i \in I}$ of X with $\bar{V}_i \subset U_i$, $i \in I$. Then every set $A \cap \bar{V}_i$ is compact in X. From what was proved above, there exists a section $f_i \in \mathscr{E}(X)$ with values between 0 and 1 so that

$$f_i | A \cap \bar{V}_i = 1 \quad \text{and} \quad f_i | X \backslash U_i = 0 \text{ (as sections)} \quad \text{for all} \quad i \in I.$$

Since the covering $\{U_i\}_{i\in I}$ is locally finite, the product

$$g := \prod_{i\in I} (1 - f_i)$$

defines a differentiable function on X. Namely if $x \in X$ is fixed, then there is a neighborhood Z of x and a finite subset T of I so that $U_i \cap Z = \emptyset$ for all $i \in I \setminus T$. Since $f_i | Z = 0$ for $i \in I \setminus T$, $g|Z$ is the *finite* product $\prod_{i\in T} (1 - f_i)$. Obviously $f := 1 - g \in \mathscr{E}(x)$ is the desired section. □

The following is an immediate consequence of Theorems 2 and 4.

Theorem 5. *If X is a differentiable manifold, then every sheaf of $\mathscr{E}^{\mathbb{R}}$-modules over X is soft.*

3. Flabby Sheaves. *A sheaf $\mathscr{S}$ over X is called flabby if, for every open set $U \subset X$, the restriction mapping $\mathscr{S}(X) \to \mathscr{S}(U)$ is surjective. On a metrizable space, every flabby sheaf is soft* (Theorem 1!). The converse is not valid. For example the structure sheaf $\mathscr{E}^{\mathbb{R}}$ on a differentiable manifold is not flabby.

Every sheaf $\mathscr{S}$ determines a flabby sheaf $\mathscr{F}(\mathscr{S})$ as follows: For every open set $U \subset X$, one defines

$$F(U) := \prod_{x\in U} \mathscr{S}_x = \{s\colon U \to \mathscr{S}_U \,|\, \pi s = \mathrm{id}\}.$$

If r_V^U are the obvious restrictions, then the system $\{F(U), r_V^U\}$ is a presheaf which satisfies $\mathscr{S}1$ and $\mathscr{S}2$. The associated sheaf, denoted by $\mathscr{F}(\mathscr{S})$, is obviously *flabby*. Its sections over $U \subset X$ are all "not necessarily continuous sections" in $\mathscr{S}$ over U. One has the canonical injection $j\colon \mathscr{S} \to \mathscr{F}(\mathscr{S})$.

Every sheaf mapping $\varphi\colon \mathscr{S}' \to \mathscr{S}$ determines a sheaf mapping $\mathscr{F}(\varphi)\colon \mathscr{F}(\mathscr{S}') \to \mathscr{F}(\mathscr{S})$ by which $\mathscr{F}$ is a *covariant functor* (Flabby Functor!). If $\mathscr{S}$ is an $\mathscr{R}$-sheaf, then $\mathscr{F}(\mathscr{S})$ is also an $\mathscr{R}$-sheaf and j is an $\mathscr{R}$-monomorphism. It is easy to show that

the functor $\mathscr{F}$ is additive and exact on the category of $\mathscr{R}$-sheaves.[3]

Flabbiness (and also softness) is not destroyed by continuous mappings:

If $f\colon X \to Y$ is continuous and $\mathscr{S}$ is a flabby (soft) sheaf on X, then the image sheaf $f_*(\mathscr{S})$ is flabby (soft) on Y.

The proof is trivial.

4. Exactness of the Functor Γ for Flabby and Soft Sheaves. The section functor Γ is left exact, but not in general exact. Thus the following is quite important.

---

[3] A functor T on a category in which the morphisms $\alpha\colon A \to A'$ and $\beta\colon A \to A'$ can be added is called *additive* if, for all such morphisms with T-images $T\alpha\colon TA \to TA'$, $T\beta\colon TA \to TA'$, it always follows that $T(\alpha + \beta) = T\alpha + T\beta$.

Exactness Lemma. *Let* $0 \to \mathcal{S}' \to \mathcal{S} \to \mathcal{S}'' \to 0$ *be an exact* $\mathcal{R}$*-sequence over* X. *Then, in the following cases, the induced sequence* $0 \to \mathcal{S}'(X) \to \mathcal{S}(X) \to \mathcal{S}''(X) \to 0$ *is exact:*

1) $\mathcal{S}'$ *is flabby.*
2) X *is paracompact and* $\mathcal{S}'$ *is soft.*

For a proof see [TF] p. 148 and 153. We reproduce here the proof of 2). Let $s'' \in \mathcal{S}''(X)$. We can always lift s'' *locally* to a section $\mathcal{S}$. Since X is paracompact, there exists a locally finite (open) cover $\{U_i\}_{i \in I}$ of X and sections $s_i \in \mathcal{S}(U_i)$ which are preimages of $s'' \mid U_i$. By the Shrinking Theorem there exists a cover $\{V_i\}_{i \in I}$ of X with $\bar{V}_i \subset U_i$ for all $i \in I$. The set E of all pairs (J, t), where $J \subset I$ and $t \in \mathcal{S}\left(\bigcup_{j \in J} \bar{V}_j\right)$ is a preimage of the section $s'' \mid \bigcup_{j \in J} V_j$, is non-empty and partially ordered: $(J_1, t_1) \leq (J_2, t_2)$ if and only if $J_1 \subset J_2$ and $t_2 \mid \bigcup_{j \in J_1} \bar{V}_j = t_1$. By Zorn's lemma there exists a maximal element (L, s) in E. Since the family $\{\bar{V}_i\}$ is locally finite, the set $V_L := \bigcup_{j \in L} \bar{V}_j$ is closed in X. Suppose there exists $i \in I \backslash L$. Then the sections $s \in \mathcal{S}(V_L)$ and $s_i \in \mathcal{S}(U_i)$ differ by a section $t' \in \mathcal{S}'(V_L \cap \bar{V}_i)$. Since $\mathcal{S}'$ is soft and $V_L \cap \bar{V}_i$ is closed, we can continue t' to a section $s'_i \in \mathcal{S}'(U_i)$. Thus s and $s_i - s'_i$ are equal on $V_L \cap \bar{V}_i$ and consequently s is continuable to $V_L \cup \bar{V}_i$. Unless $V_i = \emptyset$, this contradicts the maximality of (L, s). Thus $L = I$ and s represents s'' on all of X. □

Corollary. *For every exact* $\mathcal{R}$*-sequence* $0 \to \mathcal{S}' \to \mathcal{S} \to \mathcal{S}'' \to 0$ *the following are true:*

1) *If* $\mathcal{S}'$ *and* $\mathcal{S}$ *are flabby, then* $\mathcal{S}''$ *is flabby*
2) *If* X *is paracompact and* $\mathcal{S}'$ *and* $\mathcal{S}$ *are soft, then* $\mathcal{S}''$ *is soft.* □

One usually calls a long *exact* $\mathcal{R}$-sequence

$$0 \to \mathcal{S} \to \mathcal{S}^0 \to \mathcal{S}' \to \cdots \to \mathcal{S}^q \to \cdots$$

of $\mathcal{R}$-sheafs and $\mathcal{R}$-homomorphisms an (injective) $\mathcal{R}$*-resolution* of the $\mathcal{R}$-sheaf $\mathcal{S}$. The following theorem, which is important for cohomology, is a consequence of the above corollary.

Exactness Theorem. *Let* $0 \to \mathcal{S} \to \mathcal{S}^0 \to \cdots \to \mathcal{S}^q \to \cdots$ *be an* $\mathcal{R}$*-resolution of* $\mathcal{S}$ *over* X. *If*

1) *every sheaf* $\mathcal{S}, \mathcal{S}^q, q \geq 0$, *is flabby, or*
2) X *is paracompact and all sheaves* $\mathcal{S}, \mathcal{S}^q, q \geq 0$, *are soft, then the associated sequences of sections,* $0 \to \mathcal{S}(X) \to \mathcal{S}^0(X) \to \cdots \to \mathcal{S}^q(X) \to \cdots$, *is exact.*

Proof: We define $\mathcal{Z}^q := \mathcal{K}er(\mathcal{S}^q \to \mathcal{S}^{q+1}) = \mathcal{I}m(\mathcal{S}^{q-1} \to \mathcal{S}^q)$ and, by the assumed exactness of the long sequence, we obtain the exact $\mathcal{R}$-sequence

$0 \to \mathscr{Z}^{q-1} \to \mathscr{S}^{q-1} \to \mathscr{Z}^q \to 0$, $q \geq 1$. Since $\mathscr{Z}^0 \cong \mathscr{S}$ is flabby (soft), it follows inductively from the above corollary that *all* of the sheaves $\mathscr{Z}^q$ are flabby (soft). Thus all of the induced sequences $0 \to \mathscr{Z}^{q-1}(X) \to \mathscr{S}^{q-1}(X) \to \mathscr{Z}^q(X) \to 0$ are exact (i.e. $\mathscr{Z}^q(X) = \operatorname{Im}(\mathscr{S}^{q-1}(X) \to \mathscr{S}^q(X))$). Since $\mathscr{Z}^q(X) = \operatorname{Ker}(\mathscr{S}^q(X) \to \mathscr{S}^{q+1}(X))$, the exactness of the long sequence follows in both cases. □

Chapter B. Cohomology Theory

The cohomology groups $H^q(X, \mathscr{S})$ of a topological space X with coefficients in a sheaf of $\mathscr{R}$-modules $\mathscr{S}$ are introduced via the canonical flabby resolution of $\mathscr{S}$. Moreover the Čech and alternating Čech coholomology groups, $\check{H}^q(X, \mathscr{S})$ and $\check{H}^q_a(X, \mathscr{S})$, are studied (section 2). By means of the important Leray Theorem (section 3) it is proved, for paracompact spaces, that

$$\check{H}^q_a(X, \mathscr{S}) \xrightarrow{\sim} \check{H}^q(X, \mathscr{S}) \xrightarrow{\sim} H^q(X, \mathscr{S}), \qquad q \geq 0.$$

As standard literature, we refer to [EFV], [FAC] and [TF]. We use the same notation as in Chapter A.

§ 1. Flabby Cohomology Theory

In this paragraph we give a brief report on the basic theorems of flabby cohomology theory. In particular we show that an arbitrary *acyclic resolution* can be used to compute the cohomology groups (the formal de Rham Lemma).

1. Cohomology of Complexes. Let R be a commutative ring (in all of the applications $R := \mathscr{R}(X)$). A sequence

$$K^0 \xrightarrow{d^0} K^1 \xrightarrow{d^1} K^2 \longrightarrow \cdots \longrightarrow K^q \xrightarrow{d^q} K^{q+1} \xrightarrow{d^{q+1}} \cdots$$

of R-modules and R-homomorphisms is called a *complex* if $d^{q+1} \cdot d^q = 0$ for all q. We write $K^\bullet = (K^q, d^q)$ for such a complex. The elements of K^q are called *q-cochains* and the maps d^q are called *coboundary mappings.*

If $K'^\bullet = (K'^q, d'^q)$ is another complex, then a *homomorphism* of *complexes* $\varphi^\bullet\colon K'^\bullet \to K^\bullet$ is a sequence $\varphi^\bullet = (\varphi_q)$ of R-homomorphisms $\varphi_q\colon K'^q \to K^q$ which are *compatible* with the coboundary mappings: $d^q \varphi_q = \varphi_{q+1} d'^q$, $q \geq 0$. With these as morphisms, the *complexes form an abelian category.*

For every complex $K^\bullet$ one defines the R-modules

$$Z^q(K^\bullet) := \operatorname{Ker} d^q \quad \text{and} \quad B^q(K^\bullet) := \operatorname{Im} d^{q-1},$$

the *q-cocycles* and *q-coboundaries*. Since $d^{q+1}d^q = 0$, it follows that $Z^q(K^\bullet) \subset B^q(K^\bullet)$. Thus we may define the q-th *cohomology* module of a complex $K^\bullet$ by

$$H^0(K^\bullet) := Z^0(K^\bullet) \quad \text{and} \quad H^q(K^\bullet) := Z^q(K^\bullet)/B^q(K^\bullet), \qquad q \geq 1.$$

The elements of $H^q(K^\bullet)$ are called *cohomology classes*.

If $\varphi^\bullet\colon K'^\bullet \to K^\bullet$ is a homomorphism of complexes, then $\varphi_q(Z^q(K'^\bullet)) \subset Z^q(K^\bullet)$ and $\varphi_q(B^q(K'^\bullet)) \subset B^q(K^\bullet)$. Thus $\varphi^\bullet$ induces homomorphisms

$$H^q(\varphi^\bullet)\colon H^q(K'^\bullet) \to H^q(K^\bullet), \qquad q \geq 0,$$

of cohomology modules. Thus H^q is a *covariant additive functor* from the category of complexes with values in the category of R-modules.

A sequence $K'^\bullet \xrightarrow{\varphi^\bullet} K^\bullet \xrightarrow{\psi^\bullet} K''^\bullet$ of complexes, where $\varphi^\bullet$ and $\psi^\bullet$ are homomorphisms of complexes, is exact whenever each of the sequences $K'^q \xrightarrow{\varphi_q} K^q \xrightarrow{\psi_q} K''^q$ is exact.

The following is fundamental for the cohomology theory:

Lemma 1. *Let* $0 \to K'^\bullet \xrightarrow{\varphi^\bullet} K^\bullet \xrightarrow{\psi^\bullet} K'' \to 0$ *be a "short" exact sequence of complexes* ($0 :=$ zero complex). *Then there exists a natural connecting homomorphism* $\delta^q\colon H^q(K''^\bullet) \to H^{q+1}(K'^\bullet)$, $q \geq 0$, *which depends functorially on* $\varphi^\bullet$ *and* $\psi^\bullet$, *so that the long cohomology sequence*

$$0 \to H^0(K'^\bullet) \to \cdots \to H^q(K'^\bullet) \to H^q(K^\bullet) \to H^q(K''^\bullet) \xrightarrow{\delta^q} H^{q+1}(K'^\bullet) \to \cdots$$

is exact.

Furthermore, the long exact sequence in cohomology preserves commutativity: *Let*

$$\begin{array}{ccccccccc}
0 & \longrightarrow & K'^\bullet & \longrightarrow & K^\bullet & \longrightarrow & K'' & \longrightarrow & 0 \\
 & & \downarrow & & \downarrow & & \downarrow & & \\
0 & \longrightarrow & L'^\bullet & \longrightarrow & L^\bullet & \longrightarrow & L'' & \longrightarrow & 0
\end{array}$$

be a commutative diagram of exact sequences of complexes. Then the diagram of long exact cohomology sequences,

$$\begin{array}{ccccccccccc}
\cdots & \longrightarrow & H^q(K^\bullet) & \longrightarrow & H^q(K''^\bullet) & \xrightarrow{\delta^q} & H^{q+1}(K'^\bullet) & \longrightarrow & H^{q+1}(K^\bullet) & \longrightarrow & \cdots \\
 & & \downarrow & & \downarrow & & \downarrow & & \downarrow & & \\
\cdots & \longrightarrow & H^q(L^\bullet) & \longrightarrow & H^q(L''^\bullet) & \xrightarrow{\delta^q} & H^{q+1}(L'^\bullet) & \longrightarrow & H^{q+1}(L^\bullet) & \longrightarrow & \cdots
\end{array}$$

is everywhere commutative.

2. Flabby Cohomology Theory. Given an $\mathcal{R}$-resolution

$$0 \longrightarrow \mathcal{S} \longrightarrow \mathcal{T}^0 \xrightarrow{t^0} \mathcal{T}^1 \longrightarrow \cdots \longrightarrow \mathcal{T}^q \xrightarrow{t^q} \cdots$$

of an $\mathcal{R}$-sheaf $\mathcal{S}$, there exists the associated complex at the section level,

$$\mathcal{T}^0(X) \xrightarrow{t^0_*} \mathcal{T}^1(X) \longrightarrow \cdots \longrightarrow \mathcal{T}^q(X) \xrightarrow{t^q_*} \cdots,$$

where we always write t^q_* instead of $\Gamma(t^q)$. Hence one has the cohomology modules

$$H^0(\mathcal{T}^\bullet(\mathcal{S})) = \operatorname{Ker} t^0_* \cong \mathcal{S}(X), \quad H^q(\mathcal{T}^\bullet(\mathcal{S})) = \operatorname{Ker} t^q_* / \operatorname{Im} t^{q-1}_*, \qquad q \geq 1.$$

In section 4.3 we embedded every $\mathcal{R}$-module $\mathcal{S}$ in a functorial way in a flabby $\mathcal{R}$-sheaf $\mathcal{F}^0 := \mathcal{F}(\mathcal{S})$: $0 \longrightarrow \mathcal{S} \xrightarrow{j} \mathcal{F}^0$. This procedure can be iterated: Let $\mathcal{F}^1 := \mathcal{F}(\mathcal{F}^0/j(\mathcal{S}))$ and let $f^0: \mathcal{F}^0 \to \mathcal{F}^1$ be the composition of the quotient homomorphism $\mathcal{F}^0 \to \mathcal{F}^0/j(\mathcal{S})$ with the injection $\mathcal{F}^0/j(\mathcal{S}) \to \mathcal{F}^1$. If $0 \longrightarrow \mathcal{S} \xrightarrow{j} \mathcal{F}^0 \xrightarrow{f^0} \mathcal{F}^1 \longrightarrow \cdots \xrightarrow{f^{q-1}} \mathcal{F}^q$ is an exact $\mathcal{R}$-sequence with $\mathcal{F}^i$ flabby for $i \leq q$, then we set $\mathcal{F}^{q+1} := \mathcal{F}(\mathcal{F}^q/\mathcal{I}m\, f^{q-1})$, where $f^{-1} := j$, and we define f^q to be the composition of $\mathcal{F}^q \to \mathcal{F}^q/\mathcal{I}m\, f^{q-1}$ with $\mathcal{F}^q/\mathcal{I}m\, f^{q-1} \to \mathcal{F}^{q+1}$. It follows that

every $\mathcal{R}$-sheaf $\mathcal{S}$ possesses a canonical flabby $\mathcal{R}$-resolution

$$0 \xrightarrow{j} \mathcal{S} \xrightarrow{f^0} \mathcal{F}^0(\mathcal{S}) \longrightarrow \mathcal{F}^1(\mathcal{S}) \longrightarrow \cdots \longrightarrow \mathcal{F}^q(\mathcal{S}) \xrightarrow{f^q} \cdots.$$

We use $\mathcal{F}^\bullet(\mathcal{S})$ to denote the associated complex at the section level.

Definition (Cohomology Modules): *The $\mathcal{R}$-modules*

$$H^q(X, \mathcal{S}) := H^q(\mathcal{F}^\bullet(\mathcal{S})), \qquad q = 0, 1, ., \ldots,$$

are called the cohomology modules of the $\mathcal{R}$-sheaf $\mathcal{S}$ over X.

Since the flabby functor $\mathcal{F}$ is covariant and additive, $\mathcal{S} \rightsquigarrow \mathcal{F}^\bullet(\mathcal{S})$ is also such a functor. This has the following consequence:

I) *For every $q \geq 0$, the functor $\mathcal{S} \rightsquigarrow H^q(X, \mathcal{S})$ is additive and covariant. The functors $\mathcal{S} \rightsquigarrow \mathcal{S}(X)$ and $\mathcal{S} \rightsquigarrow H^0(X, \mathcal{S})$ are isomorphic.*

Since $\mathcal{F}$ is an exact functor, Lemma 1 implies the following:

II) *Every exact $\mathcal{R}$-sequence $0 \to \mathcal{S}' \to \mathcal{S} \to \mathcal{S}'' \to 0$ functorially determines a connecting homomorphism $\delta^q: H^q(X, \mathcal{S}'') \to H^{q+1}(X, \mathcal{S}')$, $q \geq 0$, so that the induced long cohomology sequence,*

$$0 \to H^0(X, \mathcal{S}') \to \cdots \to H^q(X, \mathcal{S}) \to H^q(X, \mathcal{S}'') \xrightarrow{\delta^q} H^{q+1}(X, \mathcal{S}') \to \cdots$$

is exact.

Remark: The above two statements are frequently used in applications (e.g. in the theory of Stein spaces) in the following manner:

Given a sheaf epimorphism $\rho\colon \mathcal{S} \to \mathcal{S}''$ *for which* $H^1(X, \mathcal{K}er\,\rho) = 0$, *then the induced homomorphism* $\mathcal{S}(X) \to \mathcal{S}''(X)$ *between modules of sections is surjective.*

It should be said here that in this book there appear to be no applications of the higher cohomology functor H^q, $q \geq 2$. However, if in the Stein theory one wants to prove vanishing theorems of the form $H^q(X, \mathcal{S}) = 0$, $q \geq 1$, (Theorem B) by a "method de descente," then they are nevertheless indispensible: One first proves the claim for every large q and gets it for all q by *decreasing induction* from q to $q-1$.

Since cohomology preserves commutative diagrams, we have the following:

II') *If*

$$\begin{array}{ccccccccc} 0 & \longrightarrow & \mathcal{S}' & \longrightarrow & \mathcal{S} & \longrightarrow & \mathcal{S}'' & \longrightarrow & 0 \\ & & \downarrow & & \downarrow & & \downarrow & & \\ 0 & \longrightarrow & \mathcal{T}' & \longrightarrow & \mathcal{T} & \longrightarrow & \mathcal{T}'' & \longrightarrow & 0 \end{array}$$

is a commutative diagram of sheaves and sheaf homomorphism whose rows are exact, then all of the induced diagrams

$$\begin{array}{ccc} H^q(X, \mathcal{S}'') & \xrightarrow{\delta^q} & H^{q+1}(X, \mathcal{S}') \\ \downarrow & & \downarrow \\ H^q(X, \mathcal{T}'') & \xrightarrow{\delta^q} & H^{q+1}(X, \mathcal{T}') \end{array}$$

are commutative.

Since the Exactness Theorem of Section 4.4 says that every flabby resolution of a flabby sheaf determines an *exact* sequence at the level of sections, we see that,

III) *for every flabby sheaf* $\mathcal{S}$ *of abelian groups over a topological space* X,

$$H^q(X, \mathcal{S}) = 0$$

for all $q \geq 1$.

We now prove the following uniqueness theorem for cohomology. *Let* $\tilde{H}^q$ *be a sequence of functors along with connecting homomorphism* $\tilde{\delta}^q$, $q \geq 0$ *having the properties in* I), II) *and* III). *Then for every* $q \geq 0$, *there exists a natural functor isomorphism* $F^q\colon \tilde{H}^q(X, S) \to H^q(X, S)$ *which is compatible with the connecting homomorphisms.*

Proof: (by induction on q) The existence of F^0 is clear from I). The exact $\mathcal{R}$-sequence $0 \to \mathcal{S} \to \mathcal{F} \to \mathcal{T} \to 0$ with the flabby sheaf $\mathcal{F} := \mathcal{F}^0(\mathcal{S})$ and

$\mathscr{T} := \mathscr{F}/\mathscr{S}$ determines, by II) and III), the following commutative diagram with exact rows (already F^0 is an isomorphism):

$$\begin{array}{ccccccccc} 0 \longrightarrow & \tilde{H}^0(X, \mathscr{S}) & \longrightarrow & \tilde{H}^0(X, \mathscr{F}) & \longrightarrow & \tilde{H}^0(X, \mathscr{T}) & \xrightarrow{\tilde{\delta}^0} & \tilde{H}^1(X, \mathscr{S}) & \longrightarrow 0 \\ & \downarrow & & \downarrow F^0 & & \downarrow F^0 & & \downarrow F^1 & \\ 0 \longrightarrow & H^0(X, \mathscr{S}) & \longrightarrow & H^0(X, \mathscr{F}) & \longrightarrow & H^0(X, \mathscr{T}) & \xrightarrow{\delta^0} & H^1(X, \mathscr{S}) & \longrightarrow 0 \end{array}$$

One sees that there is exactly one isomorphism $F^1: \tilde{H}^1(X, \mathscr{S}) \simeq H^1(X, \mathscr{S})$ which is compatible with $\tilde{\delta}^0$ and δ^0.

Let $q > 1$ and suppose that F^{q-1} has already been constructed. Since $\mathscr{F}$ is flabby, the cohomology sequences give us the following commutative diagram with exact rows:

$$\begin{array}{ccccccc} 0 & \longrightarrow & \tilde{H}^{q-1}(X, \mathscr{T}) & \xrightarrow{\tilde{\delta}^{q-1}} & \tilde{H}^q(X, \mathscr{S}) & \longrightarrow & 0 \\ & & \downarrow F^{q-1} & & \downarrow F^q & & \\ 0 & \longrightarrow & H^{q-1}(X, \mathscr{T}) & \xrightarrow{\delta^{q-1}} & H^q(X, \mathscr{S}) & \longrightarrow & 0 \end{array}$$

Again it follows immediately that there exists a unique isomorphism $F^q: \tilde{H}^q(X, \mathscr{S}) \to H^q(X, \mathscr{S})$, which is compatible with $\tilde{\delta}^{q-1}$, δ^{q-1}. □

The Exactness Theorem of Section 4.4 also implies a result which is important for applications in Chapter II.4:

IV) *If X is metrizable and $\mathscr{S}$ is a soft sheaf of abelian groups on X, then $H^q(X, \mathscr{S}) = 0$ for all $q \geq 1$.* For a proof one needs only to observe that in the flabby resolution of $\mathscr{S}$ all of the sheaves $\mathscr{F}^q(\mathscr{S})$ are now automatically soft (compare 4.3).

3. The Formal de Rham Lemma. In order to calculate the cohomology modules $H^q(X, \mathscr{S})$, one does not absolutely need the canonical flabby resolution of $\mathscr{S}$. Following the standard nomenclature, we call an $\mathscr{R}$-resolution, $0 \to \mathscr{S} \to \mathscr{T}^0 \to \cdots \to \mathscr{T}^q \to \cdots$, of $\mathscr{S}$ *acyclic* if, for all $n \geq 0$ and $q \geq 1$, $H^q(X, \mathscr{T}^n) = 0$.

Theorem (Formal de Rham Lemma). *Let*

$$0 \longrightarrow \mathscr{S} \xrightarrow{i} \mathscr{T}^0 \xrightarrow{t^0} \cdots \longrightarrow \mathscr{T}^q \xrightarrow{t^q} \cdots$$

be an acyclic resolution of $\mathscr{S}$ and $\mathscr{T}^\bullet(\mathscr{S})$ the associated complex at the section level. Then there exists natural $\mathscr{R}(X)$-isomorphisms

$$\tau_p: H^p(\mathscr{T}^\bullet(\mathscr{S})) \to H^p(X, \mathscr{S}), \qquad p = 0, 1, 2, \ldots.$$

Proof: We construct the τ_v's inductively. The existence of τ_0 is clear. Let $\mathscr{G} := \mathscr{K}er\, t^1 \cong \mathscr{T}^0/i(\mathscr{S})$. In the cohomology sequence associated to $0 \to \mathscr{S} \to$

$\mathscr{T}^0 \to \bar{\mathscr{S}} \to 0$, $\mathscr{T}^0(X) \longrightarrow \bar{\mathscr{S}}(X) \xrightarrow{\delta_0} H^1(X, \mathscr{S}) \longrightarrow H^1(X, \mathscr{T}^0) \longrightarrow \cdots$, we have $\bar{\mathscr{S}}(X) = \operatorname{Ker} t^1_*$ and $\operatorname{Im} t^0_* = \operatorname{Ker} \delta_0$. Thus δ_0 induces an $\mathscr{R}(X)$-monomorphism τ_1: $\operatorname{Ker} t^1_X/\operatorname{Im} t^0_X \to H^1(X, \mathscr{S})$ which, since $H^1(X, \mathscr{T}^0) = 0$, is also surjective.

For $\bar{\mathscr{S}}$ we have the acyclic resolution

$$0 \longrightarrow \bar{\mathscr{S}} \xrightarrow{\bar{i}} \mathscr{T}^1 \xrightarrow{t^1} \cdots \longrightarrow \mathscr{T}^q \xrightarrow{t^q} \cdots,$$

where $\bar{i}$ is determined by $\mathscr{T}^0 \to \mathscr{T}^1$ modulo $i(\mathscr{S})$. For the associated section complex $\mathscr{T}^\bullet(\bar{\mathscr{S}})$, we have

$$H^n(\mathscr{T}^\bullet(\bar{\mathscr{S}})) = H^{n+1}(\mathscr{T}^\bullet(\mathscr{S})), \qquad n \geq 1.$$

Now suppose that the existence of the isomorphisms up to index p for all acyclic resolutions has already been shown. Then, along with the isomorphisms $\tau_1, \ldots, \tau_p$, we have the isomorphism $\bar{\tau}_p$: $H^p(\mathscr{T}^\bullet(\bar{\mathscr{S}})) \cong H^p(X, \bar{\mathscr{S}})$. Since $H^p(X, \mathscr{T}^0) = H^{p+1}(X, \mathscr{T}^0) = 0$, the connecting homomorphism

$$\delta_p\colon H^p(X, \bar{\mathscr{S}}) \to H^{p+1}(X, \mathscr{S})$$

associated to $0 \to \mathscr{S} \to \mathscr{T}^0 \to \bar{\mathscr{S}} \to 0$ is bijective. Consequently,

$$\tau_{p+1} := \delta_p \bar{\tau}_p\colon H^{p+1}(\mathscr{T}^\bullet(\mathscr{S})) \to H^{p+1}(X, \mathscr{S})$$

is the desired isomorphism. □

Remark: The homomorphisms τ_p, $p \geq 0$, exist for every resolution of $\mathscr{S}$. This is easily seen by writing down the double complex behind the above argument. The acyclicity just forces all of them to be bijective.

Due to III), *every* flabby resolution of $\mathscr{S}$ is acyclic. Thus one can use *any* flabby resolution of $\mathscr{S}$ to determine its cohomology. We can make the analogous remark (using IV) for soft sheaves.

If X is metrizable and $0 \to \mathscr{S} \to \mathscr{T}^0 \to \mathscr{T}^1 \to \cdots$ is a resolution of $\mathscr{S}$ by soft $\mathscr{R}$-sheaves, then there exists natural isomorphisms $H^p(\mathscr{T}^\bullet(\mathscr{S})) \cong H^p(X, \mathscr{S})$, $p \geq 0$.

This result will be important in Chapter II.4.

§ 2. Čech Cohomology

The system $S = \{S(U), r^U_V\}$ always denotes an R-presheaf and $\mathfrak{U} = \{U_i\}$, $i \in I$, is reserved for an open cover of X. In this section we introduce the (alternating) Čech cohomology modules $H^q(\mathfrak{U}, S)$ and their limit groups $\check{H}^q(X, S)$. A vanishing theorem for compact "blocks" which is important for later applications, is proved. The theory of the long exact Čech cohomology sequence is discussed in detail. A good readable presentation of Čech theory can be found in [TAG] as well as [FAC].

If $c \in S(U)$, then we use the suggestive notation $c|V$ for $r_V^U(c)$. For every $q+1$ indices $i_0, \ldots, i_q \in I$, we write $U(i_0, \ldots, i_q) := U_{i_0} \cap \cdots \cap U_{i_q}$.

1. Čech Complexes. For every $q \geq 0$, the product

$$C^q(\mathfrak{U}, S) := \prod_{(i_0, \ldots, i_q) \in I^{q+1}} S(U(i_0, \ldots, i_q))$$

is an $R(X)$-module. Its elements (the *q-cochains*) are all functions c which associate to every $(q+1)$-tuple a value $c(i_0, \ldots, i_q) \in S(U(i_0, \ldots, i_q))$.

One defines the coboundary map $d^q\colon C^q(\mathfrak{U}, S) \to C^{q+1}(\mathfrak{U}, S)$ by

$$(d^q c)(i_0, \ldots, i_{q+1}) := \sum_{k=0}^{q+1} (-1)^k c(i_0, \ldots, \hat{i}_k, \ldots, i_{q+1}) \,|\, U(i_0, \ldots, \hat{i}_k, \ldots, i_{q+1}).$$

Obviously d^q is an $R(X)$-homomorphism. One verifies by direct calculations that $d^{q+1}d^q = 0$. Therefore $C^\bullet(\mathfrak{U}, S) := (C^q(\mathfrak{U}, S), d^q)$ is a complex of $R(X)$-modules, the so-called *Čech complex with respect to* $\mathfrak{U}$ *with values in the presheaf* S.

Every R-presheaf homomorphism $\varphi\colon S' \to S$ determines $R(X)$-homomorphisms $C^q(\mathfrak{U}, \varphi)\colon C^q(\mathfrak{U}, S') \to C^q(\mathfrak{U}, S)$, $q \geq 0$, which are compatible with the coboundary mappings. Consequently $C^\bullet(\mathfrak{U}, -) = (C^q(\mathfrak{U}, -))$ is a *covariant functor from the category of R-presheaves into the category of complexes of R(X)-modules. It is clear that*

the functor $C^\bullet(\mathfrak{U}, -)$ *is additive and exact.*

The *Čech cohomology modules of S with respect to* $\mathfrak{U}$ *are* defined to be the cohomology modules of the complex $C^\bullet(\mathfrak{U}, S)$:

$$H^q(\mathfrak{U}, S) := H^q(C^\bullet(\mathfrak{U}, S)) = Z^q(\mathfrak{U}, S)/B^q(\mathfrak{U}, S), \qquad q \geq 0.$$

Hence we have a sequence $H^q(\mathfrak{U}, -)$, $q \geq 0$, of covariant, additive functors from the category of R-presheaves to the category of $R(X)$-modules. Since $C^\bullet(\mathfrak{U}, -)$ is exact, every exact sequence $0 \to S' \to S \to S'' \to 0$ of R-presheaves determines (by Lemma 5.1) an *exact* long cohomology sequence

$$\cdots \longrightarrow H^q(\mathfrak{U}, S) \longrightarrow H^q(\mathfrak{U}, S'') \xrightarrow{\delta^q} H^{q+1}(\mathfrak{U}, S') \longrightarrow \cdots.$$

For every $\mathcal{R}$-sheaf $\mathcal{S}$, we always have the canonical presheaf $\Gamma(\mathcal{S})$. Using this, one sets

$$C^q(\mathfrak{U}, \mathcal{S}) := C^q(\mathfrak{U}, \Gamma(\mathcal{S})) = S(U(i_0, \ldots, i_q))$$

and further

$$H^q(\mathfrak{U}, \mathcal{S}) := H^q(\mathfrak{U}, \Gamma(\mathcal{S})), \qquad q \geq 0,$$

where $H^0(\mathfrak{U}, -)$ is isomorphic to the section functor $\mathscr{S} \rightsquigarrow \mathscr{S}(X)$ on the category of $\mathscr{R}$-sheaves.

2. Alternating Čech Complexes. A q-cochain $c \in C^q(\mathfrak{U}, S)$ is called *alternating* if, for every permutation π of $\{0, 1, \ldots, q\}$, it follows that $c(i_{\pi(0)}, \ldots, i_{\pi(q)}) = \operatorname{sgn} \pi \cdot c(i_0, \ldots, i_q)$ and furthermore $c(i_0, \ldots, i_q) = 0$ whenever two arguments are the same.

The set of all alternating q-cochains forms an $R(X)$-submodule $C^q_a(\mathfrak{U}, S)$ of $C^q(\mathfrak{U}, S)$. If c is alternating, then so is $d^q c$. Thus d^q induces an $R(X)$-homomorphism $d^q_a\colon C^q_a(\mathfrak{U}, S) \to C^{q+1}_a(\mathfrak{U}, S)$. Hence $C^{\bullet}_a(\mathfrak{U}, S) := (C^q_a(\mathfrak{U}, S), d^q_a)$ is a subcomplex of $C^{\bullet}(\mathfrak{U}, S)$, called the *alternating Čech complex*. Most properties of $C^{\bullet}(\mathfrak{U}, S)$ carry over immediately to $C^{\bullet}_a(\mathfrak{U}, S)$. For example, $C^{\bullet}_a(\mathfrak{U}, -)$ is a *covariant, additive, exact functor* from the category of R-presheaves to the category of $R(X)$-modules. The $R(X)$-modules

$$H^q_a(\mathfrak{U}, S) := H^q(C^{\bullet}_a(\mathfrak{U}, S)), \qquad q \geq 0,$$

are called the alternating Čech cohomology modules of $\mathfrak{U}$ *with respect* to S.

In important cases $C^q_a(\mathfrak{U}, S) = 0$. For example,

if there exists a natural number $d \geq 1$ so that, for all pair-wise different indices $i_0, \ldots, i_q \in I$, the intersections $U(i_0, \ldots, i_q)$ are empty, then, for every presheaf S of abelian groups

$$C^q_a(\mathfrak{U}, S) = 0 \quad \text{and consequently} \quad H^q_a(\mathfrak{U}, S) = 0 \quad \text{for all} \quad q \geq d.$$

Proof: Let $c \in C^q_a(\mathfrak{U}, S)$, where $q \geq d$. If the indices are pair-wise different, then $U(i_0, \ldots, i_q) = \emptyset$ and $c(i_0, \ldots, i_q) = 0$. If two indices are the same, then $c(i_0, \ldots, i_q)$ is zero by the alternating conditions. □

The injection of complexes $C^{\bullet}_a(\mathfrak{U}, S) \to C^{\bullet}(\mathfrak{U}, S)$ induces an $R(X)$-homomorphism $i_q(\mathfrak{U})\colon H^q_a(\mathfrak{U}, S) \to H^q(\mathfrak{U}, S)$, $q \geq 0$, so that, for every presheaf homomorphism, $\varphi\colon S' \to S$, the diagram

$$\begin{array}{ccc} H^q_a(\mathfrak{U}, S') & \longrightarrow & H^q_a(\mathfrak{U}, S) \\ \Big\downarrow{\scriptstyle i_q(\mathfrak{U})} & & \Big\downarrow{\scriptstyle i_q(\mathfrak{U})} \\ H^q(\mathfrak{U}, S') & \longrightarrow & H^q(\mathfrak{U}, S) \end{array}$$

is commutative. As a matter of fact one can show that the *maps $i_q(\mathfrak{U})$ are always isomorphisms.* We do not know a standard reference for this, but the reader should see [FAC], p. 214. We will never use these isomorphisms anyway.

3. Refinements and the Čech Cohomology Modules $\check{H}^q(X, S)$. Suppose that $\mathfrak{U} = \{U_i\}$, $i \in I$, and $\mathfrak{V} = \{V_j\}$, $j \in J$, are covers of X and that $\mathfrak{V}$ is a refinement of $\mathfrak{U}$: $\mathfrak{V} < \mathfrak{U}$. *Every* associated refinement mapping $\tau\colon J \to I$ determines an $R(X)$-

homomorphism $C^q(\tau): C^q(\mathfrak{U}, S) \to C^q(\mathfrak{V}, S)$, $q \geq 0$, where the q-cochain $c = \{c(i_0, \ldots, i_q)\} \in \prod S(U(i_0, \ldots, i_q))$ is mapped to the q-cochain $c' \in \prod S(V(j_0, \ldots, j_q))$ with

$$c'(j_0, \ldots, j_q) := c(\tau j_0, \ldots, \tau j_q) \mid V(j_0, \ldots, j_q).$$

(note that $V(j_0, \ldots, j_q) \subset U(\tau j_0, \ldots, \tau j_q)$).

One effortlessly verifies that all of the maps $C^q(\tau)$ are compatible with the coboundary mappings. Thus one sees that

if $\mathfrak{V} < \mathfrak{U}$, *then the refinement map* τ *induces a homomorphism of complexes* $C^\bullet(\tau): C^\bullet(\mathfrak{U}, S) \to C^\bullet(\mathfrak{V}, S)$ *and consequently we have* $R(X)$*-homomorphisms*

$$h^q(\tau): H^q(\mathfrak{U}, S) \to H^q(\mathfrak{V}, S), \qquad q \geq 0. \qquad \square$$

If $\tau': J \to I$ is another refinement map, then

$$(k^q c)(j_0, \ldots, j_{q-1}) := \sum_0^{q-1} (-1)^\nu c(\tau j_0, \ldots, \tau j_\nu, \tau' j_{\nu+1}, \ldots, \tau' j_{q-1}) \mid V(j_0, \ldots, j_{q-1})$$

defines an $R(X)$-homomorphism $k^q: C^q(\mathfrak{U}, S) \to C^{q-1}(\mathfrak{V}, S)$, $q \geq 1$, which is a so-called "homotopy" operator for the coboundary operator d. In other words,

$$dk^q + k^{q+1}d = C^q(\tau') - C^q(\tau) \quad \text{for} \quad q \geq 1,$$

and

$$k^1 d = C^0(\tau') - C^0(\tau).$$

It follows that

$$(*) \quad \begin{aligned} &(C^q(\tau') - C^q(\tau))Z^q(\mathfrak{U}, S) \subset B^q(\mathfrak{V}, S) \quad \text{for} \quad q \geq 1, \qquad \text{and} \\ &(C^0(\tau') - C^0(\tau))Z^0(\mathfrak{U}, S) = 0. \end{aligned}$$

Thus at the level of cohomology one obtains the same homomorphism for either restriction: $h^q(\tau) = h^q(\tau')$. Hence we may write $h^q(\mathfrak{U}, \mathfrak{V})$ in place of $h^q(\tau)$. We note that $h^q(\mathfrak{U}, \mathfrak{U}) = \text{id}$ and, if $\mathfrak{W} < \mathfrak{V} < \mathfrak{U}$, then $h^q(\mathfrak{W}, \mathfrak{U}) = h^q(\mathfrak{W}, \mathfrak{V})h^q(\mathfrak{V}, \mathfrak{U})$.

By observing the usual logical precautions, one can consider the "set of all open covers of X." This set is partially ordered with respect to the relation $\mathfrak{V} < \mathfrak{U}$. Every system $\{H^q(\mathfrak{U}, S), h^q(\mathfrak{V}, \mathfrak{U})\}$, $q = 0, 1, 2, \ldots$, is directed by this ordering. Thus we have the *inductive limit*

$$\check{H}^q(X, S) := \varinjlim H^q(\mathfrak{U}, S), \qquad q = 0, 1, 2, \ldots.$$

The $R(X)$-module $\check{H}^q(X, S)$ is called the *q-th Čech cohomology module of X with coefficients in the R-presheaf S*. We denote the canonical map $H^q(\mathfrak{U}, S) \to \check{H}^q(X, S)$ by $h^q(\mathfrak{U})$. Thus $h^q(\mathfrak{V})h^q(\mathfrak{V}, \mathfrak{U}) = h^q(\mathfrak{U})$, whenever $\mathfrak{V} < \mathfrak{U}$. The functor $\check{H}^q(X, -)$ is again covariant and additive.

The Čech cohomology modules for $\mathscr{R}$-sheaves $\mathscr{S}$ are defined by

$$\check{H}^q(X, \mathscr{S}) := \check{H}^q(X, \Gamma(\mathscr{S})), \qquad q \geq 0.$$

Acting on the category of $\mathscr{R}$-sequences, $\check{H}^0(X, -)$ is isomorphic to the section functor. Hence the functors $\mathscr{S} \rightsquigarrow \check{H}^q(X, \mathscr{S})$ have the property I) of Section 5.2.

4. The Alternating Čech Cohomology Modules $\check{H}^q_a(X, S)$. The considerations in the preceding paragraph can be repeated mutatis mutandis for alternating Čech complexes. Namely, by means of $C^q(\tau)$, alternating cochains are mapped to alternating cochains and thus one obtains a homomorphism of complexes $C^\bullet_a(\tau)$: $C^\bullet_a(\mathfrak{U}, S) \to C^\bullet_a(\mathfrak{V}, S)$ as well as the $R(X)$-homomorphism $h^q_a(\tau)$: $H^q_a(\mathfrak{U}, S) \to H^q_a(\mathfrak{V}, S)$, $q \geq 0$. The equations analogous to (*) show that $h^q_a(\tau)$ is likewise independent of the choice of the refinement map. Correspondingly we write $h^q_a(\mathfrak{V}, \mathfrak{U})$ instead of $h^q_a(\tau)$. Every system $\{H^q_a(\mathfrak{U}, S), h^q_a(\mathfrak{V}, \mathfrak{U})\}$ is directed, the direct limit $R(X)$-module

$$H^q_a(X, S) := \varinjlim \check{H}^q_a(\mathfrak{U}, S)$$

being called the q-th *alternating* Čech cohomology module with coefficients in the presheaf S.

For sheaves $\mathscr{S}$, one sets $\check{H}^q_a(X, \mathscr{S}) := \check{H}^q_a(X, \Gamma(\mathscr{S}))$. This is again a covariant, additive functor on the category of sheaves which has property I) of 5.2.

If $\mathfrak{V} < \mathfrak{U}$, then $i_q(\mathfrak{V}) h^q_a(\mathfrak{V}, \mathfrak{U}) = h^q(\mathfrak{V}, \mathfrak{U}) i_q(\mathfrak{U})$, where $i_q(\mathfrak{V})$ and $i_q(\mathfrak{U})$ are the natural homomorphisms from Section 2. Thus the limit homomorphism i_q: $\check{H}^q_a(X, \mathscr{S}) \to \check{H}^q(X, \mathscr{S})$ is induced, having the property that $i_q h^q_a(\mathfrak{U}) = h^q(\mathfrak{U}) i_q(\mathfrak{U})$, $q \geq 0$.

5. The Vanishing Theorem for Compact Blocks. For the proofs of Theorems A and B in Chapter III.3.2 (Theorem 1), we need the following:

Vanishing Theorem. *Let $B := \{x \in \mathbb{R}^m \mid a_\mu \leq x_\mu \leq b_\mu, 1 \leq \mu \leq m\}$ be a non-empty compact block in $\mathbb{R}^m$ and $\mathscr{S}$ a sheaf of abelian groups over B. Then*

$$H^q_a(B, \mathscr{S}) = 0 \quad \textit{for all} \quad q \geq 3^m$$

The proof will follow from a simple lemma. We denote with $|\ \ |$ the euclidian norm on $\mathbb{R}^m$ and associate to every set $M \subset \mathbb{R}^m$ its *diameter*, $d(M) := \sup_{x, y \in M} |x - y|$.

Lemma. *Let $\mathfrak{U} = \{U_i\}$ be a cover of the compact block B. Then there exists a real number $\lambda > 0$ (the so-called Lebesgue number of $\mathfrak{U}$) so that every open set V in B with $d(V) < \lambda$ lies in some U_i.*

Proof: Suppose that for every $\nu = 1, 2, \ldots$, there exists an open set $V_\nu \subset B$ with $d(V_\nu) < \nu^{-1}$ which lies in no U_i. Let $p_\nu \in V_\nu$. Since B is compact, the sequence (p_ν)

has an accumulation point $p \in B$. There exists $j \in I$ with $p \in U_j$. It is clear that, for ν large enough, $V_\nu \subset U_j$, which is the desired contradiction. $\square$

We now prove the Vanishing Theorem. For every integer $n \geq 1$ we construct as follows a cover V_n of B: Let

$$x(l_1, \ldots, l_m) := \left(a_1 + \frac{l_1}{n}(b_1 - a_1), \ldots, a_m + \frac{l_m}{n}(b_m - a_m)\right) \in \mathbb{R}^m,\ l_\mu = 0, 1, \ldots, n.$$

$$B(l_1, \ldots, l_m) := \left\{x \in B \,\middle|\, \|x_\mu - x_\mu(l_1, \ldots, l_m\| < 2\,\frac{b_\mu - a_\mu}{n}\right\}$$

$$\mathfrak{B}_n := \{B(l_1, \ldots, l_m) \mid 0 < l_\mu \leq n\}$$

Then $\mathfrak{B}_n$ is a cover of B. The sets $B(l_1, \ldots, l_m)$ and $B(l'_1, \ldots, l'_m)$ have a non-empty intersection if and only if $|l_\mu - l'_\mu| \leq 1$ for $\mu = 1, \ldots, m$. For a fixed $(l_1, \ldots, l_m) \in \mathbb{Z}^m$, there exists at most 3^m integral lattices points $(l_1, \ldots, l_m)$ which fulfill these conditions. Consequently, the intersection of every 3^{m+1} different elements of V_n is empty and it follows that $H^q_a(\mathfrak{B}_m, \mathscr{S}) = 0$ for all $q \geq 3^m$.

Since the diameter of every set $B(l_1, \ldots, l_m) \in \mathfrak{B}_m$ is smaller than $2m/n \cdot \max|b_\mu - a_\mu|$, it follows from the Lemma that, for every cover $\mathfrak{U}$ of B, there exists an index p so that $\mathfrak{B}_p$ is finer than $\mathfrak{U}$. This says that the covers of the form $\mathfrak{B}_m$ are cofinal in the direct limiting process which defines $H^q_a(X, \mathscr{S})$. Thus

$$\check{H}^q_a(X, \mathscr{S}) = \varinjlim H^q_a(\mathfrak{B}_m, \mathscr{S}) = 0. \qquad \square$$

Remark: It is clear that the bound 3^m can be greatly improved. In fact, a little work yields $m + 1$ as the best possible such bound.

6. The Long Exact Cohomology Sequence. For every exact sequence of presheaves, $0 \to S' \to S \to S'' \to 0$, and every cover $\mathfrak{U}$, we have a commutative diagram

$$\begin{array}{ccccccccc}
0 & \longrightarrow & C^\bullet_a(\mathfrak{U}, S') & \longrightarrow & C^\bullet_a(\mathfrak{U}, S) & \longrightarrow & C^\bullet_a(\mathfrak{U}, S'') & \longrightarrow & 0 \\
 & & \downarrow & & \downarrow & & \downarrow & & \\
0 & \longrightarrow & C^\bullet(\mathfrak{U}, S') & \longrightarrow & C^\bullet(\mathfrak{U}, S) & \longrightarrow & C^\bullet(\mathfrak{U}, S'') & \longrightarrow & 0
\end{array}$$

of complexes with exact rows. By the results in Paragraph 5.1 one has the commutative diagram of long exact cohomology sequences. Since direct limits of exact sequences are exact, it follows that

for every exact sequence $0 \to S' \to S \to S'' \to 0$ of R-presheaves there exists a commutative diagram of long exact cohomology sequences

$$\begin{array}{ccccccccc}
0 \to & \check{H}^0_a(X, S') & \to \cdots \longrightarrow & \check{H}^q_a(X, S) & \longrightarrow & \check{H}^q_a(X, S'') & \xrightarrow{\check{\delta}_a{}^q} & \check{H}^{q+1}_a(X, S') & \longrightarrow \cdots \\
 & \downarrow & & \downarrow & & \downarrow & & \downarrow & \\
0 \longrightarrow & \check{H}^0(X, S') & \longrightarrow \cdots \longrightarrow & \check{H}^q(X, S) & \longrightarrow & \check{H}^q(X, S'') & \xrightarrow{\check{\delta}^q} & \check{H}^{q+1}(X, S') & \longrightarrow \cdots.
\end{array}$$

For every exact sequence of $\mathcal{R}$-sheaves, $0 \to \mathcal{S}' \to \mathcal{S} \to \mathcal{S}'' \to 0$, one has (see Paragraph 1.6) the exact $\Gamma(\mathcal{R})$-presheaf sequence, $0 \to \Gamma(\mathcal{S}') \to \Gamma(\mathcal{S}) \to \tilde{S} \to 0$, where $\tilde{S} := \{\tilde{S}(U), r^U_V\}$, with $\tilde{S}(U) := \mathcal{S}(U)/\mathcal{S}'(U)$ is a presheaf whose sheaf is in fact $\mathcal{S}''$. Thus one has a commutative diagram,

$$\begin{array}{ccccccccc}
\cdots \longrightarrow & \check{H}^q_a(X, \mathcal{S}') & \longrightarrow & \check{H}^q_a(X, \mathcal{S}) & \longrightarrow & \check{H}^q_a(X, \tilde{S}) & \xrightarrow{\check{\delta}^q_a} & \check{H}^{q+1}_a(X, \mathcal{S}') & \longrightarrow \cdots \\
 & \downarrow{\scriptstyle i_q} & & \downarrow{\scriptstyle i_q} & & \downarrow{\scriptstyle i_q} & & \downarrow{\scriptstyle i_{q+1}} & \\
\cdots \longrightarrow & \check{H}^q(X, \mathcal{S}') & \longrightarrow & \check{H}^q(X, \mathcal{S}) & \longrightarrow & \check{H}^q(X, \tilde{S}) & \xrightarrow{\check{\delta}^q} & \check{H}^{q+1}(X, \mathcal{S}') & \longrightarrow \cdots,
\end{array}$$

with exact rows. The natural presheaf homomorphism $\tilde{S} \to \Gamma(\mathcal{S}'')$ induces homomorphisms $\check{H}^q(X, \tilde{S}) \to \check{H}^q(X, \mathcal{S}'')$ and $\check{H}^q_a(X, \tilde{S}) \to \check{H}^q_a(X, \mathcal{S}'')$. Thus, if we can prove that these are in fact isomorphisms, we can always replace $\tilde{S}$ by $\mathcal{S}''$ in the above diagram. With this in mind, we now show the following:

Let X be paracompact and T be a presheaf (of abelian groups) over X with associated sheaf $\mathcal{T} := \check{\Gamma}(T)$. Then the natural map $\alpha\colon T \to \Gamma(\mathcal{T})$ induces genuine isomorphisms

$$\check{H}^q(X, T) \to \check{H}^q(X, \mathcal{T}) \quad \text{and} \quad \check{H}^q_a(X, T) \to \check{H}^q_a(X, \mathcal{T}), \qquad q \geq 0.$$

Proof: Assuming the statement in the case $\mathcal{T} = 0$, the general case can be derived as follows: The map α yields $0 \longrightarrow \operatorname{Ker} \alpha \longrightarrow T \xrightarrow{\bar{\alpha}} \operatorname{Im} \alpha \longrightarrow 0$, $0 \longrightarrow \operatorname{Im} \alpha \xrightarrow{j} \Gamma(\mathcal{T}) \longrightarrow \tilde{T} \longrightarrow 0$, where $\tilde{T} := \Gamma(\mathcal{T})/\operatorname{Im} \alpha$, as exact sequences of presheaves. The associated long exact cohomology sequences are

$$\cdots \to \check{H}^q(X, \operatorname{Ker} \alpha) \to \check{H}^q(X, T) \xrightarrow{\bar{\alpha}_*} \check{H}^q(X, \operatorname{Im} \alpha) \to \check{H}^{q+1}(X, \operatorname{Ker} \alpha) \to \cdots$$

and

$$\cdots \to \check{H}^{q-1}(X, \tilde{T}) \to \check{H}^q(X, \operatorname{Im} \alpha) \xrightarrow{j_*} \check{H}^q(X, \mathcal{T}) \to \check{H}^q(X, \tilde{T}) \to \cdots$$

Since the functor $\check{\Gamma}$ is exact, $\check{\Gamma}(\operatorname{Ker} \alpha) = 0 = \check{\Gamma}(\tilde{T})$. Assuming the result for the zero sheaf, all homomorphisms

$$\alpha_*\colon \check{H}^q(X, T) \to \check{H}^q(X, \operatorname{Im} \alpha), \qquad j_*\colon \check{H}^q(X, \operatorname{Im} \alpha) \to \check{H}^q(X, \mathcal{T}), \qquad q \geq 0,$$

are bijective.

It remains to show that, in the case that $\mathcal{T} = 0$, it follows that $\check{H}^q(X, \mathcal{T}) = \check{H}^q_a(X, \mathcal{T}) = 0$. This is obviously contained in the following:

Lemma. *Let $\mathfrak{U} = \{U_i\}$, $i \in I$, be a locally-finite cover of X. Then, assuming $\check{\Gamma}(T) = 0$, given a cochain $c \in C^q(\mathfrak{U}, T)$, there exists a refinement $\mathfrak{W} = \{W_j\}, j \in J$, of $\mathfrak{U}$ (with refinement map $\tau\colon J \to I$) so that $C^q(\tau)c = 0$.*

Proof: Let $\mathfrak{V} = \{V_i\}$, $i \in I$, be a cover of X with $\bar{V}_i \subset U_i$ (Shrinking Theorem). Let $J := X$ and take $\tau\colon J \to I$ so that $x \in V_{\tau x}$. Since $\mathfrak{U}$ is locally finite, every point x

possesses a neighborhood W_x which has non-empty intersection with only finitely many U_i's. By shrinking if necessary, we may assume that

$$1) \qquad W_x \subset U_i \quad \text{for} \quad x \in U_i \quad \text{and} \quad W_x \subset V_i \quad \text{for} \quad x \in V_i.$$

Then $\mathfrak{W} := \{W_x\}$, $x \in X$, is a refinement of both $\mathfrak{B}$ and $\mathfrak{U}$ with refinement map τ. Since $V_i \subset U_i$, we may assume that

$$2) \qquad \text{if} \quad W_x \cap V_i \neq \emptyset \quad \text{then} \quad x \in U_i.$$

Finally, since $\check{\Gamma}(T) = 0$, W_x can be chosen so that

$$3) \quad c(i_0, \ldots, i_q) \mid W_x = 0 \quad \text{for all} \quad i_0, \ldots, i_q \in I \quad \text{and all} \quad x \in U(i_0, \ldots, i_q).$$

For any $(q+1)$-points $x_0, \ldots, x_q$ with $W_{x_0} \cap \cdots \cap W_{x_q} \neq \emptyset$, we now consider $C^q(\tau)c(x_0, \ldots, x_q) = c(\tau x_0, \ldots, \tau x_q) \mid W(x_0, \ldots, x_q)$. Since $W_{x_0} \cap W_{x_k} \neq \emptyset$ for $k = 0, \ldots, q$, W_{x_0} must have a non-trivial intersection with every $V_{\tau x_k}$. Thus, by 2), $x_0 \in U_{\tau x_k}$ for all k (i.e. $x_0 \in U(\tau x_0, \ldots, \tau x_q)$). Now 3) implies that $c(\tau x_0, \ldots, \tau x_q) \mid W_{x_0} = 0$ and thus $c(\tau x_0, \ldots, \tau x_q) \mid W(x_0, \ldots, x_q) = 0$. □

Thus, in summary, we have shown that

if X is paracompact, then the cohomology functors $\check{H}^q$ and $\check{H}^q_a$ have the property II) *of paragraph* 5.2.

We will verify property III) in the next section.

§ 3. The Leray Theorem and the Isomorphism Theorems $\check{H}_{al}(X, \mathscr{S}) \simeq \check{H}^q(X, \mathscr{S}) \simeq H^q(X, \mathscr{S})$

The Leray Theorem is fundamental for later applications in the Stein theory as well as being the key to proving that all of the cohomologies are the same. It states that, for special covers $\mathfrak{U}$ of X, the groups $\check{H}^q(X, \mathscr{S})$ are isomorphic to $H^q(\mathfrak{U}, \mathscr{S})$, $q \geq 0$. By means of a canonical resolution of $\mathscr{S}$ relative to $\mathfrak{U}$, we reduce this theorem to the formal de Rham Lemma. As an application of the Leray theorem, we show that on paracompact spaces the Čech cohomology for flabby sheaves $\mathscr{F}$ vanishes. As a consequence the maps i_q: $\check{H}^q_a(X, \mathscr{S}) \simeq H^q(X, \mathscr{S})$ and $\check{H}^q(X, \mathscr{S}) \simeq H^q(X, \mathscr{S})$ are isomorphisms.

1. The Canonical Resolution of a Sheaf Relative to a Cover. If $\mathscr{S}$ is an $\mathscr{R}$-sheaf over X, then for every open set $Y \subset X$, we use $S\langle Y \rangle$ to denote that $\mathscr{R}$-sheaf which is the trivial extension of $\mathscr{S} \mid Y$ to X. Thus $\mathscr{S}\langle Y \rangle = i_*(\mathscr{S} \mid Y)$, where i: $Y \to X$ is the injection. For all open $U \subset X$, we have $\mathscr{S}\langle Y \rangle(U) = S(U \cap Y)$.

Now let $\mathfrak{U} = \{U_i\}$, $i \in I$, be an open cover of X. For every $\mathscr{R}$-sheaf $\mathscr{S}$ we form the $\mathscr{R}$-sheaves $\mathscr{S}(i_0, \ldots, i_p) := \mathscr{S}\langle U(i_0, \ldots, i_p) \rangle$, $i_0, \ldots, i_p \in I$, and further (see paragraph 0.7) we have the direct product sheaf

(0) $$\mathscr{S}^p := \prod_{(i_0, \dots, i_p) \in I^{p+1}} \mathscr{S}(i_0, \dots, i_p), \qquad p = 0, 1, \dots.$$

These product sheaves are obviously $\mathscr{R}$-sheaves as well. If one introduces the open cover $\mathfrak{U} \mid U = \{U_i'\}$, $U_i' := U_i \cap U$, $i \in I$, of U, then $\mathscr{S}(U(i_0, \dots, i_p) \cap U) = (\mathscr{S} \mid U)(U'(i_0, \dots, i_p))$. Thus we may write:

(1) $$\mathscr{S}^p(U) = C^p(\mathfrak{U} \mid U, \mathscr{S} \mid U), \qquad p \geq 0.$$

Hence $\mathscr{S}^p(U)$ is the p-cochain module of $\mathscr{S} \mid U$ with respect to the cover $\mathfrak{U} \mid U$, and one has the cochain complex

$$C^\bullet(\mathfrak{U} \mid U, \mathscr{S} \mid U) = (\mathscr{S}^p(U), d_U^p).$$

If one associates to every section $s \in \mathscr{S}(U)$ the o-cochain $c(i) := (s \mid U_i \cap U)$, then one has an $\mathscr{R}(U)$-homomorphism j_U: $\mathscr{S}(U) \to \mathscr{S}^0(U)$ such that $d_U^0 j_U = 0$. For all U, V with $V \subset U$ the restrictions $S^p(U) \to S^p(V)$ are compatible with d_U^p and d_V^p. Thus, taking the direct limit of the above complexes, we have an $\mathscr{R}$-sheaf sequence

$$\mathscr{S} \xrightarrow{j} \mathscr{S}^0 \xrightarrow{d^0} \mathscr{S}^1 \longrightarrow \cdots \longrightarrow \mathscr{S}^p \xrightarrow{d^p} \mathscr{S}^{p+1} \longrightarrow,$$

where $d^0 j = 0$ and $d^{p+1} d^p = 0$. We claim that,

for every $\mathscr{R}$-sheaf $\mathscr{S}$ over X and every cover $\mathfrak{U}$ of X,

(R) $$0 \longrightarrow \mathscr{S} \xrightarrow{j} \mathscr{S}^0 \xrightarrow{d^0} \mathscr{S}^1 \longrightarrow \cdots \longrightarrow \mathscr{S}^p \xrightarrow{d^p} \mathscr{S}^{p+1} \longrightarrow \cdots$$

is an $\mathscr{R}$-resolution of $\mathscr{S}$.

The equations $\operatorname{Ker} j_x = 0$ and $\operatorname{Im} j_x = \operatorname{Ker} d_x^0$, $x \in X$, follow trivially. It thus remains to verify the inclusion $\operatorname{Ker} d_x^p \subset \operatorname{Im} d_x^{p-1}$, $p \geq 1$. For that we need the following Lemma:

Lemma. *Let $\mathscr{T}$ be a sheaf of abelian groups over a topological space M and let $\mathfrak{W} = \{W_j\}$, $j \in J$, be an open cover of M. Suppose that $W_k = M$ for some $k \in J$. Then $H^q(\mathfrak{W}, \mathscr{T}) = 0$ for all $q \geq 1$ (i.e. every cocycle $c = \{c(j_0, \dots, j_q)\} \in Z^q(W, T)$, $q \geq 1$, is a coboundary.)*

Proof: Let $c \in Z^q(\mathfrak{W}, \mathscr{T})$. It is always the case that $W(j_0, \dots, j_{q-1}) = W(k, j_0, \dots, j_{q-1})$. Thus $b := \{b(j_0, \dots, j_{q-1})\}$, where $b(j_0, \dots, j_{q-1}) := c(k, j_0, \dots, j_{q-1}) \in \mathscr{T}(W(j_0, \dots, j_{q-1}))$, is a $(q-1)$-cochain in $C^{q-1}(\mathfrak{W}, T)$. By the definition of the coboundary map,

$$(d^{q-1} b)(j_0, \dots, j_q) = \sum_{\nu=1}^{q} (-1)^\nu c(k, j_0, \dots, \hat{j}_\nu, \dots, j_q) \mid W(j_0, \dots, \hat{j}_\nu, \dots, j_q).$$

However, by assumption,

$$0 = (d^q c)(k, j_0, \ldots, j_q) = c(j_0, \ldots, j_q) - \sum_{\nu=0}^{q} (-1)^\nu c(k, j_0, \ldots, \hat{j}_\nu, \ldots, j_q).$$

Thus

$$(d^{q-1}b)(j_0, \ldots, j_q) = c(j_0, \ldots, j_q). \qquad \square$$

Now we can easily prove the claimed inclusions $\operatorname{Ker} d_x^p \subset \operatorname{Im} d_x^{p-1}$, $p \geq 1$.

Let $s_x \in \operatorname{Ker} d_x^p$. Then there is a neighborhood M of x and a representation $s \in \operatorname{Ker} d_M^p = Z^p(\mathfrak{U} | M, \mathcal{S} | M)$. We can choose M so small that it is contained in a set U_k. Thus M is a member of the cover $\mathfrak{U} | M$ and, by the above Lemma (with $\mathcal{T} := \mathcal{S} | M$), there exists an element $t \in C^{p-1}(\mathfrak{U} | M, \mathcal{S} | M) = \mathcal{S}^{p-1}(M)$ with $d_M^p t = s$. Letting $t_x \in S_x^{p-1}$ be the germ determined by t, it follows that $d_x^p t_x = s_x$. $\square$

We call the just constructed resolution (R) of $\mathcal{S}$ the *canonical resolution of $\mathcal{S}$ relative to the cover* $\mathfrak{U}$. We emphasize that it is in general not acyclic.

2. Acyclic Covers. For every flabby sheaf $\mathcal{F}$ over X, the sheaves $\mathcal{F}\langle U \rangle$ are also flabby over X. A flabby resolution $0 \to \mathcal{S} \to \mathcal{F}^0 \to \mathcal{F}^1 \to \cdots$ of $\mathcal{S}$ induces the flabby resolution $0 \to \mathcal{S}\langle U \rangle \to \mathcal{F}^0\langle U \rangle \to \mathcal{F}^1\langle U \rangle \to \cdots$ of $\mathcal{S}\langle U \rangle$ over X. Thus the groups $H^q(X, \mathcal{S}\langle U \rangle)$ are the cohomology groups of the section complex $\mathcal{F}^0\langle U \rangle(X) \to \mathcal{F}^1\langle U \rangle(X) \to \cdots$. Since $0 \to \mathcal{S} | U \to \mathcal{F}^0 | U \to \mathcal{F}^1 | U \to \cdots$ is a flabby resolution of $\mathcal{S} | U$ over U and since $\mathcal{F}^i\langle U \rangle(X) = \mathcal{F}^i(U)$, we obtain for every $\mathcal{R}$-sheaf $\mathcal{S}$ the equations

$$(2) \qquad H^q(X, \mathcal{S}\langle U \rangle) = H^q(U, \mathcal{S}), \qquad U = \mathring{U} \subset X, \qquad q \geq 0.$$

In order to compute the cohomology groups $H^q(X, \mathcal{S}^p)$, where $\mathcal{S}^p = \prod_{i_0, \ldots, i_p} \mathcal{S}(i_0, \ldots, i_p)$, we note the following:

Lemma. *Let $(\mathcal{T}_j)$, $j \in J$, be a locally finite family of $\mathcal{R}$-sheaves (i.e. every point $x \in X$ has a neighborhood V so that all but finitely many $\mathcal{T}_j | V$ are the zero sheaf over V). Then, defining $\mathcal{T} := \prod_{j \in J} \mathcal{T}_j$, we have the canonical isomorphism*

$$H^q(X, \mathcal{T}) \cong \prod_{j \in J} H^q(X, \mathcal{T}_j), \qquad q \geq 0.$$

The reader can find a simple proof in [TF], p. 175.

We now assume that the given cover $\mathfrak{U}$ is locally finite. Thus, for every $p \geq 0$, the family $\{\mathcal{S}(i_0, \ldots, i_p)\}$, $i_0, \ldots, i_p \in I$, is locally finite. Hence the Lemma, together with equation (0), implies that

$$H^q(X, \mathcal{S}^p) \cong \prod_{i_0, \ldots, i_p} H^q(X, \mathcal{S}(i_0, \ldots, i_p)).$$

Since $\mathscr{S}(i_0, \ldots, i_p) = \mathscr{S}\langle U(i_0, \ldots, i_p)\rangle$, equation (2) implies that, if $\mathfrak{U}$ is a locally finite cover of X, then there is a natural isomorphism

$$(3) \qquad H^q(X, \mathscr{S}^p) = \prod_{i_0, \ldots, i_p} H^q(U(i_0, \ldots, i_p), S), \qquad p, q \geq 0.$$

It is now easy to give a sufficient condition for the acyclicity of the canonical resolution (R). One calls an open cover $\mathfrak{U}$ of X *acyclic with respect to* $\mathscr{S}$ if $H^q(U(i_0, \ldots, i_p), \mathscr{S}) = 0$ for all $p \geq 0$, $q \geq 1$. Thus we can say that,

if $\mathfrak{U}$ is a locally finite cover of X which is acyclic with respect to $\mathscr{S}$, then the canonical resolution of $\mathscr{S}$ relative to $\mathfrak{U}$ is an acyclic resolution of $\mathscr{S}$.

3. The Leray Theorem. The following is now easy to show:

Theorem (Leray Theorem). *If $\mathfrak{U}$ is a locally finite cover of X which is acyclic with respect to the $\mathscr{R}$-sheaf $\mathscr{S}$, then there exists natural $\mathscr{R}(X)$-isomorphisms*

$$H^p(\mathfrak{U}, \mathscr{S}) \simeq H^p(X, \mathscr{S}), \qquad p = 0, 1, 2, \ldots.$$

Proof: We now know that the canonical resolution of $\mathscr{S}$ relative to $\mathfrak{U}$, $0 \to \mathscr{S} \to \mathscr{S}^0 \to \mathscr{S}^1 \to \cdots$, is acyclic. Thus, by the formal de Rham Lemma, there is a canonical $\mathscr{R}(X)$-isomorphism of the p-th cohomology module of the section complex $(\mathscr{S}^i(X), d^i_X)$ onto $H^p(X, \mathscr{S})$, $p \geq 1$. Now, by equation (1), $(\mathscr{S}^i(X), d^i_X) = C^\bullet(\mathfrak{U}, \mathscr{S})$. Since $H^p(C^\bullet(\mathfrak{U}, \mathscr{S})) = H^p(\mathfrak{U}, \mathscr{S})$ by definition, the proof is finished. □

Remark: The assumption that $\mathfrak{U}$ is locally finite is quite important in the above proof (e.g. equation (3) is used). However, with other methods one can show that it is superfluous. In all of the applications in several complex variables one can get along with locally finite covers anyway (see Chapter 4).

4. The Isomorphism Theorem $\check{H}^q_a(X, \mathscr{S}) \cong \check{H}^q(X, \mathscr{S}) \cong H^q(X, \mathscr{S})$. If $\mathscr{F}$ is a flabby sheaf over X, then *every* covering $\mathfrak{U}$ is acyclic with respect to $\mathscr{F}$. Since $H^p(X, \mathscr{F}) = 0$ for all $p \geq 1$, the Leray Theorem implies that $H^p(\mathfrak{U}, \mathscr{F}) = 0$ for all $p \geq 1$ whenever $\mathfrak{U}$ is a locally finite cover. Since on paracompact spaces every cover has a locally finite refinement, we now have the following:

If X is paracompact and $\mathscr{F}$ is a flabby sheaf of abelian groups over X, then $\check{H}^p(X, \mathscr{F}) = 0$ for all $p \geq 1$.

Thus on paracompact spaces, we now know that the functors $\check{H}^q$ (together with the connecting homomorphisms $\check{\delta}^q$) satisfy properties I)–III) of paragraph 5.2. Thus over such spaces *the flabby and Čech cohomology theories are isomorphic:*

If X is paracompact and $\mathscr{S}$ is an $\mathscr{R}$-sheaf, then there are natural isomorphisms

$$\check{H}^q(X, \mathscr{S}) \simeq H^q(X, \mathscr{S}), \qquad q \geq 0.$$

In closing we want to make it clear that all of the above remarks for $\check{H}^q$ go through for $\check{H}^q_a$. The "alternating" $\mathscr{R}(U)$-module $\mathscr{S}^p_a(U) := C^p_a(\mathfrak{U} \mid U, \mathscr{S} \mid U)$ is contained in the $\mathscr{R}(U)$-module $\mathscr{S}^p(U) = C^p(\mathfrak{U} \mid U, \mathscr{S} \mid U)$. These modules along with the natural restriction mappings form canonical presheaves for the sheaves of $\mathscr{R}$-modules $\mathscr{S}^p_a$ over X. If one "well orders" the index set I, then

$$(0') \qquad \mathscr{S}^p_a = \prod_{i_0 < \cdots < i_p} \mathscr{S}(i_0, \ldots, i_p),$$

where the product is taken over *all* increasing $(p+1)$-tuples in I^{p+1}. One further verifies that $d^p(\mathscr{S}^p_a) \subset \mathscr{S}^{p+1}_a$ and that the induced $\mathscr{R}$-sequence,

$$(R') \qquad 0 \longrightarrow \mathscr{S} \xrightarrow{\ j\ } \mathscr{S}^0_a \xrightarrow{d_a{}^0} \mathscr{S}^1_a \xrightarrow{d_a} \cdots,$$

is as before an $\mathscr{R}$-resolution of $\mathscr{S}$. In fact the construction in the proof of the Lemma shows that every alternating cocycle is an alternating coboundary. We refer to this as the *canonical alternating resolution of* $\mathscr{S}$ *relative to* $\mathfrak{U}$. It exists for *every* open cover.

If $\mathfrak{U}$ is now locally finite, then (from (0')) we obtain

$$(3') \qquad H^q(X, \mathscr{S}^p_a) \cong \prod_{i_0 < \cdots < i_p} H^q(U(i_0, \ldots, i_p), \mathscr{S}).$$

From this one infers that, for every locally finite cover $\mathfrak{U}$ which is acyclic with respect to $\mathscr{S}$, the canonical alternating resolution (R') is acyclic. Consequently in this case (by the formal de Rham Lemma) the cohomology module $H^p(X, \mathscr{S})$ is isomorphic to the p-th cohomology module of the section complex $(\mathscr{S}^i_a(X), d^i_{a*})$. But $(\mathscr{S}^i_a(X), d^i_{a*}) = C^\bullet_a(\mathfrak{U}, \mathscr{S})$. So

$$H^p(X, \mathscr{S}) \cong H^p(C^\bullet_a(\mathfrak{U}, \mathscr{S})) = H^p_a(\mathfrak{U}, \mathscr{S}), \qquad p = 0, 1, \ldots.$$

If $\mathscr{S} = \mathscr{F}$ is flabby, then, as at the beginning of this section, it follows that $H^p_a(\mathfrak{U}, \mathscr{F}) = 0$ for every locally finite covering $\mathfrak{U}$ of X. Hence, in the paracompact case, the functors $\check{H}^q_a$ likewise have property III). Therefore the flabby and alternating cohomology theories are isomorphic. We summarize now as follows:

If X is paracompact, then, for every $\mathscr{R}$-sheaf $\mathscr{S}$, there exists natural isomorphisms

$$\check{H}^q_a(X, \mathscr{S}) \xrightarrow{\sim} \check{H}^q(X, \mathscr{S}) \xrightarrow{\sim} H^q(X, \mathscr{S}), \qquad q \geq 0.$$

The first isomorphism is as a matter of fact the map i_q introduced in Section 2.4. To see this note that the homomorphism from $\check{H}^q_a(X, S)$ to $\check{H}^q(X, S)$ is induced by the canonical homomorphism from $\mathscr{S}^q_a$ to $\mathscr{S}^q$. Since the sections in $\mathscr{S}^q_a$ (resp. $\mathscr{S}^q$) are the alternating q-cochains (resp. q-cochains), the map $\mathscr{S}^q_a \to \mathscr{S}^q$ is induced by the injection $C^q_a(\mathfrak{U}, \mathscr{S}) \to C^q(\mathfrak{U}, \mathscr{S})$. The latter determines the map i_q. □

As a side result we have the Vanishing Theorem of Paragraph 2.5 for compact blocks B for the flabby cohomology as well:

$$H^q(B, \mathscr{S}) = 0 \text{ for all large } q.$$

Chapter I. Coherence Theory for Finite Holomorphic Maps

In this chapter it is shown that for every finite holomorphic map $f: X \to Y$ the image functor f_* is exact in the category of coherent $\mathcal{O}_X$-sheaves. It is further proved that if $\mathcal{S}$ is a coherent $\mathcal{O}_X$-sheaf, then $f_*(\mathcal{S})$ is coherent. These two theorems are important ingredients in the proof of Theorem B (see Chapter IV, 1.2).

§ 1. Finite Maps and Image Sheaves

The main result of this section is Theorem 4 which is concerned with the exactness of the image functor f_*. Several elementary facts about finite maps are needed for its proof. We begin by assembling these. The reader can find a detailed presentation of these foundations in [CAS].

1. Closed and Finite Maps. All topological spaces which occur, X, Y, ..., are assumed to be Hausdorff.

Definition 1. (Closed finite maps): *Let $f: X \to Y$ be a map. One says that f is closed if the image by f of every closed set in X is again closed in Y. One calls f finite if it is first continuous and closed and if, for all $y \in Y$, the fiber $f^{-1}(y)$ is finite.*

The following lemma is extremely useful:

Lemma 2. *Let $f: X \to Y$ be closed and $U \subset X$ an open neighborhood of a fiber $f^{-1}(y)$ for some $y \in Y$. Then there exists a neighborhood V of y in Y so that $f^{-1}(V) \subset U$.*

Proof: Since U is open, $f(X \backslash U)$ is closed in Y. The set $V := Y \backslash f(X \backslash U)$ is open and contains y. Clearly $f^{-1}(V) \subset U$. □

From here on we consider only finite mappings. The identity id: $X \to X$ is trivially finite. If $f: X \to Y$ and $g: Y \to Z$ are finite then the composition $gf: X \to Z$ is likewise finite. Suppose X' is closed in X. Then the induced map $f': X' \to Y$, defined by $f' := f \mid X'$, is a finite map. Let V be open in Y and let $U := f^{-1}(V)$. If we define f_U by $f_U := f \mid U$, then it is again a finite map.

2. The Bijection $f_*(\mathcal{S})_y \to \prod_{i=1}^{t} \mathcal{S}_{x_i}$. Let $f\colon X \to Y$ be finite. Take y to be an arbitrary point in Y and $x_1, \ldots, x_t \in X$ its different preimages in X. Let $\mathcal{S}$ be a sheaf on X. Then for every open set $V \subset X$ we associated $\mathcal{S}(f^{-1}(V))$. This functor is a canonical presheaf whose sheaf is called the direct image sheaf, $f_*(\mathcal{S})$. Every germ $\sigma_y \in f_*(\mathcal{S})_y$ is represented in a neighborhood V_0 of y by a section $s \in \mathcal{S}(f^{-1}(V_0))$. Since $f^{-1}(V_0)$ is a neighborhood of every x_i, the section s determines germs $s_{x_i} \in \mathcal{S}_{x_i}$, $i = 1, \ldots, t$. These germs are determined by σ_y independent of the choice of the representations s. Thus the map

$$\check{f}\colon f_*(\mathcal{S})_y \to \prod_{1}^{t} \mathcal{S}_{x_i}, \qquad \sigma_y \mapsto (s_{x_1}, \ldots, s_{x_t})$$

of the stalk $f_*(\mathcal{S})_y$ into the cartesian product of the stalks $\mathcal{S}_{x_i}$ is well-defined. In fact $s_{x_i} = \hat{f}_{x_i}(\sigma_y)$, where $\hat{f}_{x_i}\colon f_*(\mathcal{S})_y \to \mathcal{S}_{x_i}$ is the map considered in Chapter A.

Theorem 3. *The map $\check{f}$ is bijective.*

Proof: Injectivity: Let $\sigma_y, \sigma'_y \in f_*(\mathcal{S})_y$. There exists a neighborhood V' of y such that σ_y and σ'_y have representatives $s, s' \in \mathcal{S}(f^{-1}(V))$. If $\check{f}(\sigma_y) = \check{f}(\sigma'_y)$ then, for all $i = 1, \ldots, t$, there exists a neighborhood of x_i, $U_i \subset f^{-1}(V')$, so that $s \mid U_i = s' \mid U_i$. Since $\bigcup_{i=1}^{t} U_i$ is a neighborhood of $f^{-1}(y)$, Lemma 2 implies that there exists a neighborhood V of y with $V \subset V'$ and $f^{-1}(V) \subset \bigcup_{i=1}^{t} U_i$. Thus $s \mid f^{-1}(V) = s' \mid f^{-1}(V)$ and consequently $\sigma_y = \sigma'_y$.

Surjectivity: Let $(s_{x_1}, \ldots, s_{x_t})$ be an arbitrary point in $\prod_{1}^{t} \mathcal{S}_{x_i}$. For each i we choose a neighborhood U_i of x_i and a section $s_i \in \mathcal{S}(U_i)$ which represents $\mathcal{S}_{x_i}$. Since X is Hausdorff, one can assume that the U_i's are pairwise disjoint. Define $U := \bigcup_{i=1}^{t} U_i$, we thus have a section $s \in \mathcal{S}(U)$ which is defined by $s \mid U_i = s_i$, $i = 1, \ldots, t$. Again using Lemma 2, since U is a neighborhood of $f^{-1}(y)$, there exists a neighborhood V of y such that $f^{-1}(V) \subset U$. Certainly $s \mid f^{-1}(V) \in \mathcal{S}(f^{-1}(V))$ and its germ, $\sigma_y \in f_*(S)_y$ satisfies $\check{f}(\sigma_y) = (s_{x_1}, \ldots, s_{x_t})$. □

3. The Exactness of the Functor f_*. Again let $f\colon X \to Y$ be finite and as above let $x_1, \ldots, x_t \in X$ be the distinct preimages of $y \in Y$. Take $\mathcal{S}$ and $\mathcal{S}'$ to be sheaves over X and $\varphi\colon \mathcal{S}' \to \mathcal{S}$ a sheaf mapping. Thus we have the induced mapping of sheaves $f_*(\varphi)\colon f_*(\mathcal{S}') \to f_*(\mathcal{S})$. Now φ gives us maps of the stalks $\varphi_{x_i}\colon \mathcal{S}'_{x_i} \to \mathcal{S}_{x_i}$. Hence we have the diagram

$$\text{(D)} \qquad \begin{array}{ccc} f_*(\mathcal{S}')_y & \xrightarrow{f_*(\varphi)_y} & f_*(\mathcal{S})_y \\ \Big\downarrow \check{f}' & & \Big\downarrow \check{f} \\ \prod_{1}^{t} \mathcal{S}'_{x_i} & \xrightarrow{\prod_{1}^{t} \varphi_{x_i}} & \prod_{1}^{t} \mathcal{S}_{x_i} \end{array}$$

where $\check{f}'$ and $\check{f}$ are the bijections associated to $\mathscr{S}'$ and $\mathscr{S}$ respectively (see Theorem 3). It follows immediately that *diagram* (D) *is commutative.*

Let $\mathscr{S}'$ and $\mathscr{S}$ be sheaves of abelian groups and suppose that $\varphi\colon \mathscr{S}' \to \mathscr{S}$ is a sheaf homomorphism in the category of such sheaves. Then $f_*(\mathscr{S}')$ and $f_*(\mathscr{S})$ are sheaves of abelian groups, the mappings $\check{f}'$ and $\check{f}$ are group isomorphisms and $f_*(\varphi)$ is a homomorphism of sheaves of abelian groups. This immediately implies that all of the maps in diagram (D) are group homomorphisms. Hence the following is evident.

Theorem 4 (*Exactness Theorem*): *Let $f\colon X \to Y$ be a finite map and suppose that $\mathscr{S}' \to \mathscr{S} \to \mathscr{S}''$ is an exact sequence of sheaves of abelian groups on X. Then the sequence of image sheaves*

$$f_*(\mathscr{S}') \to f_*(\mathscr{S}) \to f_*(\mathscr{S}''),$$

is likewise exact.

Proof: By the remarks above, the two sequences,

$$\prod_1^t \mathscr{S}'_{x_i} \to \prod_1^t \mathscr{S}_{x_i} \to \prod_1^t \mathscr{S}''_{x_i}$$

and

$$f_*(\mathscr{S}')_y \to f_*(\mathscr{S})_y \to f_*(\mathscr{S}'')_y,$$

are isomorphic. Therefore the exactness of the first implies the exactness of the second. □

4. The Isomorphisms $H^q(X, \mathscr{S}) = H^q(Y, f_*(\mathscr{S}))$. The following direct application of Theorem 4 is needed in Chapter IV.

Theorem 5. *Let $f\colon X \to Y$ be a finite map. Then for every sheaf of abelian groups ($\mathbb{C}$-vector spaces) on X, $\mathscr{S}$, there exist natural group ($\mathbb{C}$-vector space) isomorphisms*

$$H^q(X, \mathscr{S}) \cong H^q(Y, f_*(\mathscr{S})), \qquad q \geq 0.$$

Proof: Let $0 \longrightarrow \mathscr{S} \xrightarrow{\ j\ } \mathscr{S}^0 \xrightarrow{\ d^0\ } \mathscr{S}^1 \cdots$ be the canonical flabby resolution of $\mathscr{S}$ on X. By Theorem 4 the sequence of image sheaves,

$$(*) \qquad 0 \longrightarrow f_*(\mathscr{S}) \xrightarrow{f_*(j)} f_*(\mathscr{S}^0) \xrightarrow{f_*(d^0)} f_*(\mathscr{S}^1) \longrightarrow \cdots$$

is exact. Furthermore $f_*(\mathscr{S}^i)$, $i \geq 0$, is flabby (see A.4.3). Hence $(*)$ is a flabby resolution of $f_*(\mathscr{S})$ and as a result

$$H^q(Y, f_*(\mathscr{S})) \cong \mathrm{Ker}[f_*(\mathscr{S}^q)(Y) \to f_*(\mathscr{S}^{q+1})(Y)]/\mathrm{Im}[f_*(\mathscr{S}^{q-1}(Y)) \to f_*(\mathscr{S}^q)(Y)].$$

But the canonical isomorphisms between $f_*(\mathscr{S}^q)(Y)$ and $\mathscr{S}^q(X)$ are compatible with the homomorphisms $f_*(d^q)$ and d^q. Consequently

$$\operatorname{Ker}[f_*(\mathscr{S}^q)(Y) \to f_*(\mathscr{S}^{q+1})(Y)] \cong \operatorname{Ker}[\mathscr{S}^q(X) \to \mathscr{S}^{q+1}(X)], \qquad q \geq 0.$$

Since the analogous remark holds for the image groups,

$$H^q(Y, f_*(\mathscr{S})) = \operatorname{Ker}[\mathscr{S}^q(X) \to \mathscr{S}^{q+1}(X)]/\operatorname{Im}[\mathscr{S}^{q-1}(X) \to \mathscr{S}^q(X)] \cong H^q(X, \mathscr{S})$$

for all $q \geq 0$.

5. The $\mathcal{O}_y$-Module Isomorphism $\tilde{f}: f_*(\mathscr{S})_y \to \prod_1^t \mathscr{S}_{x_i}$. Now let X, Y be complex spaces and $f: X \to Y$ a finite *holomorphic* mapping. Thus if $\mathscr{S}$ is an $\mathcal{O}_X$-sheaf, then $f_*(\mathscr{S})$ is $\mathcal{O}_Y$-sheaf. Let $y \in Y$ and $\{x_1, \ldots, x_t\} = f^{-1}(y)$. Then we can associate to f the *lifting* homomorphisms $f_i^*: \mathcal{O}_y \to \mathcal{O}_{x_i}$, $1 \leq i \leq t$. Every $\mathcal{O}_{x_i}$-module, $\mathscr{S}_{x_i}$, is therefore in the obvious way an $\mathcal{O}_y$-module. Consequently $\prod_1^t \mathscr{S}_{x_i}$ is an $\mathcal{O}_y$-module. Likewise the stalk $f_*(\mathscr{S})_y$ is an $\mathcal{O}_y$-module. The multiplication is defined by taking $h \in \mathcal{O}_Y(V)$, $s \in \mathscr{S}(f^{-1}(V))$ and letting $h \cdot s := h \circ f \cdot s \in \mathscr{S}(f^{-1}(V))$, where $h \circ f \in \mathcal{O}_X(f^{-1}(V))$ is the lift of h by f. This has the following consequence (also see [CAS]):

If $f: X \to Y$ is a finite holomorphic map, then every bijection

$$\tilde{f}: f_*(\mathscr{S})_y \to \prod_1^t \mathscr{S}_{x_i}, \qquad y \in Y, \qquad \{x_1, \ldots, x_t\} = f^{-1}(y),$$

is an $\mathcal{O}_y$-module isomorphism.

§ 2. The General Weierstrass Division Theorem and the Weierstrass Isomorphism

The key tools of this section are the Weierstrass division theorem and Hensel's lemma for convergent power series. The reader can find these in [AS] (Theorem I.4.2, p. 34/35, and Theorem I.5.6, p. 49/50). The main result here is that for certain finite holomorphic maps f, the direct image map f_* preserves coherence.

1. Continuity of Roots. Let B be a domain in $\mathbb{C}^n$. We use $z = (z_1, \ldots, z_n)$ for the coordinates in $\mathbb{C}^n$ and consider a monic polynomial,

$$\omega = \omega(w, z) := w^b + a_1(z)w^{b-1} + \cdots + a_b(z) \in \mathcal{O}(B)[w],$$

in another variable $w \in \mathbb{C}$. The degree of ω is b, $1 \leq b \leq \infty$, and the coefficients are holomorphic on B: $a_j \in \mathcal{O}(B)$ for all j. The zero set of ω,

$$A := \{(w, z) \in \mathbb{C} \times B \,|\, \omega(w, z) = 0\},$$

is an analytic set in $\mathbb{C} \times B$. The projection $\mathbb{C} \times B \to B$ induces by restriction a *continuous, surjective* map $\pi: A \to B$. Each fiber, $\pi^{-1}(z)$, $z \in B$, contains at most b different points. Furthermore we have the following

Theorem 1 (Continuity of the Roots). *The map $\pi: A \to B$ is closed and therefore finite.*

Proof: Let M be closed in A and $y \in B$ be an accumulation point of $\pi(M)$. Thus there is a sequence $(c_\nu, y_\nu) \in M$ so that the y_ν's converge to y. We note that $c_\nu^b = -(a_1(y_\nu)c_\nu^{b-1} + \cdots + a_b(y_\nu))$. Hence, if $|c_\nu| \geq 1$, then $|c_\nu| \leq |c_\nu^b| \leq |a_1(y_\nu)| + \cdots + |a_b(y_\nu)|$. Therefore

$$(*) \qquad |c_\nu| \leq \max\{1, |a_1(y_\nu)| + \cdots + |a_b(y_\nu)|\}.$$

Since (y_ν) converges to y, each sequence $a_j(y_\nu)$, $1 \leq j \leq b$, converges and is in particular bounded. Thus the sequence $\{c_\nu\} \subset \mathbb{C}$ is bounded and has a subsequence $\{c_{\nu_k}\}$ which converges to $c \in \mathbb{C}$. Hence $\{(c_{\nu_k}, y_{\nu_k})\}$ converges to $(c, y) \in A$. By assumption M is closed. Thus $(c, y) \in M$ and consequently $y = \pi(c, y) \in \pi(M)$.

2. The General Weierstrass Division Theorem. We retain the notation used in the above paragraph. Let $y \in B$ be fixed and $x_i = (w_i, y)$ be preimages of y under the map π. For the sake of brevity we write $\mathcal{O}_{x_i}$ (resp. $\mathcal{O}_y$) instead of $\mathcal{O}_{\mathbb{C} \times B, x_i}$ (resp. $\mathcal{O}_{B,y}$). It follows that $\mathcal{O}_y[w] \subset \mathcal{O}_y\langle w - w_i\rangle = \mathcal{O}_{x_i}$. Thus every polynomial $p \in \mathcal{O}_y[w]$ determines at each x_i a germ $p_{x_i} \in \mathcal{O}_{x_i}$.

Theorem 2 (Weierstrass Division Theorem). *Let t germs $f_i \in \mathcal{O}_{x_i}$, $1 \leq i \leq t$, be arbitrarily prescribed. Then there is a polynomial $r \in \mathcal{O}_y[w]$ with* deg *$r < b$ and t germs $q_i \in \mathcal{O}_{x_i}$, $1 \leq i \leq t$, such that*

$$f_i = q_i \omega_{x_i} + r_i$$

for all i. The polynomial r and the germs $q_1, \ldots, q_t$ are uniquely determined by $f_1, \ldots, f_t$ and w.

Remark: If $t = 1$ then ω_{x_1} is a Weierstrass polynomial in $(w - w_1)$ and Theorem 2 is the usual Weierstrass division theorem.

Proof: Let $y = 0$. Then

$$\omega(w, 0) = (w - w_1)^{b_1} \cdots (w - w_t)^{b_t},$$

where $b_i \geq 1$ and $\sum_1^t b_i = b$. By Hensel's lemma there exist monic, pairwise relatively prime polynomials $\omega_i(w, z) \in \mathcal{O}_y[w]$ so that, for the germ $\omega_y \in \mathcal{O}_y[w]$ determined by $\omega \in \mathcal{O}(B)[w]$,

$$\omega_y = \omega_1 \cdots \omega_t,$$

where $\omega_i(w, 0) = (w - w_i)^{b_i}$, $1 \le i \le t$. Every induced germ $\omega_{ix_i} \in \mathcal{O}_{x_i} \cong \mathcal{O}_y\langle w - w_i\rangle$ is distinguished in $w - w_i$ and of degree b_i. Furthermore each polynomial $e_i := \prod_{j \ne i} \omega_j \in \mathcal{O}_y[w]$ induces a unit $e_{ix_i} \in \mathcal{O}_{x_i}$. This follows because $e_i(x_i) = \prod_{j \ne i} \omega_j(w_i, 0) = \prod_{j \ne i} (w_i - w_j)^{b_j} \ne 0$. Clearly $\omega_y = e_i \omega_i$, $1 \le i \le t$. □

After the above preparations, the existence and uniqueness statements now follow in a few lines:

Existence: It is enough to prove existence statement for t-tuples of the special form $(0, \ldots, 0, f_i, 0, \ldots, 0)$, $f_i \in \mathcal{O}_{x_i}$. The more general result will follow immediately by addition. By the usual Weierstrass Theorem, given $f_i e_{ix_i}^{-1} \in \mathcal{O}_{x_i}$, we have the decomposition

$$f_i e_{ix_i}^{-1} = q_i \cdot \omega_{ix_i} + r'_{x_i},$$

where $q_i \in \mathcal{O}_{x_i}$, $r' \in \mathcal{O}_y[w - w_i]$ and $\deg r' < b_i$.

Let $r := r' e_i$. Then $r \in \mathcal{O}_y[w]$ and with $\deg r < b_i + \deg e_i = b$. Thus, since $\omega_y = e_i \omega_i$,

$$f_i = q_i \omega_{x_i} + r_{x_i}.$$

If $j \ne i$, then, since $\omega_i(x_j) = (w_j - w_i)^{b_i} \ne 0$, every germ $\omega_{ix_j} \in \mathcal{O}_{x_j}$ is a unit. Thus one can determine $q_j := -r'_{x_j} \omega_{ix_j}^{-1} \in \mathcal{O}_{x_j}$ for all $j \ne i$. It follows that, for such j,

$$0 = q_j \omega_{x_j} + r_{x_j}.$$

Uniqueness: It is enough to show that if we have t equations $q_i \omega_{x_i} + r_{x_i} = 0$, where $q_{x_i} \in \mathcal{O}_{x_i}$ and $r \in \mathcal{O}_y[w]$ with $\deg r < b$, $i = 1, \ldots, t$, then $r = 0$ and $q_1 = \cdots = q_t = 0$. Since ω_{ix_i} is distinguished in $(w - w_i)$ and since $\omega_{x_i} = e_{ix_i} \omega_{ix_i}$, it follows that $-r_{x_i} = (q_i e_{ix_i}) \omega_{ix_i}$ is the Weierstrass decomposition of $-r_{x_i} \in \mathcal{O}_y[w - w_i]$ in $\mathcal{O}_{x_i} \cong \mathcal{O}_y\langle w - w_i\rangle$ with respect to ω_{ix_i}, $1 \le i \le t$. Thus $q_i e_{ix_i}$ is a polynomial in w. In other words, ω_i divides r in $\mathcal{O}_y[w]$, $i = 1, \ldots, t$. Since $\omega_1, \ldots, \omega_t$ are pairwise relatively prime, $\omega_y = \omega_1 \cdots \omega_t$ divides r in $\mathcal{O}_y[w]$. But $\deg r < \deg \omega$. Thus $r = 0$ and consequently $q_1 = \cdots = q_t = 0$.

3. The Weierstrass Homomorphism $\mathcal{O}_B^b \xrightarrow{\sim} \pi_*(\mathcal{O}_A)$. We maintain the above notation. For short we write $\mathcal{O}$ instead of $\mathcal{O}_{\mathbb{C} \times B}$. The polynomial ω generates a sheaf of ideals $\mathcal{J} := \omega\mathcal{O}$. The quotient sheaf $\mathcal{O}/\mathcal{J}$ is "supported" on A. We let $\mathcal{O}_A$ denote the restriction of $\mathcal{O}/\mathcal{J}$ to A. Thus A is a complex space with structure sheaf $\mathcal{O}_A$. The projection $\pi\colon A \to B$ is holomorphic and (by Theorem 1) is finite.

We now construct an $\mathcal{O}_B$-sheaf homomorphism $\mathring{\pi}\colon \mathcal{O}_B^b \to \pi_*(\mathcal{O}_A)$. It will be shown that it is in fact an isomorphism. Let U be open in B and let $s = (r_0, \ldots, r_{b-1}) \in \mathcal{O}_B^b$ be a section. The polynomial $r := \sum_{\beta=0}^{b-1} r_\beta w^\beta \in \mathcal{O}(\mathbb{C} \times U)$ determines (module $\mathscr{I}$) a section $\bar{s} \in (\mathcal{O}/\mathscr{I})(\mathbb{C} \times U)$. By the canonical isomorphisms, $(\mathcal{O}/\mathscr{I})(\mathbb{C} \times U) \cong \mathcal{O}_A(\pi^{-1}(U)) = \pi_*(\mathcal{O}_A)(U)$. Thus we may think of $\bar{s}$ as a section $\mathring{s} \in \pi_*(\mathcal{O}_A)(U)$. The map

$$\mathring{\pi}_U\colon \mathcal{O}_B^b(U) \to \pi_*(\mathcal{O}_A)(U), \qquad s \to \mathring{s},$$

is obviously an $\mathcal{O}_B(U)$-module homomorphism. The collection of these homomorphism is compatible with the restriction maps. Hence we obtain an $\mathcal{O}_B$-sheaf homomorphism $\mathring{\pi}\colon \mathcal{O}_B^b \to \pi_*(\mathcal{O}_A)$. We call $\mathring{\pi}$ the *Weierstrass homomorphism* with respect to ω.

Theorem 3. *The Weierstrass homomorphism $\mathring{\pi}\colon \mathcal{O}_B^b \to \pi_*(\mathcal{O}_A)$ is an $\mathcal{O}_B$-isomorphism.*

Proof: It is enough to show that every homomorphism of germs, $\mathring{\pi}_z\colon \mathcal{O}_{B,z}^b \to \pi_*(\mathcal{O}_A)_z$, $z \in B$, is bijective. Let $x_1, \ldots, x_t \in A$ be the preimages of z under the map π. We identify $\mathcal{O}_{x_i}/\mathscr{I}_{x_i}$ with $(\mathcal{O}/\mathscr{I})_{x_i}$ and bring into play the $\mathcal{O}_{B,z}$-isomorphism

$$\check{\pi}_z\colon \pi_*(\mathcal{O}_A)_z \xrightarrow{\sim} \prod_1^t (\mathcal{O}/\mathscr{I})_{x_i} = \prod_1^t (\mathcal{O}_{x_i}/\mathscr{I}_{x_i}).$$

Thus we have the commutative diagram

$$\begin{array}{ccc} \mathcal{O}_{B,z}^b & \xrightarrow{\mathring{\pi}_z} & \pi_*(\mathcal{O}_A)_z \\ & \searrow{\scriptstyle \tau_z} & \downarrow{\scriptstyle \check{\pi}_z} \\ & & \prod_1^t (\mathcal{O}_{x_i}/\mathscr{I}_{x_i}) \end{array},$$

where $\tau_z(r_{0z}, \ldots, r_{b-1,z}) = (r_{x_1} \bmod \mathscr{I}_{x_1}, \ldots, r_{x_t} \bmod \mathscr{I}_{x_t})$ and $r := \sum_{\beta=0}^{b-1} r_{\beta z} w^\beta$. The bijectivity of $\mathring{\pi}_z$ is equivalent to the bijectivity of τ_z. As one easily confirms, the latter is exactly the content of Theorem 2. □

It should be mentioned in passing that up to this point we have not used the coherence of the sheaf $\mathcal{O}$. Using Theorem 3 one can carry out by induction an elegant coherence proof for this structure sheaf. The reader can find this and related details in [CAS].

4. The Coherence of the Direct Image Functor π_*. Let $z \in B$ be arbitrary and let $x_1, \ldots, x_t \in A$ be its preimages under π. If $\mathscr{S}$ is a coherent $\mathcal{O}_A$-sheaf, then for each i there exists an open neighborhood U_i of x_i and a sequence $\mathcal{O}_{U_i}^{p_i} \to \mathcal{O}_{U_i}^{q_i} \to \mathscr{S}_{U_i} \to 0$, $p_i, q_i \geq 1$, which is exact on U_i. We may assume that $p_i = p$ and $q_i = q$ independent of i and that the U_i's are pairwise disjoint. Then $U := \bigcup_1^t U_i$ is a neighborhood of the fiber $\pi^{-1}(z)$ and we have the exact sequence

(*) $$\mathcal{O}_U^p \to \mathcal{O}_U^q \to \mathscr{S}_U \to 0.$$

By Lemma 1.2 we may take U smaller so that $U = \pi^{-1}(V)$ where V is a neighborhood of z in B. Then the induced map $\pi_U\colon U \to V$ is finite and the functor $(\pi_U)_*$ carries the sequence $(*)$ over to an exact sequence on V (see Theorem 1.4):

$$(\pi_U)_*(\mathcal{O}_U^p) \to (\pi_U)_*(\mathcal{O}_U^q) \to (\pi_U)_*(\mathcal{S}_U) \to 0.$$

Of course $(\pi_U)_*(\mathcal{S}_U) = \pi_*(\mathcal{S})_V$. Moreover, using the polynomial ω_V which is the restriction of ω to V, Theorem 3 yields an $\mathcal{O}_V$-isomorphism

$$(\pi_U)_*(\mathcal{O}_U^r) = \pi_*(\mathcal{O}_A^r)_V \cong \mathcal{O}_V^{br}.$$

Thus we have the sequence

$$\mathcal{O}_V^{bp} \to \mathcal{O}_V^{bq} \to \pi_*(\mathcal{S})_V \to 0$$

which is exact on V. This proves that $\pi_*(\mathcal{S})_V$ is coherent. (Here the coherence of $\mathcal{O}_V$ is used in an essential way!) For future reference, we state this result explicitely:

Coherence Theorem 4. *If $\mathcal{S}$ is a coherent $\mathcal{O}_A$-sheaf, then $\pi_*(\mathcal{S})$ is a coherent $\mathcal{O}_B$-sheaf.*

§ 3. The Coherence Theorem for Finite Holomorphic Maps

The purpose of this section is to prove that the direct image of a coherent sheaf by a finite map is coherent. This is proved in three steps. First, Theorem 2.4 is applied in the situation where the sheaf is coherent on some neighborhood of the origin in $\mathbb{C} \times \mathbb{C}^n$ and the finite map is the restriction of the projective $\mathbb{C} \times \mathbb{C}^n \to \mathbb{C}^n$ to the support of the sheaf. This is immediately generalized by induction to the case of $\mathbb{C}^m \times \mathbb{C}^n \to \mathbb{C}^n$. Then we show that any finite map can be locally realized as the restriction of such a projection and consequently a local version of the desired coherence theorem is proved. Finally, since coherence is a local notion, the global result (Theorem 3) follows in a few lines.

1. The Projection Theorem. Let $w = (w_1, \dots, w_m)$(resp. $z = (z_1, \dots, z_n)$) be the coordinates in $\mathbb{C}^m$ (resp. $\mathbb{C}^n$). The objective of this paragraph is to prove the following:

Theorem 1. *Let $\mathcal{S}$ be a coherent analytic sheaf on some neighborhood U of the origin $(0, 0) \in \mathbb{C}^m \times \mathbb{C}^n$. Suppose that $(0, 0)$ is isolated in $\operatorname{supp} \mathcal{S} \cap (\mathbb{C}^m \times \{0\})$. Then there exist open neighborhoods W and Z of $0 \in \mathbb{C}^m$, $0 \in \mathbb{C}^n$ with $W \times Z \subset U$ such that the following hold for the projection $\varphi\colon W \times Z \to Z$:*

1) *The restriction map $\varphi \,|\operatorname{supp} \mathcal{S} \cap (W \times Z)$ is finite.*
2) *The image sheaf $\varphi_*(S_{W \times Z})$ is a coherent sheaf of $\mathcal{O}_Z$-modules.*

Proof: a) We may assume that supp $\mathscr{S}$ intersects the plane $\mathbb{C}^m \times \{0\}$ at the origin $x := (0, 0)$. The annihilator sheaf $\mathrm{An}(\mathscr{S}) \subset \mathcal{O}_U$ is coherent on U and vanishes exactly on $\mathrm{supp}(\mathscr{S})$. Thus there is a germ $f \in \mathrm{An}(\mathscr{S})_x$ with $f(w, 0) \neq 0$. We may assume that f is distinguished in w_1. In order to do this we need only to make a linear change of variables in $w_1, \ldots, w_m$. Thus by the preparation theorem, there exists a unit $e \in \mathcal{O}_x$ and a polynomial $\omega_x = w_1^b + \sum_{i=1}^{b} a_{ix} w_1^{b-i}$ whose coefficients a_{ix} are holomorphic germs in $w := (w_2, \ldots, w_n)$ and z and which vanishes at $0 \in \mathbb{C}^{m+n-1}$ with $f = e\omega_x$. Since e is a unit, ω_x is also in $\mathrm{An}(S)_x$. Now we can find a neighborhood W_1 of the origin in the w_1-axes and a neighborhood $T \subset \mathbb{C}^{m+n-1}$ of the origin in the (w', z)-space with $W_1 \times T \subset U$ and such that every germ a_{ix} has a holomorphic representative $a_i \in \mathcal{O}(T)$. Thus the polynomial $\omega := w_1^b + \sum_{i=1}^{b} a_i w_1^{b-i} \in \mathcal{O}(T)[w_1]$ is a section of $\mathrm{An}(\mathscr{S})$ on $W_1 \times T$. We let A denote the zero set of ω in $\mathbb{C} \times T$.

Since the equation $\omega(w_1, 0, \ldots, 0) = 0$ is solved only by $w_1 = 0$, we can take T to be small enough so that $A \subset W_1 \times T$. Now let $\psi\colon W_1 \times T \to T$ be projection on the second factor and define $\pi\colon A \to T$ by $\pi := \psi \,|\, A$. By Theorem 2.1, if one equips A with the structure sheaf $\mathcal{O}_A := \mathcal{O}/\omega\mathcal{O} \,|\, A$, π is a finite map. Since $\mathscr{S}_{(W_1 \times T)\setminus A} = 0$, we may consider $\mathscr{S}_{W_1 \times T}$ to be a coherent sheaf of $\mathcal{O}_A$-modules and $\psi_*(\mathscr{S}_{W_1 \times T}) = \pi_*(\mathscr{S}_{W_1 \times T})$. Furthermore, by Theorem 2.4, $\mathscr{S}' := \psi_*(\mathscr{S}_{W_1 \times T})$ is a coherent sheaf of $\mathcal{O}_T$-modules. Since supp $\mathscr{S}$ is a closed analytic subset of A, $\psi \,|\, \mathrm{supp}\, \mathscr{S} \cap (W_1 \times T)\colon \mathrm{supp}\, \mathscr{S} \cap (W_1 \times T) \to T$ is likewise finite. We note that all of the above remains valid if T is taken to be a smaller neighborhood of the origin in $\mathbb{C}^{m-1} \times \mathbb{C}^n$.

b) The proof of Theorem 1 is now quickly finished by an argument involving induction on m. If $m = 1$, then we take $W := W_1$, $Z := T$, $\varphi := \psi$ and the above arguments yield the desired result. If $m > 1$, then we consider the coherent image sheaf of $\mathcal{O}_T$-modules, $\mathscr{S}'$, on T. Since $\{(0, 0)\} = \mathrm{supp}\, \mathscr{S} \cap (W_1 \times T)$, the origin is isolated in the intersection of the plane $\mathbb{C}^{m-1} \times \{0\}$ with $\mathrm{supp}\, \mathscr{S}' = \psi(\mathrm{supp}\, \mathscr{S}_{W_1 \times T})$. Hence we may apply the induction assumption: There exists a neighborhood W' of $0 \in \mathbb{C}^{m-1}$ and a neighborhood Z of $0 \in \mathbb{C}^n$ with $W' \times Z \subset T$ so that the projection $\chi\colon W' \times Z \to Z$ induces a finite map $\mathrm{supp}\, \mathscr{S}' \cap (W' \times Z) \to Z$ and the image sheaf $\chi_*(\mathscr{S}'_{W' \times Z})$ is a coherent sheaf of $\mathcal{O}_Z$-modules. We now set $W := W_1 \times W'$ and redefine T to be the (possibly) smaller set $W' \times Z$. Thus the projection $\varphi\colon W \times Z \to Z$ factors through $W' \times Z$: $W \times Z \xrightarrow{\psi} W' \times Z \xrightarrow{\chi} Z$. Since $\varphi_*(\mathscr{S}_{W \times Z}) \cong \chi_*(\psi_*(\mathscr{S}_{W \times Z}))$ and $\psi_*(\mathscr{S}_{W \times Z}) = \mathscr{S}'$ is coherent, the coherence statement 2) has been proved. Moreover, the restriction of φ to $\mathrm{supp}\, \mathscr{S} \cap (W \times Z)$ also factors into two finite maps:

$$\mathrm{supp}\, \mathscr{S} \cap (W \times Z) \to \mathrm{supp}\, \mathscr{S}' \cap (W' \times Z) \to Z.$$

Thus we have the finiteness of $\varphi \,|\, \mathrm{supp}\, \mathscr{S} \cap (W \times Z)$.

2. Finite Holomorphic Maps (Local Case). Let X and Y be complex spaces and $f\colon X \to Y$ a holomorphic map. Let $\mathscr{S}$ be a coherent sheaf of $\mathcal{O}_X$-modules.

Theorem 2. *Suppose that x_0 is an isolated point of* $\operatorname{supp} \mathcal{S} \cap f^{-1}(f(x_0))$ *in X. Let U_0 be any open neighborhood of x_0 in X. Then there exist neighborhoods U, V of $x_0 \in X$ (resp. $f(x_0) \in Y$) with $U \subset U_0$ and $f(U) \subset V$ such that the following hold:*

1) *The restriction of the induced map f_U: $U \to V$ to* $\operatorname{supp} \mathcal{S} \cap U$ *is finite.*
2) *The image sheaf $(f_U)_*(\mathcal{S}_U)$ is a coherent sheaf of $\mathcal{O}_V$-modules.*

Proof. a) We first consider the case where Y is a domain in $\mathbb{C}^n$. We choose a neighborhood $U' \subset U_0$ of x_0 in X so that there is a biholomorphic embedding i: $U' \to G$ of U' into a domain $G \subset \mathbb{C}^m$. Defining $f' := f \mid U'$, the "product map" $i \times f'$: $U' \to G \times Y$ is therefore a biholomorphic embedding of U' into $G \times Y \subset \mathbb{C}^m \times \mathbb{C}^n$. Now the trivial continuation of the image sheaf $(i \times f')_*(\mathcal{S}_{U'})$ to all of $G \times Y$ is a coherent sheaf of $\mathcal{O}_{G \times Y}$-modules, $\mathcal{S}^*$. Moreover $\operatorname{supp} \mathcal{S}^* \cap (G \times \{y_0\}) = (i \times f')(x_0)$, where $y_0 := f(x_0)$. Thus we are in a position to apply Theorem 1: There exist neighborhoods W and Z of $i(x_0) \in G$ and $y_0 \in Y$ with $W \times Z \subset G \times Y$ such that, if φ: $W \times Z \to Z$ is the projection, the following hold:

1) The restriction of φ to $\operatorname{supp} \mathcal{S}^* \cap (W \times Z)$ is finite.
2) The sheaf $\varphi_*(\mathcal{S}^*_{W \times Z})$ is a coherent sheaf of $\mathcal{O}_Z$-modules.

Let $V := Z$ and $U := (i \times f')^{-1}(W \times Z)$. Then, since it is the composition of two finite maps, $f_U | \operatorname{supp} \mathcal{S} \cap U$: $\operatorname{supp} \mathcal{S} \cap U \to V$ is finite and the image sheaf $(f_U)_*(\mathcal{S}_U) = \varphi_*(\mathcal{S}^*_{W \times Z})$ is coherent on V.

b) Now let Y be arbitrary. Since the conclusions of the theorem are local in nature, we may assume that Y is a complex subspace of a domain B in $\mathbb{C}^n$. Taking ι: $Y \to B$, we define the finite map $\tilde{f}$: $X \to B$ by $\tilde{f} := \iota \circ f$. Applying part a) to $\tilde{f}$, there exist neighborhoods U and $\tilde{V}$ of $x_0 \in X$ and $y_0 \in B$ so that $\tilde{f}_U$: $U \to \tilde{V}$ induces a finite map $\operatorname{supp} \mathcal{S} \cap U \to \tilde{V}$ and such that $(\tilde{f}_U)_*(\mathcal{S}_U)$ is a coherent sheaf of $\mathcal{O}_{\tilde{V}}$-modules. We define $V := \tilde{V} \cap Y$ as a neighborhood of y_0 in Y. Since $\operatorname{Im} \tilde{f}_U \subset V$, it follows that f_U: $U \to V$ induces a finite holomorphic map $\operatorname{supp} \mathcal{S} \cap U \to V$. Furthermore $(\tilde{f}_U)_*(\mathcal{S}_U)$ is just the trivial continuation of $(f_U)_*(\mathcal{S}_U)$ to B. So $(f_U)_*(\mathcal{S}_U)$ is a coherent $\mathcal{O}_V$-sheaf. □

Since $\operatorname{supp} \mathcal{O}_X = X$, the following case is an immediate application of Theorem 2 in the case where $\mathcal{S} = \mathcal{O}_X$.

Corollary. *Let f: $X \to Y$ be a holomorphic map such that x_0 is an isolated point of the fiber $f^{-1}(f(x_0))$. Then there exist neighborhoods U and V of $x_0 \in X$ and $f(x_0) \in Y$ with $f(U) \subset V$ so that the restriction f_U: $U \to V$ is finite.*

3. Finite Holomorphic Maps and Coherence. The theorem below is the main result of this section. It is now just an easy consequence of Theorem 2.

Theorem 3 (The Direct Theorem for Finite Maps). *Let f: $X \to Y$ be a finite holomorphic map and $\mathcal{S}$ be a coherent sheaf of $\mathcal{O}_X$-modules. Then $f_*(\mathcal{S})$ is a coherent sheaf of $\mathcal{O}_Y$-modules.*

Proof: Let $y \in Y$ be arbitrary and $x_1, \ldots, x_t \in X$ be the different f-preimages of y. By Theorem 2 there are neighborhoods U_i and V_i of $x_i \in X$ and $y \in Y$ with $f(U_i) \subset V_i$ so that the induced map $f_{U_i}\colon U_i \to V_i$ is finite and $f_{U_i*}(\mathscr{S}_{U_i})$ is coherent on V_i, $1 \leq i \leq t$. We may assume that the U_i's are pairwise disjoint. Let V be a neighborhood of $y \in Y$ which is contained in $\bigcup_1^t V_i$. Then for $1 \leq i \leq t$ we have the finite maps $f_i\colon W_i \to V$, where $W_i := U_i \cap f^{-1}(V)$ and $f_i := f \mid W_i$. Furthermore

$$f_{i*}(\mathscr{S}_{W_i}) = f_{U_i*}(\mathscr{S}_{U_i})_V$$

and thus $f_{i*}(\mathscr{S}_{W_i})$ is a coherent $\mathcal{O}_V$-sheaf, $i = 1, \ldots, t$.

We now choose V so small that $U := f^{-1}(V) = \bigcup_1^t W_i$ (this is possible by Lemma 1.2) and we consider the restriction of f to U, $f_U\colon U \to V$. Certainly $f_{U*}(\mathscr{S}_U) = f_*(\mathscr{S})_V$. Since $W_i \cap W_j = \emptyset$ for $i \neq j$,

$$f_*(\mathscr{S})(V') = \prod_1^t \mathscr{S}(W_i \cap f^{-1}(V'))$$

for every open set $V' \subset V$. But each $f_{i*}(\mathscr{S}_{W_i})$ is coherent. Thus $f_*(\mathscr{S})_V$ is a coherent sheaf of $\mathcal{O}_V$-modules. □

Chapter II. Differential Forms and Dolbeault Theory

In this chapter Dolbeault cohomology theory is presented. One of the basic tools is the $\bar{\partial}$-integration lemma for closed (p, q)-forms (Theorem 4.1). The proof of this lemma is based on the existence of bounded solutions of the inhomogeneous Cauchy-Riemann differential equation $\partial g/\partial \bar{z} = f$. This solution is constructed in Paragraph 3 by means of the classical integral operator

$$Tf(z, u) = \frac{1}{2\pi i} \iint_B \frac{f(\zeta, u)}{\zeta - z} \, d\zeta \wedge d\bar{\zeta}.$$

The Dolbeault theory yields among other things the fact that $H^q(Q, \mathcal{O}) = 0, q \geq 1$, for compact blocks $Q \subset \mathbb{C}^m$ (Theorem 4.4). This vanishing theorem is needed in Chapter III in order to prove Theorem B for compact blocks.

In Section 1.2 we collect in more detail than is really necessary the general facts about differential forms on manifolds. We always use X to denote a differentiable manifold of real dimension m and $\mathscr{E}^{\mathbb{R}}$ its real structure sheaf. For short we write $\mathscr{E}$ for the sheaf $\mathscr{E}^{\mathbb{R}} := \mathscr{E}^{\mathbb{R}} + i\mathscr{E}^{\mathbb{R}}$. Beginning in Section 2, X has the additional structure of a complex manifold, with complex structure sheaf $\mathcal{O} \subset \mathscr{E}$. The symbols U, V are reserved for open sets in X. Real local coordinates on X are denoted by $u_1, \ldots, u_m$.

§ 1. Complex Valued Differential Forms on Differentiable Manifolds

1. Tangent Vectors. We need the following representation theorem for germs of functions.

Theorem 1. *Let $u_1, \ldots, u_m \in \mathscr{E}^{\mathbb{C}}(U)$ be local coordinates on $U \subset X$. Then every germ $f_x \in \mathscr{E}^{\mathbb{R}}_x$, $x \in U$, can be written as*

$$f_x = f_x(x) + \sum_{\mu=1}^{m} (u_{\mu x} - u_\mu(x)) g_{\mu x}, \qquad g_{\mu x} \in E^{\mathbb{R}}_x.$$

The values $g_{\mu x}(x) \in \mathbb{R}$ *of* $g_{\mu x}$ *at* x *are uniquely determined by* f_x:

$$g_{\mu x}(x) = \frac{\partial f_x}{\partial u_\mu}(x), \qquad \mu = 1, \ldots, m.$$

Proof: Without loss of generality we may take U to be an open set in $\mathbb{R}^m$ (the space of real m-tuples $(u_1, \ldots, u_m)$) with x the origin. For every $f_x \in \mathscr{E}_x^{\mathbb{R}}$ there exists a ball $B \subset U$ about $x = 0$ and a representative $f \in \mathscr{E}^{\mathbb{R}}(B)$ of f_x. In B we have

$$\begin{aligned} f(u_1, \ldots, u_m) - f(0) &= \sum_{\mu=1}^{m} (f(0, \ldots, 0, u_\mu, \ldots, u_m) \\ &\qquad - f(0, \ldots, 0, u_{\mu+1}, \ldots, u_m)) \\ &= \sum_{\mu=1}^{m} \int_0^{u_\mu} \frac{\partial f}{\partial y}(0, \ldots, 0, y, u_{\mu+1}, \ldots, u_m)\, dy. \end{aligned}$$

If one sets $g_\mu(u_1, \ldots, u_m) := \int_0^1 \partial f/\partial t\, (0, \ldots, 0, tu_\mu, u_{\mu+1}, \ldots, u_m)\, dt$, then $g_\mu \in \mathscr{E}^{\mathbb{R}}(B)$ and, by substituting $y = tu_\mu$,

$$u_\mu g_\mu = \int_0^{u_\mu} \frac{\partial f}{\partial y}(0, \ldots, 0, y, u_{\mu+1}, \ldots, u_m)\, dy.$$

Thus one has $f = f(0) + u_1 g_1 + \cdots + u_m g_m$ on B. Since $x = 0$ and $u_\mu(x) = 0$, this yields the desired equation for f_x. The determination of $g_{\mu x}(x)$ follows trivially by differentiation. □

Definition 2. (Tangent Vector). *An* $\mathbb{R}$*-linear mapping* $\xi: \mathscr{E}_x^{\mathbb{R}} \to \mathbb{R}$ *is called a tangent vector at* $x \in X$ *if it satisfies the product rule:*

$$\xi(f_x g_x) = \xi(f_x) g_x(x) + f_x(x) \xi(g_x), \qquad f_x, g_x \in \mathscr{E}_x^{\mathbb{R}}.$$

Thus $\xi(r) = 0$ for all constant germs $r \in \mathbb{R}$.

The set of all tangent vectors at $x \in X$ is an $\mathbb{R}$-vector space which we denote by $T(x)$ and call the *tangent space of* X *at* x.

Theorem 3. *Let* $u_1, \ldots, u_m$ *be coordinates on* $U \subset X$. *Then the* m *partial derivatives*

$$\left.\frac{\partial}{\partial u_\mu}\right|_x : \mathscr{E}_x^{\mathbb{R}} \to \mathbb{R}, \qquad f_x \mapsto \frac{\partial f_x}{\partial u_\mu}(x), \qquad \mu = 1, \ldots, m,$$

form a basis for $T(x)$. For every $\xi \in T(x)$ it follows that

$$\xi = \sum_{\mu=1}^{m} \xi(u_{\mu x}) \left.\frac{\partial}{\partial u_\mu}\right|_x.$$

Proof: It is clear that the partial derivative maps are tangent vectors. Since

$$\frac{\partial u_{ix}}{\partial u_j}(x) = \delta_{ij}, \qquad 1 \le i, j \le m,$$

these vectors are linearly independent. Thus it remains to verify the equation of ξ. By Theorem 1, we have

$$f_x = f_x(x) + \sum_{\mu=1}^{m} (u_{\mu x} - u_\mu(x)) g_{\mu x}, \qquad g_{\mu x} \in \mathscr{E}_x^{\mathbb{R}}, \qquad g_{\mu x}(x) = \frac{\partial f_x}{\partial u_\mu}(x),$$

for every $f_x \in \mathscr{E}_x^{\mathbb{R}}$. Since ξ vanishes on the constant germs, the product rule implies that

$$\xi f_x = \sum_{\mu=1}^{m} \xi(u_{\mu x}) g_{\mu x}(x) = \sum_{\mu=1}^{m} \xi(u_{\mu x}) \frac{\partial f_x}{\partial u_\mu}(x). \qquad \square$$

If in general R is a commutative K-algebra over a field K, and M is an R-module, then any K-linear mapping $\alpha: R \to M$ which satisfies the product rule, $\alpha(fg) = \alpha(f)g + f\alpha(g), f, g \in R$, is called a *derivation* of R with values in M. The set $\mathscr{D}(R, M)$ of all such derivations is itself an R-module. By means of the map $\mathscr{E}_x^{\mathbb{R}} \to \mathbb{R}, f_x \to f_x(x)$, the field $\mathbb{R}$ is an $\mathscr{E}_x^{\mathbb{R}}$-module. In this way the *tangent space* $T(x)$ is the module of derivations $\mathscr{D}(\mathscr{E}_x^{\mathbb{R}}, \mathbb{R})$.

2. Vector Fields: If ξ is a map which associates to each point $x \in U \subset X$ a tangent vector $\xi_x \in T(X)$, then one calls ξ a *vector field* on U. If $V \subset U$, then ξ associates to each $f \in \mathscr{E}^{\mathbb{R}}(V)$ a real valued function $\xi(f)$:

$$\xi(f): V \to \mathbb{R}, \qquad x \mapsto \xi(f)(x) := \xi_x(f_x).$$

Definition 4. (Differentiable Vector Fields): *A vector field ξ on $U \subset X$ is called differentiable if, for every open $V \subset U$ and every $f \in \mathscr{E}^{\mathbb{R}}(V)$, the function $\xi(f)$ is itself differentiable on V.*

The set of all differentiable vector fields on $U \subset X$ form an $\mathscr{E}_U^{\mathbb{R}}(U)$-module, $T(U)$. If $V \subset U$, then one has the natural restriction r_V^U: $T(U) \to T(V)$, $\xi \mapsto \xi \mid V$. Hence

the system $\{T(U), r_V^U\}$ is a presheaf on X which satisfies the axioms S1 and S2. The associated $\mathscr{E}^{\mathbb{R}}$-sheaf is denoted by $\mathscr{T}$ and, for every $x \in X$, the stalk $\mathscr{T}_x$ is the $\mathscr{E}_x^{\mathbb{R}}$-module of all derivations of $\mathscr{E}_x^{\mathbb{R}}$ into itself. The sheaf $\mathscr{T}$ is called the sheaf of germs of differentiable vector fields on X.

We immediately have the following analog to Theorem 3:

Theorem 5. *If $u_1, \ldots, u_m$ are coordinates on $U \subset X$, then the partial derivatives*

$$\frac{\partial}{\partial u_\mu}: \mathscr{E}^{\mathbb{R}}(U) \to \mathscr{E}^{\mathbb{R}}(U), \qquad f \mapsto \frac{\partial f}{\partial u_\mu}, \qquad \mu = 1, \ldots, m,$$

form a basis for the $\mathscr{E}^{\mathbb{R}}(U)$-module $\mathscr{T}(U)$. For every $\xi \in \mathscr{T}(U)$,

$$\xi = \sum_{\mu=1}^{m} \xi(u_\mu) \frac{\partial}{\partial u_\mu}, \qquad \xi(u_\mu) \in \mathscr{E}^{\mathbb{R}}(U).$$

Thus, by means of

$$\mathscr{T}(U) \to \mathscr{E}^{\mathbb{R}}(U)^m, \qquad \xi \mapsto (\xi(u_1), \ldots, \xi(u_m)),$$

every coordinate system $u_1, \ldots, u_m$ on $U \subset X$ determines an $\mathscr{E}^{\mathbb{R}}(U)$-module isomorphism. Thus $\mathscr{T}(U)$ is free of rank m. Hence it follows that

the $\mathscr{E}^{\mathbb{R}}$-sheaf $\mathscr{T}$ is locally free of rank m.

3. Complex r-vectors. Since $\mathbb{C}$ and all of the tangent spaces $T(x)$, $x \in X$, are real vector spaces, we can make the following definition for all $r \geq 1$:

Definition 6 (r-vector). *A (complex) r-vector φ at a point $x \in X$ is an r-fold, $\mathbb{R}$-linear, alternating map φ: $T(x) \times \cdots \times T(x) \to \mathbb{C}$.*

The r-vectors at x form a complex vector space $A^r(x)$. One sets $A^0(x) := \mathbb{C}$. The Grassmann product $\wedge$ (wedge product) is defined in the direct sum

$$A(x) := \bigoplus_{r=0}^{\infty} A^r(x)$$

as follows: For $\varphi \in A^r(x)$ and $\psi \in A^s(x)$, $\varphi \wedge \psi \in A^{r+s}(x)$ with $\varphi \wedge \psi(\xi_1, \ldots, \xi_{r+s}) := ((r+s)!)/(r! \cdot s!) \sum \delta(\iota_1, \ldots, \iota_{r+s}) \varphi(\xi_{\iota_1}, \ldots, \xi_{\iota_r}) \varphi(\xi_{\iota_{r+1}}, \ldots, \xi_{\iota_{r+s}})$ for $\xi_\iota \in T(x)$.[1]

It is easy to check that $\wedge$ endows $A(x)$ with the structure of an *associative $\mathbb{C}$-algebra with* 1 *and that*

$$\varphi \wedge \psi = (-1)^{rs} \psi \wedge \varphi, \quad \text{for} \quad \varphi \in A^r(x) \quad \text{and} \quad \psi \in A^s(x)$$

[1] Here the sum is taken over all $\iota_1, \ldots, \iota_{r+s}$ from 1 to $r+s$ and $\delta(\iota_1, \ldots, \iota_{r+s})$ is Kronecher symbol representing the sign of the permutation

$$\begin{pmatrix} 1 & 2 & \cdots & m \\ i_1 & i_2 & \cdots & i_m \end{pmatrix}.$$

Now let $u_1, \ldots, u_m$ be local coordinates on $U \subset X$. For a given $x \in U$ we let $du_1, \ldots, du_m$ denote the *dual* basis of $\partial/\partial u_1|_x, \ldots, \partial/\partial u_m|_x$ for the dual *real* vector space $T(x)^*$. (We should really use the notation $du_1|_x, \ldots, du_m|_x$, but it is too cumbersome). Then $du_1, \ldots, du_m$ form a basis for the *complex* vector space $A^1(x)$. Further, as a basis for $A^r(x)$, we have

$$\{du_{\iota_1} \wedge \cdots \wedge du_{\iota_r} \,|\, 1 \le \iota_1 < \cdots < \iota_r < m\}.$$

In particular dim $A^r(x) = \binom{m}{r}$ and $A^r(x) = 0$ for $r > m$,

4. Lifting *r*-vectors. Suppose that X and Y are differentiable manifolds and $f\colon X \to Y$ is a differentiable map. If $x \in X$ and $y := f(x) \in Y$, then every germ $g_y \in \mathscr{E}_y^{\mathbb{R}}$ is lifted via f to the germ $(g_y \circ f)_x \in \mathscr{E}_x^{\mathbb{R}}$. Thus every tangent vector $\xi \in T(x)$ determines the tangent vector

$$f_\#\xi\colon \mathscr{E}_y^{\mathbb{R}} \to \mathbb{R}, \qquad g_y \mapsto \xi(g_y \circ f)_x,$$

at $y \in Y$. The map $f_\#\colon T(x) \to T(y)$ is $\mathbb{R}$-linear. Associated to $f_\#$, we have the $\mathbb{C}$-algebra homomorphism

$$f^*\colon A(y) \to A(x) \quad \text{with} \quad f^*(A^r(y)) \subset A^r(x), \qquad r \ge 0,$$

defined by

$$(f^*\varphi)(\xi_1, \ldots, \xi_r) := \varphi(f_\#(\xi_1), \ldots, f_\#(\xi_r)),$$

where $\varphi \in A(y)$ and $\xi_1, \ldots, \xi_r \in T(x)$. One also uses the suggestive notation of $\varphi \circ f$ instead of f^*. Obviously * is a *contravariant functor* and $_\#$ is *covariant*.

We now write $f_\#$ and f^* in terms of local coordinates. Suppose that $n := \dim_{\mathbb{R}} Y$ and that $w_1, \ldots, w_n$ (resp. $u_1, \ldots, u_m$) are local coordinates on $W \subset Y$ (resp. $U \subset X$). We assume that $f(U) \subset W$. Thus $f\,|\,U\colon U \to W$ is represented by n functions $w_\nu = f_\nu(u_1, \ldots, u_m) \subset \mathscr{E}^{\mathbb{R}}(U)$, $1 \le \nu \le m$. In this language

$$f_\#\left(\frac{\partial}{\partial u_i}\right) = \sum_{\nu=1}^{n} \frac{\partial f_\nu}{\partial u_i}(x)\frac{\partial}{\partial w_\nu} \quad \text{and} \quad f^*(dw_j) = \sum_{\mu=1}^{m} \frac{\partial f_j}{\partial u_\mu}(x)\, du_\mu, \qquad \begin{matrix} 1 \le i, \\ j \le m,\, n. \end{matrix}$$

Since f^* is linear, it is therefore determined for all r-vectors:

$$f^*\left(\sum_{1 \le \iota_1 < \cdots < \iota_r \le n} a_{\iota_1 \cdots \iota_r}\, dw_{\iota_1} \wedge \cdots \wedge dw_{\iota_r}\right) = \sum_{1 \le \iota_1 < \cdots < \iota_r \le n} a_{\iota_1 \cdots \iota_r}\, f^*(dw_{\iota_1}) \wedge \cdots \wedge f^*(dw_{\iota_r})$$

5. Complex Valued Differential Forms. If the map φ associates to each point $x \in U \subset X$ an r-vector $\varphi_x \in A^r(x)$, then one calls φ an r-vector over U. If $\xi_1, \ldots, \xi_r \in T(V)$ are differentiable vector fields on $V \subset U$ and $\xi_{ix} \in T(x)$ denotes the tangent vector of ξ_i at x, then, to every r-vector φ over U, we have a *complex valued* function

$$\varphi(\xi_1, \ldots, \xi_r)\colon V \to \mathbb{C}, \qquad x \to \varphi(\xi_1, \ldots, \xi_r)(x) := \varphi_x(\xi_{1x}, \ldots, \xi_{rx})$$

Definition 7 (*Differential forms of degree r*). *An r-vector φ on $U \subset X$ is called a (complex valued) differential r-form (of degree r) on U, if for every open set $V \subset U$ and every system $\xi_1, \ldots, \xi_r \in T(V)$, the function $\varphi(\xi_1, \ldots, \xi_r)$ is differentiable on V.*

The set of all r-forms on U is an $\mathscr{E}(U)$-module $A^r(U)$. If $V \subset U$, then one has the natural restriction $r_V^U\colon A^r(U) \to A^r(V)$, $\varphi \mapsto \varphi \mid V$. Thus one sees that

the system $\{A^r(U), r_V^U\}$ is a canonical presheaf whose sheaf $\mathscr{A}^r$ is a sheaf of $\mathscr{E}$-modules called the sheaf of germs of (complex valued) r-forms on X.

If $u_1, \ldots, u_m$ are local coordinates on U, then $du_1, \ldots, du_m \in A^1(U)$. For every vector field $\xi = \sum_1^m f_i \frac{\partial}{\partial u_i}$, $f_i \in \mathscr{E}^{\mathbb{R}}(U)$, and every $\mu = 1, \ldots, m$, it is clear that $du_\mu(\xi) = f_\mu$. More generally we have the following:

Theorem 8. *If $u_1, \ldots, u_m$ are local coordinates on U, then the family $\{du_{\iota_1} \wedge \cdots \wedge du_{\iota_r} \mid 1 \le \iota_1 \le \cdots \le \iota_r \le m\}$ is a basis of the $\mathscr{E}(U)$-module $A^r(U)$. For every r-form $\varphi \in A^r(U)$,*

$$\varphi = \sum_{1 \le \iota_1 < \cdots < \iota_r \le m} a_{\iota_1 \cdots \iota_r}\, du_{\iota_1} \wedge \cdots \wedge du_{\iota_r},$$

where $a_{\iota_1 \cdots \iota_r} = \varphi(\partial/\partial x_{\iota_1}, \ldots, \partial/\partial x_{\iota_r}) \in \mathscr{E}(U)$.

The proof is trivial. □

By the above, every local coordinate system $u_1, \ldots, u_m$ on U induces an $\mathscr{E}(U)$-module isomorphism,

$$A^r(U) \to \mathscr{E}(U)^{\binom{m}{r}}, \qquad \varphi \mapsto \left(\varphi\left(\frac{\partial}{\partial u_{\iota_1}}, \ldots, \frac{\partial}{\partial u_{\iota_r}}\right)\right)$$

Thus $A^r(U)$ is free of rank $\binom{m}{r}$ and hence

the $\mathscr{E}$-sheaf $\mathscr{A}^r$ is locally free of rank $\binom{m}{r}$.

One should note that $\mathscr{A}^0 = \mathscr{E}$. The sheaf $\mathscr{A}^1$ is often called the sheaf of germs of *Pfaffian forms on X*. Another way of looking at $\mathscr{A}^1$ is $\mathscr{A}^1 = \mathscr{H}om_{\mathbb{R}}(\mathscr{T}, \mathbb{C})$, where $\mathbb{R}$ and $\mathbb{C}$ denote the obvious constant sheaves. □

The $\mathscr{E}$-sheaf $\mathscr{A}^r$ is canonically isomorphic to the r-fold Grassmann product of the $\mathscr{E}$-sheaf $\mathscr{A}^1$ with itself over the sheaf of rings $\mathscr{E}$: $\mathscr{A}^r \cong \bigwedge^r \mathscr{A}^1 = \mathscr{A}^1 \wedge \cdots \wedge \mathscr{A}^1$ (r-times). The $\mathscr{E}$-sheaf $A := \bigoplus_0^m A^r$ of germs of *all* differential forms on X is locally free of rank 2^m.

The reader should note that we always consider *complex differential forms*, while in the theory of differentiable manifolds one usually studies only real differential forms.

Conjugation in the field $\mathbb{C}$, $z \mapsto \bar{z}$, induces a *conjugation* $\bar{\ }: \mathscr{E} \to \mathscr{E}$ which associates the function $\bar{f}$, defined by $x \mapsto \overline{f(x)}$, to each $f \in \mathscr{E}(U)$. Furthermore one has a natural conjugation which preserves the degree: If $\varphi \in A^r(U)$ is given is local coordinates by $\sum a_{\iota_1 \cdots \iota_r} du_{\iota_1} \wedge \cdots \wedge du_{\iota_r}$, then $\bar{\varphi} := \sum \bar{a}_{\iota_1 \cdots \iota_r} du_{\iota_1} \wedge \cdots \wedge du_{\iota_r}$ is the complex conjugate differential form to φ. For $a \in \mathbb{C}$ and $\varphi, \psi \in A(U)$, one has the following elementary rules:

$$\overline{\varphi + \psi} = \bar{\varphi} + \bar{\psi}, \qquad \overline{a\varphi} = \bar{a}\bar{\varphi}, \qquad \overline{\varphi \wedge \psi} = \bar{\varphi} \wedge \bar{\psi}.$$

6. Exterior Derivative. The differential operator $d: \mathscr{A} \to \mathscr{A}$ which maps $\mathscr{A}^r$ into $\mathscr{A}^{r+1}$ plays a fundamental role in the study of differential forms. The following theorem describes the action of d on modules of sections on open sets U.

Theorem 9. *There exists a unique $\mathbb{C}$-linear map $d: A(U) \to A(U)$ having the following properties:*

1) *d maps $A^r(U)$ into $A^{r+1}(U)$, $r \geq 0$.*
2) *For every differentiable function $f \in \mathscr{E}(U) = A^0(U)$ and every vector field $\xi \in T(U)$, $df(\xi) = \xi(f)$.*
3) *$d\bar{\varphi} = \overline{d\varphi}$ for all $\varphi \in A(U)$.*
4) *$d(\varphi \wedge \psi) = d\varphi \wedge \psi + (-1)^r \varphi \wedge d\psi$, for $\varphi \in A^r(U)$.*
5) *$d \circ d = 0$*

For a proof see [ARC] as well as [DF].

One calls d the *exterior* or *total* derivative. If $u_1, \ldots, u_m$ are local coordinates in U and $f \in \mathscr{E}(U)$, then 2) immediately implies that

$$df = \sum_{\mu=1}^{m} \frac{\partial f}{\partial u_\mu} du_\mu$$

Thus, as is suggested by the notation, the exterior derivative of u_i is du_i. In general if

$$\varphi = \sum a_{\iota_1 \cdots \iota_r} du_{\iota_1} \wedge \cdots \wedge du_{\iota_r} \in A^r(U),$$

then

$$\begin{aligned} d\varphi &= \sum (da_{\iota_1 \cdots \iota_r}) \wedge du_{\iota_1} \wedge \cdots \wedge du_{\iota_r} \\ &= \sum \frac{\partial a_{\iota_1 \cdots \iota_r}}{\partial u_\iota} du_\iota \wedge du_{\iota_1} \wedge \cdots \wedge du_{\iota_r}. \end{aligned}$$

7. Lifting Differential Forms. Let X, Y be differentiable manifolds and $f: X \to Y$ a differentiable map. If $W \subset Y$ is open, then the differential forms on W

can be lifted via f to differential forms on $f^{-1}(W)$: There exists a $\mathbb{C}$-algebra homomorphism

$$f^*: A^r(W) \to A^r(f^{-1}(W)), \qquad \varphi \mapsto \varphi \circ f.$$

Applying the notation and equations of Section 4, the action of f^* is easily computed in local coordinates:

Let $u_1, \ldots, u_m$ (resp. $w_1, \ldots, w_n$) be local coordinates on $U \subset f^{-1}(W)$ (resp. W). Suppose that the map $f|U: U \to W$ is given by $w_\nu = f_\nu(u_1, \ldots, u_m) \in \mathscr{E}^{\mathbb{R}}(U)$, $1 \leq \nu \leq n$. If φ is an r-form

$$\varphi = \sum a_{\iota_1 \ldots \iota_r} dw_{\iota_1} \wedge \cdots \wedge dw_{\iota_r} \in A^r(W), \qquad a_{\iota_1 \ldots \iota_r} \in \mathscr{E}(W),$$

then

$$f^*(\varphi)|U := (\varphi \circ f)|U = \sum (a_{\iota_1 \ldots \iota_r} \circ f)\, df_{\iota_1} \wedge \cdots \wedge df_{\iota_r} \in A^r(U),$$

where $a_{\iota_1 \ldots \iota_r} \circ f \in \mathscr{E}(U)$ denotes the lift of the function $a_{\iota_1 \ldots \iota_r}$.

It is clear that

$$\overline{f^*(\varphi)} = f^*(\bar{\varphi})$$

Less trivial, but nevertheless true, is

$$f^*(d\varphi) = d(f^*\varphi) \quad \textit{for all} \quad \varphi \in A(W).$$

In other words, *exterior differentiation commutes with lifting.* For proofs of all of these statements we refer the reader to [DF].

Remark: The lifting of differential forms is never used in this book.

8. The de Rham Cohomology Groups. Using the sheaves $\mathscr{A}^r$ and exterior differentiation d, one forms on X the $\mathbb{C}$-sheaf sequence

$$\mathscr{E} = \mathscr{A}^0 \xrightarrow{d} \mathscr{A}^1 \xrightarrow{d} \cdots \longrightarrow \mathscr{A}^r \xrightarrow{d} \mathscr{A}^{r+1} \xrightarrow{d} \cdots \xrightarrow{d} \mathscr{A}^m \longrightarrow 0$$

Since $d \circ d = 0$, this is a *complex* with $\operatorname{Ker} d|\mathscr{A}^0 = \mathbb{C}$.

Theorem 10. *Let i denote the embedding of the constant sheaf $\mathbb{C}$ in $\mathscr{A}^0$.* Then the sequence

$$(*) \qquad 0 \longrightarrow \mathbb{C} \xrightarrow{i} \mathscr{A}^0 \xrightarrow{d} \mathscr{A}^1 \xrightarrow{d} \cdots \xrightarrow{d} \mathscr{A}^m \longrightarrow 0$$

is exact and consequently a resolution of the sheaf $\mathbb{C}$.

This local theorem is an immediate consequence of the famous

Lemma of Poincaré. Let D be a *convex domain in* $\mathbb{R}^m$ *and* $\varphi \in A^r(D)$, $r \geq 1$. *Assume that* φ *is closed* (*i.e.* $d\varphi = 0$). *Then* φ *is exact* (*i.e. there exists an* $(r-1)$*-form* $\psi \in A^{r-1}(D)$ *such that* $d\psi = \varphi$).

For a proof we refer the reader to [DF].

Since all sheaves $\mathscr{A}^r$ are *soft*, the sequence $(*)$ is an *acyclic* resolution of $\mathbb{C}$ over X. The formal de Rham Lemma (Chapter B.1.3) therefore yields the classical Theorem of de Rham: *Let X be a differentiable manifold and $\mathscr{A}^r$, $r \geq 0$, the sheaf of germs of (complex valued) differentiable r-forms on X. Then there exist the following natural isomorphisms:*

$$H^0(X, \mathbb{C}) \cong \operatorname{Ker}(d \,|\, \mathscr{A}^0(X))$$

and

$$H^q(X, \mathbb{C}) \cong \operatorname{Ker}(d \,|\, \mathscr{A}^q(X))/d\mathscr{A}^{q-1}(X), \qquad q \geq 1.$$

One usually calls the groups on the right sides of these isomorphisms the *de Rham cohomology groups of* X (with complex coefficients). Although they are obtained via the differentiable structure of X, they are *topological invariants* of the manifold X, being isomorphic to "singular" cohomology groups.

The above considerations for the d-operator will be used again for the $\bar{\partial}$-operator in later sections. This will lead to the Dolbeault cohomology groups (see Section 4). At the point where the Poincaré Lemma enters for the d-operator, we will use the Lemma of Dolbeault which is proved in Section 3.

§ 2. Differential Forms on Complex Manifolds

In this section X always denotes a complex manifold of dimension m with structure sheaf $\mathcal{O}$. The inclusion $\mathcal{O} \subset \mathscr{E}$ together with the conjugation $\bar{\ }: A \to A$ gives rise to a double gradation of A which contains important information about X.

1. The Sheaves $\mathscr{A}^{1,0}$, $\mathscr{A}^{0,1}$ and Ω^1. If $z_1 = x_1 + iy_1, \ldots, z_m = x_m + iy_m \in \mathcal{O}(U)$ are complex coordinates in $U \subset X$, then, since d is $\mathbb{C}$-linear, $dz_\mu = dx_\mu + idy_\mu$, $d\bar{z}_\mu = dx_\mu - idy_\mu$, $1 \leq \mu \leq m$. Furthermore

$$dx_\mu = \tfrac{1}{2}(dz_\mu + d\bar{z}_\mu), \qquad dy_\mu = \frac{1}{2i}(dz_\mu - d\bar{z}_\mu), \qquad \mu = 1, \ldots, m.$$

Thus the family $\{dz_1, d\bar{z}_1, \ldots, dz_m, d\bar{z}_m\}$ is a basis for $\mathscr{A}^1(U)$ as an $\mathscr{E}(U)$-module. Hence the following is immediate:

Theorem 1. *If $z_1, \ldots, z_m$ are local complex coordinates on $U \subset X$, then for every $r \geq 0$, an $\mathscr{E}(U)$-basis of $\mathscr{A}^1(U)$ is given by the $\binom{2m}{r}$ forms*

$$\{dz_{i_1} \wedge \cdots \wedge dz_{i_p} \wedge d\bar{z}_{j_1} \wedge \cdots \wedge d\bar{z}_{j_q} \,|\, 1 \leq i_1 < \cdots < i_p \leq m,$$
$$1 \leq j_1 < \cdots < j_1 \leq m, \; p + q = r\}.$$

Proof: These $\binom{2m}{r}$ forms generate $\mathscr{A}^r(U)$. Since $\mathscr{A}^r(U)$ is a free $\mathscr{E}(U)$-module of rank $\binom{2m}{r}$, they form a basis. □

For every function $f \in \mathscr{E}(U)$ we have

$$df = \sum_1^m \left(\frac{\partial f}{\partial x_\mu} dx_\mu + \frac{\partial f}{\partial y_\mu} dy_\mu \right).$$

If as usual one sets

$$\frac{\partial f}{\partial z_\mu} := \frac{1}{2}\left(\frac{\partial f}{\partial x_\mu} - i \frac{\partial f}{\partial y_\mu} \right) \quad \text{and} \quad \frac{\partial f}{\partial \bar{z}_\mu} := \frac{1}{2}\left(\frac{\partial f}{\partial x_\mu} + i \frac{\partial f}{\partial y_\mu} \right), \qquad \mu = 1, \ldots, m,$$

then one obtains

$$df = \sum_1^m \left(\frac{\partial f}{\partial z_\mu} dz_\mu + \frac{\partial f}{\partial \bar{z}_\mu} d\bar{z}_\mu \right) \quad \text{for all} \quad f \in \mathscr{E}(U).$$

A function $f \in \mathscr{E}(U)$ is holomorphic on U if it satisfies the *Cauchy-Riemann differential equations*:

$$\frac{\partial f}{\partial \bar{z}_\mu} = 0, \qquad \mu = 1, \ldots, m,$$

or equivalently when

$$df = \sum_1^m \frac{\partial f}{\partial z_\mu} dz_\mu.$$

We now introduce three important subsheaves of $\mathscr{A}^1$. We start with the inclusion $d\mathcal{O} \subset \mathscr{A}^1$ and denote by $\mathscr{A}^{1,0}$ the $\mathscr{E}$-subsheaf of $\mathscr{A}^1$ generated by $d\mathcal{O}$. One calls $\mathscr{A}^{1,0}$ *the sheaf of germs of differentiable* (1, 0)*-forms on* X. Explicitly,

$$\mathscr{A}_x^{1,0} = \left\{ \sum_{i=1}^{<\infty} g_{ix}\, df_{ix} \,\middle|\, g_{ix} \in \mathscr{E}_x, f_{ix} \in \mathcal{O}_x \right\}, \qquad x \in X.$$

Along with $\mathscr{A}^{1,0}$ we consider its conjugate $\mathscr{E}$-sheaf $\mathscr{A}^{0,1} := \bar{\mathscr{A}}^{1,0}$. Since d is interchangable with conjugation,

$$\mathscr{A}_x^{0,1} = \left\{ \sum_{i=1}^{\infty} g_{ix}\, d\bar{f}_{ix} \,\middle|\, g_{ix} \in \mathscr{E}_x, f_{ix} \in \mathcal{O}_x \right\}, \qquad x \in X.$$

In other words, $\mathscr{A}^{0,1}$ can be described as the $\mathscr{E}$-subsheaf of $\mathscr{A}^1$ which is generated by $d\bar{\mathcal{O}}$. One calls $\mathscr{A}^{0,1}$ *the sheaf of germs of differentiable* (0, 1)*-forms on* X.

Finally we let Ω^1 denote the $\mathcal{O}$-subsheaf of $\mathscr{A}^1$ which is generated by $d\mathcal{O}$. Thus $\Omega^1 \subset \mathscr{A}^{1,0}$ and

$$\Omega_x^1 = \left\{ \sum_{i=1}^{<\infty} g_{ix}\, df_{ix} \,\middle|\, g_{ix}, f_{ix} \in \mathcal{O}_x \right\}, \qquad x \in X.$$

One calls Ω^1 the *sheaf of germs of holomorphic* 1-*forms on* X.

Hence three sheaves $\mathscr{A}^{1,0}$, $\mathscr{A}^{0,1}$, and Ω^1 are *uniquely* determined by the complex structure on X. In local coordinates, they are described as follows:

Theorem 2. *Let $z_1, \ldots, z_m$ be local complex coordinates on $U \subset X$. Thus*

1) *The 1-forms $dz_1, \ldots, dz_m$ form a basis of $\mathscr{A}^{1,0}(U)$ as an $\mathscr{E}(U)$-module and a basis of $\Omega^1(U)$ as on $\mathcal{O}(U)$-module.*
2) *The 1-forms $d\bar{z}_1, \ldots, d\bar{z}_m$ form a basis of $\mathscr{A}^{0,1}(U)$ as an $\mathscr{E}(U)$-module.*

Proof: Certainly $dz_\mu \in \mathscr{A}^{1,0}(U)$ and $d\bar{z}_\mu \in \mathscr{A}^{0,1}(U)$, $1 \le \mu \le m$. If $x \in U$ and $f \in \mathcal{O}_x$ then $df_x = \sum_1^m \left(\frac{\partial f}{\partial z_\mu}\right)_x (dz_\mu)_x$. Hence $dz_{1x}, \ldots, dz_{mx}$ generate $\mathscr{A}_x^{1,0}$ as an $\mathscr{E}_x$-module and thus the sections $dz_1, \ldots, dz_m$ generate the $\mathscr{E}(U)$-module $\mathscr{A}^{1,0}(U)$. Analogously it follows that $\mathscr{A}^{0,1}(U)$ is generated over $\mathscr{E}(U)$ by $d\bar{z}_1, \ldots, d\bar{z}_n$ and that $dz_1, \ldots, dz_m$ form an $\mathcal{O}(U)$ generating system for $\Omega^1(U)$. Since $dz_1, d\bar{z}_1, \ldots, dz_m, d\bar{z}_m$ are linearly independent over $\mathscr{E}(U)$ (see Theorem 1), the claim follows.

2. The Sheaves $\mathscr{A}^{p,q}$ and Ω^p. Using the Grassmann product $\wedge$ which is defined for arbitrary sheaves of rings, we now construct new sheaves from the $\mathscr{E}$-sheaves $\mathscr{A}^{1,0}$, $\mathscr{A}^{0,1}$ and the $\mathcal{O}$-sheaf Ω^1. For natural numbers p, q we form (over $\mathscr{E}$) the Grassmann product

$$\mathscr{A}^{p,q} := \underbrace{\mathscr{A}^{1,0} \wedge \cdots \wedge \mathscr{A}^{1,0}}_{p\text{-times}} \wedge \underbrace{\mathscr{A}^{0,1} \wedge \cdots \wedge \mathscr{A}^{0,1}}_{q\text{-times}},$$

and (over $\mathcal{O}$) the Grassmann product

$$\Omega^p := \underbrace{\Omega^1 \wedge \cdots \wedge \Omega^1}_{p\text{-times}},$$

where we agree that $\mathscr{A}^{0,0} := \mathscr{A}^0 = \mathscr{E}$ and $\Omega^0 := \mathcal{O}$. Obviously $\mathscr{A}^{p,q}$ is an $\mathscr{E}$-subsheaf of $\mathscr{A}^{p+q}$ and Ω^p is an $\mathcal{O}$-subsheaf of $\mathscr{A}^p$.

Definition 3. *The $\mathscr{E}$-sheaf $\mathscr{A}^{p,q}$ is called the sheaf of germs of differentiable (p, q)-forms on X. The $\mathcal{O}$-sheaf Ω^p is called the sheaf of germs of holomorphic p-forms on X.*

The (p, q)-forms are rather easily described in local coordinates:

Theorem 4. *Let $z_1, \ldots, z_m$ be local complex coordinates on $U \subset X$. Then*

1) *The family*

   $\{dz_{i_1} \wedge \cdots \wedge dz_{i_p} \wedge d\bar{z}_{j_1} \wedge \cdots \wedge d\bar{z}_{j_q}, 1 \le i_q < \cdots < i_p \le m, 1 \le j_1 < \cdots < j_q \le m\}$

   is an $\mathscr{E}(U)$-basis for $\mathscr{A}^{p,q}(U)$.

2) $\mathscr{A}^r(U) = \bigoplus_{p+q=r} \mathscr{A}^{p,q}(U)$ *as an $\mathscr{E}(U)$-module.*

3) *The family $\{dz_{i_1} \wedge \cdots \wedge dz_{i_p}, \; 1 \le i_1 < \cdots < i_p \le m\}$, is an $\mathscr{O}(U)$-basis for $\Omega^p(U)$.*

Proof: The definition of $\mathscr{A}^{p,q}$ implies that the family in 1) always generates $\mathscr{A}^{p,q}(U)$ as an $\mathscr{E}(U)$-module. Since the union of such families over all p, q with $p + q = r$ is a basis for $\mathscr{A}^r(U)$ (see Theorem 1), each family is a basis of $\mathscr{A}^{p,q}(U)$. The same reasoning proves the validity of equation 2). Statement 3) now follows trivially. □

Corollary. *For all $r \ge 0$*

$$\mathscr{A}^r = \bigoplus_{p+q=r} \mathscr{A}^{p,q} \quad \text{(as } \mathscr{E}\text{-modules)}.$$

In particular $\mathscr{A}^1 = \mathscr{A}^{1,0} \oplus \mathscr{A}^{0,1}$. The $\mathscr{O}$-sheaf Ω^p is locally free of rank $\binom{m}{p}$ and is therefore coherent. Clearly $\Omega^p = 0$ for all $p > m = \dim X$.

From 1) of Theorem 4 it also follows that

$$\overline{\mathscr{A}^{p,q}} = \mathscr{A}^{q,p}.$$

3. The Derivatives ∂ and $\bar{\partial}$. If one composes $d\colon \mathscr{E} \to \mathscr{A}^1$ with the projections on the first and second summands of the decomposition $\mathscr{A}^1 = \mathscr{A}^{1,0} \oplus \mathscr{A}^{0,1}$, then one has two $\mathbb{C}$-linear maps

$$\partial\colon \mathscr{E} \to \mathscr{A}^{1,0} \quad \text{and} \quad \bar{\partial}\colon \mathscr{E} \to \mathscr{A}^{0,1}$$

with $d = \partial + \bar{\partial}$. If $z_1, \ldots, z_m$ are local coordinates on $U \subset X$, then

$$\partial f = \sum_1^m \frac{\partial f}{\partial z_\mu} dz_\mu, \qquad \bar{\partial} f = \sum_1^m \frac{\partial f}{\partial \bar{z}_\mu} d\bar{z}_\mu, \qquad \text{for } f \in \mathscr{E}(U).$$

Thus f is holomorphic if and only if $\bar{\partial} f = 0$. In other words

$$\mathscr{O} = \mathscr{K}er\, \bar{\partial}.$$

These considerations can be immediately generalized: For every (p, q)-form

$$\omega = \sum_{i,j} a_{i_1 \cdots i_p j_1 \cdots j_p}\, dz_{i_1} \wedge \cdots \wedge dz_{i_p} \wedge d\bar{z}_{j_q} \wedge \cdots \wedge d\bar{z}_{j_q} \in \mathscr{A}^{p,q}(U)$$

we set

$$\partial\omega := \sum_{\mu=1}^{m}\left(\sum_{i,j}\frac{\partial a_{i_1\cdots j_q}}{\partial z_\mu}\,dz_\mu\wedge dz_{i_1}\wedge\cdots\wedge dz_{i_p}\wedge d\bar z_{j_1}\wedge\cdots\wedge d\bar z_{j_q}\right)\in A^{p+1,q}(U)$$

and

$$\bar\partial\omega := \sum_{\mu=1}^{m}\left(\sum_{i,j}\frac{\partial a_{i_1\cdots j_q}}{\partial \bar z_\mu}\,d\bar z_\mu\wedge dz_{i_1}\wedge\cdots\wedge dz_{i_p}\wedge d\bar z_{j_1}\wedge\cdots\wedge d\bar z_{j_q}\right)\in A^{p,q+1}(U).^2$$

Thus we have defined two $\mathbb{C}$-linear maps

$$\partial\colon \mathscr{A}^{p,q}\to\mathscr{A}^{p+1,q} \quad\text{and}\quad \bar\partial\colon \mathscr{A}^{p,q}\to\mathscr{A}^{p,q+1}.$$

Since $d\omega = \sum_{i,j} da_{i_1\cdots j_q}\wedge dz_1\wedge\cdots\wedge d\bar z_{j_q}$ and, for functions

$$da = \sum_{\mu=1}^{m}\frac{\partial a}{\partial z_\mu}\,dz_\mu + \sum_{\mu=1}^{m}\frac{\partial a}{\partial \bar z_\mu}\,d\bar z_\mu,$$

it follows that

$$d\omega = \partial\omega + \bar\partial\omega \quad\text{for all}\quad \omega\in\mathscr{A}^{p,q}$$

In particular

$$d\mathscr{A}^{p,q}\subset\mathscr{A}^{p+1,q}\oplus\mathscr{A}^{p,q+1}.$$

Both maps ∂, $\bar\partial$ are in the defined obvious way as $\mathbb{C}$-linear maps

$$\partial\colon\mathscr{A}\to\mathscr{A}\quad\text{and}\quad\bar\partial\colon\mathscr{A}\to\mathscr{A}.$$

These operators, along with the total differential operator d, play important roles in the theory of complex manifolds. Their formal properties are summarized in the following:

Theorem 5. *For ∂ and $\bar\partial$ defined as above, we have the following identities:*

$$d=\partial+\bar\partial,\qquad \partial\circ\partial=0,\qquad \bar\partial\circ\bar\partial=0,\qquad \partial\circ\bar\partial+\bar\partial\circ\partial=0,$$
$$\overline{\partial\varphi}=\bar\partial\bar\varphi\quad\text{for all}\quad\varphi\in A,$$
$$\partial(\varphi\wedge\psi)=\partial\varphi\wedge\psi+(-1)^{p+q}\varphi\wedge\partial\psi\quad\text{for}\quad\varphi\in\mathscr{A}^{p,q},$$

[2] In this case $\sum_{i,j}$ means that the sum is taken over all index combinations with $1\le i_1<\cdots<i_p\le m$ and $1\le j_1<\cdots<j_q\le m$.

and

$$\bar\partial(\varphi \wedge \psi) = \bar\partial\varphi \wedge \psi + (-1)^{p+q}\varphi \wedge \bar\partial\psi \quad \text{for} \quad \varphi \in \mathscr{A}^{p,q}.$$

Proof: By definition

$$0 = d \circ d = (\partial + \bar\partial) \circ (\partial + \bar\partial) = \partial \circ \partial + \bar\partial \circ \bar\partial + (\partial \circ \bar\partial + \bar\partial \circ \partial).$$

The equations in the first row now follow, since, for every $\varphi \in \mathscr{A}^{p,q}$, $\partial \circ \partial\varphi \in \mathscr{A}^{p+2,q}$, $\bar\partial \circ \bar\partial\varphi \in \mathscr{A}^{p,q+2}$, and $(\partial\bar\partial + \bar\partial\partial)\varphi \in \mathscr{A}^{p+1,q+1}$. The equation $\overline{\partial\varphi} = \bar\partial\bar\varphi$ follows immediately if one writes φ, $\partial\varphi$ and $\bar\partial\varphi$ in local coordinates and notes that for functions f it is always the case that

$$\overline{\frac{\partial f}{\partial z_\mu}} = \frac{\partial \bar f}{\partial \bar z_\mu}, \qquad \mu = 1, \ldots, m.$$

Because of this it is also only necessary to verify the remaining equation which involves ∂ and this is again just a calculation in local coordinates. □

The operators ∂ and $\bar\partial$ are "created equal". However, since the sheaf $\mathcal{O}$ is usually given priority over $\bar{\mathcal{O}}$ in complex analysis and since $\mathcal{O} = \operatorname{Ker} \bar\partial \,|\, \mathscr{E}$, the operator $\bar\partial$ is more in the foreground of interest.

Theorem 6. *A $(p, 0)$-form $\varphi \in A^{p,0}(U)$, $p \geq 0$, is holomorphic if and only if it is $\bar\partial$-closed (i.e. $\bar\partial = 0$). Thus*

$$\Omega^p = \operatorname{Ker} \bar\partial \,|\, \mathscr{A}^{p,0} \quad \textit{for all} \quad p \geq 0.$$

Proof: If $\varphi = \sum_i a_{i_1 \cdots i_p}\, dz_{i_1} \wedge \cdots \wedge dz_{i_p}$ in local coordinates, then

$$\bar\partial\varphi = \sum_{\mu=1}^{m} \sum_i \left(\frac{\partial a_{i_1 \cdots i_p}}{\partial \bar z_\mu}\, d\bar z_\mu \wedge dz_{i_1} \wedge \cdots \wedge dz_{i_p}\right) \in \mathscr{A}^{p,1}(U).$$

Since the forms $d\bar z_\mu \wedge dz_{i_1} \wedge \cdots \wedge dz_{i_p}$ form a basis for $\mathscr{A}^{p,1}(U)$, $\bar\partial\varphi = 0$ if and only if for all μ and all $i_1, \ldots, i_p$

$$\frac{\partial a_{i_1 \cdots i_p}}{\partial \bar z_\mu} = 0.$$

In other words, $a_{i_1 \cdots i_p} \in \mathcal{O}(U)$ or, equivalently, $\varphi \in \Omega^p(U)$. □

Remark: Since $\overline{\partial\varphi} = \bar\partial\bar\varphi$, the form φ is in Ker ∂ if and only if $\bar\varphi \in \operatorname{Ker} \bar\partial$. Hence, since $A^{0,p} = \overline{A^{p,0}}$, we have the analog to Theorem 6:

$$\overline{\Omega^p} = \operatorname{Ker} \partial \,|\, A^{0,p}.$$

One calls $\bar{\Omega}^p$ the sheaf of germs of *antiholomorphic* p-forms on X. The results which we just obtained for ∂ and $\bar{\partial}$ are suggestively illustrated by the following diagram of $\mathbb{C}$-homomorphisms (i always denotes the natural injection):

$$
\begin{array}{ccccccccccccccc}
 & & 0 & & 0 & & 0 & & 0 & & & & 0 & & \\
 & & \downarrow & & \downarrow & & \downarrow & & \downarrow & & & & \downarrow & & \\
0 & \longrightarrow & \mathbb{C} & \xrightarrow{i} & \bar{\mathcal{O}} & \xrightarrow{d=\bar{\partial}} & \bar{\Omega}^1 & \xrightarrow{d} & \bar{\Omega}^2 & \xrightarrow{d} & \cdots & \xrightarrow{d} & \bar{\Omega}^m & \longrightarrow & 0 \\
 & & \downarrow i & & \downarrow i & & \downarrow i & & \downarrow i & & & & \downarrow i & & \\
0 & \longrightarrow & \mathcal{O} & \xrightarrow{i} & \mathcal{E} & \xrightarrow{\bar{\partial}} & \mathcal{A}^{0,1} & \xrightarrow{\bar{\partial}} & \mathcal{A}^{0,2} & \xrightarrow{\bar{\partial}} & \cdots & \xrightarrow{\bar{\partial}} & \mathcal{A}^{0,m} & \longrightarrow & 0 \\
 & & \downarrow d=\partial & & \downarrow \partial & & \downarrow \partial & & \downarrow \partial & & & & \downarrow \partial & & \\
0 & \longrightarrow & \Omega^1 & \xrightarrow{i} & \mathcal{A}^{1,0} & \xrightarrow{\bar{\partial}} & \mathcal{A}^{1,1} & \xrightarrow{\bar{\partial}} & \mathcal{A}^{2,2} & \xrightarrow{\bar{\partial}} & \cdots & \xrightarrow{\bar{\partial}} & \mathcal{A}^{1,m} & \longrightarrow & 0 \\
 & & \downarrow d & & \downarrow \partial & & \downarrow \partial & & \downarrow \partial & & & & \downarrow \partial & & \\
0 & \longrightarrow & \Omega^2 & \xrightarrow{i} & \mathcal{A}^{2,0} & \xrightarrow{\bar{\partial}} & \mathcal{A}^{2,1} & \xrightarrow{\bar{\partial}} & \mathcal{A}^{2,2} & \xrightarrow{\bar{\partial}} & \cdots & \xrightarrow{\bar{\partial}} & \mathcal{A}^{2,m} & \longrightarrow & 0 \\
 & & d\downarrow & & \partial\downarrow & & \partial\downarrow & & \partial\downarrow & & & & \partial\downarrow & & \\
 & & \vdots & & \vdots & & \vdots & & \vdots & & & & \vdots & & \\
 & & d\downarrow & & \partial\downarrow & & \partial\downarrow & & \partial\downarrow & & & & \partial\downarrow & & \\
0 & \longrightarrow & \Omega^m & \xrightarrow{i} & \mathcal{A}^{m,0} & \xrightarrow{\bar{\partial}} & \mathcal{A}^{m,1} & \xrightarrow{\bar{\partial}} & \mathcal{A}^{m,2} & \xrightarrow{\bar{\partial}} & \cdots & \xrightarrow{\bar{\partial}} & \mathcal{A}^{m,m} & \longrightarrow & 0 \\
 & & \downarrow & & \downarrow & & \downarrow & & \downarrow & & & & \downarrow & & \\
 & & 0 & & 0 & & 0 & & 0 & & & & 0 & &
\end{array}
$$

This diagram is *anticommutative* in every square. Every row and every column is a complex of sheaves. The family $\{\mathcal{A}^{p,q}, \partial, \bar{\partial}\}$ is thus a *double complex*. The associated *simple complex* is $\{\mathcal{A}^r, d\}$. The main result of the next section is that *all of the rows and columns in the above diagram are exact.*

Remark: Frequently one sees d' and d'' in the literature instead of ∂ and $\bar{\partial}$.

4. Holomorphic Liftings of (p, q)**-forms.** In Paragraph 1.7 we saw that a differentiable map $f: X \to Y$ induces on any open set $W \subset Y$ a $\mathbb{C}$-algebra homomorphism

$$f^*: A(W) \to A(f^{-1}(W)).$$

This maps $A^r(W)$ into $A^r(f^{-1}(W))$ for all r, is interchangable with conjugation, and with the exterior differential operator:

$$\overline{f^*(\varphi)} = f^*(\bar{\varphi}), \qquad f^* \circ d = d \circ f^*.$$

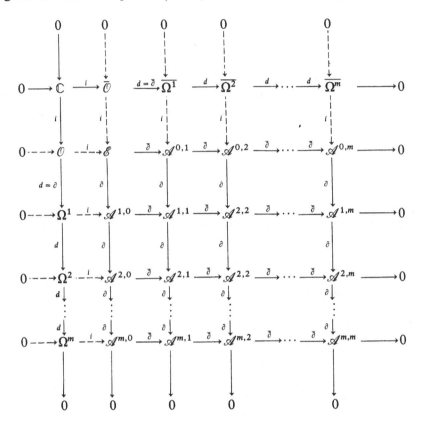

In the complex analytic case f^* has further properties.

Theorem 7. *Let X, Y be complex manifolds and $f: X \to Y$ a holomorphic map. Then*

$$f^*(A^{p,q}(W)) \subset A^{p,q}(f^{-1}(W)), \qquad f^*(\Omega^p(W)) \subset \Omega^p(f^{-1}(W))$$
$$f^* \circ \partial = \partial \circ f^*, \qquad f^* \circ \bar{\partial} = \bar{\partial} \circ f^*.$$

Proof: All of these statements are local. So let $z_1, \ldots, z_m$ (resp. $w_1, \ldots, w_n$) be local complex coordinates on $U \subset f^{-1}(W)$ (resp. W). Let $f \mid U: U \to W$ be given by the functions $w_\nu = f_\nu(z_1, \ldots, z_m) \in \mathcal{O}(U)$, $\nu = 1, \ldots, n$. Every (p, q)-form φ on W can be written

$$\varphi = \sum_{i,j} a_{i_1 \ldots j_q}\, dw_{i_1} \wedge \cdots \wedge dw_{i_p} \wedge d\bar{w}_{j_1} \wedge \cdots \wedge d\bar{w}_{j_q}.$$

Then by the result in Section 1.7

$$f^*(\varphi) \mid U = \sum_{i,j} (a_{i_1 \ldots j_q} \circ f)\, df_{i_1} \wedge \cdots \wedge df_{i_p} \wedge d\bar{f}_{j_1} \wedge \cdots \wedge d\bar{f}_{j_q} \in \mathscr{A}^{p,q}(U).$$

The fact that $f^*(\varphi) \in \mathscr{A}^{p,q}$ follows since the holomorphicity of f_i implies that $df_i \in \mathscr{A}^{1,0}$ and consequently $d\bar{f}_i \in \mathscr{A}^{0,1}$. If φ is a holomorphic p-form, then $f^*(\varphi) = \sum_i a_{i_1 \ldots i_p} \circ f\, df_{i_1} \wedge \cdots \wedge df_{i_p}$. Since f and $a_{i_1 \ldots i_p}$, are holomorphic, the coefficients $a_{i_1 \ldots i_p} \circ f$ are holomorphic. Thus $f^*(\varphi)$ is in the p-fold exterior product of $\mathcal{O}(U)$-module spanned by $d\mathcal{O}$ and is thus in $\Omega^p(U)$.

In order to see the fact that f^* commutes with ∂ and $\bar{\partial}$, we begin with the equation $f^* \circ d = d \circ f^*$. Since $d = \partial + \bar{\partial}$ and f^* is additive,

$$f^*(\partial\varphi) + f^*(\bar{\partial}\varphi) = \partial f^*\varphi + \bar{\partial} f^*\varphi \quad \text{for every } (p, q)\text{-form } \varphi.$$

Now

$$f^*(\partial\varphi),\ \partial(f^*\varphi) \in \mathscr{A}^{p+1,q} \quad \text{and} \quad f^*(\bar{\partial}\varphi),\ \bar{\partial}(f^*\varphi) \in \mathscr{A}^{p,q+1}.$$

Thus, since f^* preserves the double gradation of $\mathscr{A}$, we have the desired equalities.

§ 3. The Lemma of Grothendieck

In this section preparations are made for the proof of the fact that both the rows and columns of the diagram on p. 70 are exact. It is necessary to solve the differential equation $\partial g/\partial \bar{z} = f$, where the given function f depends on a parameter as well as z. For this we need theorems from Lebesgue integration theory as well as the Stokes' theorem in the form of a generalized Cauchy integral formula.

The letter z always denotes a complex variable and $u_1, \ldots, u_d$ are real variables. We reserve ζ and z for complex variables of integration.

1. Area Integrals and the Operator T. Let B always be a *bounded domain* in the z-plane and U an open set in the parameter space $\mathbb{R}^d$ of d-tuples $(u_1, \ldots, u_d)$. We consider complex valued continuous functions f on $B \times U$ for which the integral

$$(1) \qquad (Tf)(x, u) := \frac{1}{2\pi i} \iint_B \frac{f(\zeta, u)}{\zeta - z}\, d\zeta \wedge d\bar{\zeta}$$

exists for all points $(z, u) \in B \times U$. From the Lebesgue integration theory we know that every continuous function from $B \times U$ which is bounded on $B \times \{u\}$, $u \in U$, has this property. Note that $\iint_B \frac{d\zeta \wedge d\bar{\zeta}}{\zeta - z}$ exists and $\iint_B \frac{d\zeta \wedge d\bar{\zeta}}{(\zeta - z)^2}$ doesn't!

By the Lebesgue dominated convergence theorem,

$$(2) \qquad (Tf)(z, u) = \lim_{\varepsilon \to 0} \frac{1}{2\pi i} \iint_{G/K_\varepsilon} \frac{f(\zeta, u)}{\zeta - z}\, d\zeta \wedge d\bar{\zeta}, \qquad (z, u) \in B \times U,$$

where K_ε is the closed disk of radius $\varepsilon > 0$ about z.

The function $Tf\colon B \times U \to \mathbb{C}$ has different behavior with respect to u than it does with respect to z. In order to study the behavior of u in equation (1), it is convenient to free the denominator of the integrand from the variable z by making the substitution $\eta\colon \zeta - z$. Thus one obtains

$$(3) \qquad (Tf)(z, u) = \frac{1}{2\pi i} \iint_{B_z} \frac{f(z + \eta, u)}{\eta}\, d\eta \wedge d\bar{\eta}, \qquad (z, u) \in B \times U,$$

where $B_z := \{\eta \in \mathbb{C} \mid z + \eta \in B\}$ is the set B translated by z. One gets rid of the dependency of the domain of integration on z by extending f trivially (by 0) to all of $\mathbb{C} \times U$, calling the extended function $\tilde{f}$. In this way we have

$$(3') \qquad (Tf)(z, u) = \frac{1}{2\pi i} \iint \frac{\tilde{f}(z + \eta, u)}{\eta}\, d\eta \wedge d\bar{\eta}, \qquad (z, u) \in B \times U.$$

In polar coordinates, $\eta = re^{i\varphi}$, (3) can be written as follows:

$$(Tf)(z, u) = \frac{1}{\pi} \iint_{B_z} e^{-i\varphi} \cdot f(z + re^{i\varphi}, u)\, d\varphi \wedge dr, \qquad (z, u) \in B \times U.$$

In this way we see that if $\rho := \sup_{v,w} |v - w| < \infty$ is the diameter of B, then, since f is bounded on every $B \times \{u\}$,

Tf is bounded on $B \times \{u\}$, $u \in U$, with

$$|Tf|_{B \times \{u\}} \leq 2\rho |f|_{B \times \{u\}}.$$

We note here the trivially shown (nevertheless important) fact that T is a $\mathbb{C}$-*linear operator.*

2. The Commutivity of T with Partial Differentiation. The situation for the parameters u_μ is relatively simple:

Lemma 1. *Suppose Tf exists on $B \times U$ and that $(\partial f/\partial u_\mu)(\zeta, u)$ is continuous on $B \times U$ and is bounded on $B \times K$ for all compact K contained in U. Then*

$$\frac{\partial}{\partial u_\mu}(Tf) = T\frac{\partial f}{\partial u_\mu}.$$

Proof: The integrand of T is uniformly bounded by a Lebesgue integrable function. Thus one may interchange integration and differentiation by u_μ. □

Due to the singularity of the integrand, the situation for z and $\bar{z}$ is more complicated and requires stronger assumptions.

Lemma 2. *Let f be continuous on $B \times U$ and suppose that* $\operatorname{supp} f \cap (B \times K)$ *is compact for every compact K in U. Then Tf is continuous on $B \times U$. If in addition $\partial f/\partial z$, $\partial f/\partial \bar{z}$ exist and are continuous on $B \times U$, then,*

$$\frac{\partial}{\partial z}(Tf) = T\frac{\partial f}{\partial z}, \qquad \frac{\partial}{\partial \bar{z}}(Tf) = T\frac{\partial f}{\partial \bar{z}}.$$

Proof: We represent Tf by (3′):

$$Tf = \frac{1}{2\pi i}\iint \frac{f(z+\eta, u)}{\eta}\, d\eta \wedge d\bar{\eta}.$$

By the assumption on $\operatorname{supp} f$, the integrand is continuous on $\mathbb{C} \times U$, and on $\mathbb{C} \times B \times U$ (where $\eta \in \mathbb{C}$, $z \in B$, $u \in U$) it is uniformly bounded by a Lebesgue integrable function. Thus Tf is continuous in $B \times U$.

The additional assumptions imply that the integrand has z and $\bar{z}$ derivatives which are integrable over $\mathbb{C}$. Thus the uniform boundedness implies that $\partial/\partial z$ and $\partial/\partial \bar{z}$ commute with the integral operator T. □

Applying the above lemmas in the case where f is smooth, we have the following:

Lemma 3. *Let $f \in \mathscr{E}(B \times U)$ and suppose that* $\operatorname{supp} f \cap (B \times K)$ *is compact for every compact $K \subset U$. Then*

$$Tf \in \mathscr{E}(B \times U),$$

$$\frac{\partial}{\partial z}(Tf) = T\frac{\partial f}{\partial z}, \qquad \frac{\partial}{\partial \bar{z}}(Tf) = T\frac{\partial f}{\partial \bar{z}}, \qquad \frac{\partial}{\partial u_\mu}(Tf) = T\frac{\partial f}{\partial u_\mu}, \qquad 1 \le \mu \le d.$$

Proof: For *any* (also higher) partial derivative g of f, the support of g has compact intersection with every $B \times K$. Thus by the above lemmas, we have the desired interchangability of T with the various differential operators. At the same time all of the first partial derivatives of Tf are continuous on $B \times U$. By an iterated application of this, $Tf \in \mathscr{E}(B \times U)$.

3. The Cauchy Integral Formula and the Equation $(\partial/\partial\bar{z})(Tf) = f$. Functions of the form Tf appear as "correction terms" in the generalized Cauchy integral formula. If G is a bounded domain in $\mathbb{C}$ and ∂G is piece-wise smooth, then, for every complex valued function h which is continuously differentiable on $\bar{G}$, Stokes' theorem is equivalent to the equation

$$\int_{\partial G} h(\xi)\, d\xi = \iint_G dh(\xi) \wedge d\xi = \iint_G \frac{dh}{d\bar{\xi}}(\xi)\, d\bar{\xi} \wedge d\xi.$$

For integrands of the form

$$h(\xi) = \frac{f(\xi)}{\xi - z}, \qquad z \in G, \qquad f \text{ continuously differentiable on } \bar{G},$$

this equation is no longer valid. This is due to the fact that the singularity at z yields an additional contribution, which is in fact quite simple to compute. In order to derive a formula for h one puts a closed disk of radius ε, $K_\varepsilon(z) \subset G$, around z. Since h is continuously differentiable on $\overline{G \backslash K_\varepsilon(z)}$, Stokes' theorem can be applied on $G \backslash K_\varepsilon(z)$ instead of G.

Now $(\xi - z)^{-1}$ is holomorphic on $G \backslash \{z\}$. So it follows that $(\partial h/\partial\bar{\xi}) = (\partial f/\partial\bar{\xi}) \cdot (\xi - z)^{-1}$ and therefore

$$(*) \qquad \iint_{G\backslash K_\varepsilon} \frac{\partial f}{\partial\bar{\xi}}(\xi) \cdot (\xi - z)^{-1}\, d\bar{\xi} \wedge d\xi = \int_{\partial G} \frac{f(\xi)}{\xi - z}\, d\xi - \int_{\partial K_\varepsilon(z)} \frac{f(\xi)}{\xi - z}\, d\xi.$$

The following is just a consequence of letting $\varepsilon \to 0$ in $(*)$:

Theorem 4 (Cauchy Integral Formula). *Let f be a continuously differentiable function on $\bar{G}$. Then*

$$\frac{1}{2\pi i} \iint_G \frac{\partial f}{\partial\bar{\xi}}(\xi) \cdot (\xi - z)^{-1}\, d\xi \wedge d\bar{\xi} = f(z) - \frac{1}{2\pi i} \int_{\partial G} \frac{f(\xi)}{\xi - z}\, d\xi, \qquad z \in G.$$

Proof: In polar coordinates $\xi - z = \varepsilon e^{i\varphi}$,

$$\int_{\partial K_\varepsilon(z)} \frac{f(\xi)}{\xi - z}\, d\xi = i \int_0^{2\pi} f(z + \varepsilon e^{i\varphi})\, d\varphi;$$

and thus

$$\lim_{\varepsilon \to 0} \int_{\partial K_\varepsilon(z)} \frac{f(\xi)}{\xi - z}\, d\xi = 2\pi i f(z).$$

Substituting this in (∗) and using $d\bar{\xi} \wedge d\xi = -d\xi \wedge d\bar{\xi}$, the result follows. □

Remark: If f is holomorphic on $\bar{G}$, then $(\partial f/\partial \bar{z}) = 0$ and the above formula becomes the Cauchy integral formula of classical function theory.

We now use Theorem 4 to show that, for functions with compact support, the operator T is a left-inverse of the differential operator $(\partial/\partial \bar{z})$.

Lemma 5. *Let f be a continuously differentiable function on a bounded domain B and suppose that* $\operatorname{supp} f$ *is compact in B. Then*

$$T \frac{\partial f}{\partial \bar{z}} = f \quad \text{on} \quad B.$$

Proof: Let $z \in B$ be fixed. Since $\operatorname{supp} f$ is compact in B, we may extend f trivially (smoothly) to all of $\mathbb{C}$ and, letting G be a domain containing B,

$$T \frac{\partial f}{\partial \bar{z}}(z) = \frac{1}{2\pi i} \iint_G \frac{\partial f}{\partial \bar{\xi}}(\xi)(\xi - z)^{-1}\, d\xi \wedge d\bar{\xi}.$$

If one chooses G with piece-wise smooth boundary, then, using the fact that $f \equiv 0$ on ∂G, the Cauchy integral formula yields the desired result. □

4. A Lemma of Grothendieck. It can now be shown that for all "reasonable" functions, T is a right inverse for $\partial/\partial \bar{z}$. In other words, Tf solves the differential equation $(\partial g/\partial \bar{z}) = f$.

Theorem 6 (Grothendieck). *Let $f \in \mathscr{E}(B \times U)$ and suppose that every partial derivative of f is bounded on every $B \times K$, where K is compact in U. Then*

1) $|Tf|_{B \times \{u\}} \leq 2\rho |f|_{B \times \{u\}}$, $u \in U$, *where* $\rho =$ *diameter of* B,

2) $\dfrac{\partial}{\partial u_\mu}(Tf) = T \dfrac{\partial f}{\partial u_\mu}$ *on* $B \times U$, $\mu = 1, \ldots, d$,

3) $\dfrac{\partial}{\partial \bar{z}}(Tf) = f$ *on* $B \times U$,

4) $Tf \in \mathscr{E}(B \times U)$.

Proof: The statement 1) was proved in paragraph 1. Statement 2) follows from Lemma 1. In order to prove statements 3) and 4), let $z_0 \in B$ be fixed and choose a

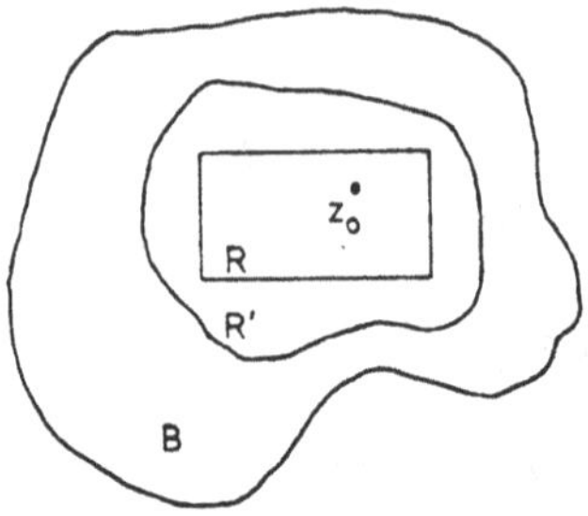

compact rectangle $R \subset B$ with $z_0 \in \mathring{R}$. There exists a function $r \in \mathscr{E}(\mathbb{C})$ which is identically 1 on R and vanishes outside of a neighborhood R' of R which is relatively compact in B (see Chapter A.4.2). We set

$$f_1 := (1 - r)f, \qquad f_2 := rf.$$

Then $f_1, f_2 \in \mathscr{E}(B \times U)$ and $f = f_1 + f_2$. Since f is bounded on every $B \times K$, so are f_1 and f_2. Thus Tf_1 and Tf_2 exist on $B \times U$ and

$$Tf = Tf_1 + Tf_2.$$

Now, by the choice of r, the function f_1 vanishes identically on $R \times U$. Thus

$$(Tf_1)(z, u) = \frac{1}{2\pi i} \iint_{B \setminus R} \frac{f_1(\xi, u)}{\xi - z}\, d\xi \wedge d\bar{\xi}, \qquad (z, u) \in B \times U.$$

As a function of z, this integral behaves quite well as long as z varies in $\mathring{R}$ and stays away from the boundary ∂R. In this case the integrand is infinitely often differentiable in z, $\bar{z}$ and the u_μ's and all derivatives are bounded. It follows that $Tf_1 \mid \mathring{R} \times U \in \mathscr{E}(\mathring{R} \times U)$ and one can interchange differentiation and integration. Since the integrand is holomorphic in z,

$$\frac{\partial}{\partial \bar{z}}(Tf_1) = 0 \quad \text{on} \quad \mathring{R} \times U.$$

We consider the integral Tf_2 whose integrand has a singularity at z. By the choice of r, $\operatorname{supp} f_2 \cap (B \times K)$ is compact for every compact $K \subset U$. Thus, by Lemma 3, $Tf_2 \in \mathscr{E}(B \times U)$ and $(\partial/\partial\bar{z})(Tf_2) = T(\partial f_2/\partial\bar{z})$. It follows from Lemma 4 that $(\partial/\partial\bar{z})(Tf_2) = f_2$ on $B \times U$ and, since $f_2 = f$ on $\mathring{R} \times U$, $(\partial/\partial\bar{z})(Tf_2) = f$ on $\mathring{R} \times U$. Since $Tf = Tf_1 + Tf_2$, statements 3) and 4) now follow from the just derived properties of Tf_1 and Tf_2. □

In the applications of Theorem 6 in the next sections, U is always an open set in the complex vector space $\mathbb{C}^d$ with variables $w_1, \ldots, w_d$. The interchangability equations 2) remain valid for the real and imaginary parts of each w_i. Thus it follows that

$$\frac{\partial}{\partial \bar{w}_i}(Tf) = T\frac{\partial f}{\partial \bar{w}_i}.$$

Hence we can rephrase Theorem 6 as follows:

Corollary. *Let U be an open set in $\mathbb{C}^d$ with variables (global coordinates) $w_1, \dots, w_d$ and suppose $f \in \mathscr{E}(B \times U)$ satisfies the same conditions as in Theorem 5. Then, if f is holomorphic in the variables w_i, it follows that Tf is also holomorphic in the w_i's.*

§ 4. Dolbeault Cohomology Theory

Associated to every closed set M in a complex manifold X are its Dolbeault cohomology groups $\mathrm{Dolb}^{p,q}(M)$. By using the results of Section 3, we will show that these groups vanish for every compact product set in $\mathbb{C}^n$. Among other things this implies the fundamental Dolbeault isomorphism $\mathrm{Dolb}^{p,q}(M) \simeq H^q(M, \Omega^p)$.

1. The Solution of the $\bar\partial$-problem for Compact Product Sets. If M is a subset of X and $\varphi \in A(M)$ is a differential form, then φ is said to be *$\bar\partial$-closed* (over M) if $\bar\partial\varphi = 0$. It is said to be $\bar\partial$-exact if there exists $\psi \in A(M)$ with $\bar\partial\psi = \varphi$. Since $\bar\partial \cdot \bar\partial = 0$, every $\bar\partial$-exact form is $\bar\partial$-closed. The $\bar\partial$-problem consists of characterizing the subsets M of X on which every $\bar\partial$-closed (p, q)-form is $\bar\partial$-exact. The following solves this problem for compact product sets:

Theorem 1 (Dolbeault–Grothendieck). *Let $K_1, \dots, K_m$ be compact sets in $\mathbb{C}$ and $K := K_1 \times \cdots \times K_m$ their product in $\mathbb{C}^m$. Then every $\bar\partial$-closed (p, q)-form, $q \geq 1$, on K is $\bar\partial$-exact.*

Proof: We let Γ_e denote the $\mathbb{C}$-vector space of all (p, q)-forms on K which, with respect to the basis dz_i, $d\bar{z}_j$, are free of $d\bar{z}_{e+1}, \dots, d\bar{z}_m$, $0 \leq e \leq m$. Note that $\Gamma_m = \mathscr{A}^{p,q}(K)$. We will show by induction on e that every $\varphi \in \Gamma_e$ satisfying $\bar\partial\varphi = 0$ has a $\bar\partial$-preimage $\psi \in A^{p,q-1}(K)$. If $e = 0$ then $\varphi = 0$ (since $q \geq 1$) and thus $\psi := 0$ does the job.

Let $e \geq 1$ and take $\varphi \in \Gamma_e$. Collecting all of the terms in φ which contain $d\bar{z}_e$, we obtain

$$\varphi = \alpha \wedge d\bar{z}_e + \beta,$$

where $\alpha \in \mathscr{A}^{p,q-1}(K)$ and $\beta \in \mathscr{A}^{p,q}(K)$ are both free of $d\bar{z}_e, \dots, d\bar{z}_m$. In particular

$$\beta \in \Gamma_{e-1}.$$

By assumption

$$0 = \bar\partial\varphi = \bar\partial\alpha \wedge d\bar{z}_e + \bar\partial\beta.$$

From this, since $d\bar{z}_e, \ldots, d\bar{z}_m$ are not found in α, it follows via an easy calculation that, for every coefficient $f \in \mathscr{E}(K)$ of α,

$$\frac{\partial f}{\partial \bar{z}_\mu} = 0 \quad \text{for} \quad \mu = e+1, \ldots, m.$$

Hence each such coefficient f is holomorphic in $z_{e+1}, \ldots, z_m$. We now apply the Lemma of Grothendieck (in the form of the Corollary) to an open neighborhood $B \times U$ of K with $B \supset K_e$ and $U \supset \prod_{\mu \neq e} K_\mu$ in which f and all of its derivatives is bounded and holomorphic in $z_{e+1}, \ldots, z_m$. For the Lemma, $z := z_e$. Hence, for every coefficient f of α, we obtain a function $\tilde{f} \in \mathscr{E}(B \times U)$ such that $(\partial \tilde{f}/\partial \bar{z}_e) = f$ and $(\partial \tilde{f}/\partial \bar{z}_\mu) = 0$, $\mu = e+1, \ldots, m$, on $B \times U$. If one replaces every such f in α with such an $\tilde{f}$, leaving α otherwise unchanged, an easy calculation shows that one has a form $\tilde{\alpha} \in \mathscr{A}^{p,q-1}(K)$ such that

$$\bar{\partial}\tilde{\alpha} = \alpha \wedge d\bar{z}_e + \gamma, \quad \text{where} \quad \gamma \in \Gamma_{e-1}.$$

Now the (p, q)-form $\delta := -\bar{\partial}\tilde{\alpha} = \beta - \gamma \in \Gamma_{e-1}$ is $\bar{\partial}$-closed. Thus, by the induction assumption, there exists $\tilde{\delta} \in \mathscr{A}^{p,q-1}(K)$ with $\bar{\partial}\tilde{\delta} = \delta$. Consequently, $\psi := \tilde{\alpha} + \tilde{\delta} \in \mathscr{A}^{p,q-1}$ is a $\bar{\partial}$-preimage of φ on K. Since $\Gamma_m = \mathscr{A}^{p,q}(K)$, the theorem is proved. □

Corollary to Theorem 1. *For every compact product set $K \subset \mathbb{C}^m$, the sequence*

$$0 \longrightarrow \Omega^p(K) \xrightarrow{i} \mathscr{A}^{p,0}(K) \xrightarrow{\bar{\partial}} \mathscr{A}^{p,1}(K) \xrightarrow{\bar{\partial}} \cdots \xrightarrow{\bar{\partial}} \mathscr{A}^{p,m}(K) \longrightarrow 0$$

is exact for all $p \geq 0$.

Proof: The exactness at the point $\mathscr{A}^{p,0}(K)$ follows from Theorem 2.6 and the exactness at all other places follows from Theorem 1. □

The proof of Theorem 1 which we reproduced above and which makes use of the integral operator T is due to Grothendieck, being communicated by Serre on 5/15/1954 ([ENS$_2$], Exposé XVIII).

Since $\overline{\partial\psi} = \bar{\partial}\bar{\psi}$, it is clear that all sequences

$$0 \longrightarrow \bar{\Omega}^p(K) \xrightarrow{i} \mathscr{A}^{0,p}(K) \xrightarrow{\partial} \mathscr{A}^{1,p}(K) \xrightarrow{\partial} \cdots \xrightarrow{\partial} \mathscr{A}^{m,p}(K) \longrightarrow 0$$

are exact. Thus the previously announced exactness of the diagram on p. 70 has now been shown.

At this point it should be said that Theorem 1 is also valid for *open* product sets: One can exhaust such domains by compact product sets, solve the $\bar{\partial}$-problem on the compact sets, and take the limit of appropriately chosen solutions as the exhaustion converges to the open set. Such a limiting process is carried out later in the much more general context of Stein spaces (Chapter IV.4). Among other things it will be shown at that time that Theorem 1 is valid for every Stein manifold.

2. The Dolbeault Cohomology Groups. On every m-dimensional complex manifold X and for every natural number $p \geq 0$, we have the following sequence of sheaves (see 2.3):

$$(*) \qquad 0 \longrightarrow \Omega^p \xrightarrow{\ i\ } \mathscr{A}^{p,0} \xrightarrow{\ \bar\partial\ } \mathscr{A}^{p,1} \xrightarrow{\ \bar\partial\ } \cdots \xrightarrow{\ \bar\partial\ } \mathscr{A}^{p,m} \longrightarrow 0.$$

Further, for every *closed* set M in X, we have the $\mathbb{C}$-complex $A^{p,\bullet}(M)$ of sections

$$\mathscr{A}^{p,0}(M) \xrightarrow{\ \bar\partial\ } \mathscr{A}^{p,1}(M) \xrightarrow{\ \bar\partial\ } \cdots \xrightarrow{\ \bar\partial\ } \mathscr{A}^{p,m}(M) \longrightarrow 0.$$

Definition 2 (Dolbeault Cohomology Groups). *The cohomology groups $H^i(\mathscr{A}^{p,\bullet}(M))$ of the complex $\mathscr{A}^{p,\bullet}(M)$ are called the Dolbeault cohomology groups of M in X. They are denoted by* $\mathrm{Dolb}^{p,i}(M)$.

$$\mathrm{Dolb}^{p,0}(M) = \mathrm{Ker}(\bar\partial\colon \mathscr{A}^{p,0}(M) \to \mathscr{A}^{p,1}(M)),$$
$$\mathrm{Dolb}^{p,q}(M) = \mathrm{Ker}(\bar\partial\colon \mathscr{A}^{p,q}(M) \to \mathscr{A}^{p,q+1}(M))/\mathrm{Im}(\bar\partial\colon \mathscr{A}^{p,q-1}(M) \to \mathscr{A}^{p,q}(M)).$$

In particular, if every $\bar\partial$-closed (p, q)-form on M is $\bar\partial$-exact, $\mathrm{Dolb}^{p,q}(M) = 0$ for $q \geq 1$.

Theorem 3. *For every $p \geq 0$, the complex $(*)$ is a resolution of the sheaf of germs of holomorphic p-forms on X. This resolution is acyclic over every closed set M in X and thus there is the natural $\mathbb{C}$-isomorphism*

$$H^i(M, \Omega^p) \cong \mathrm{Dolb}^{p,i}(M), \qquad i \geq 0$$

Proof: Since every point $x \in X$ has a neighborhood basis consisting of relatively compact product domains, the corollary to Theorem 1 shows that $(*)$ is a resolution of Ω^p. Since all sheaves $\mathscr{A}^{p,q}$ are soft on every closed set $M \subset X$, the acyclicity follows. The isomorphism is thus a consequence of the formal de Rham Theorem. □

The corollary to Theorem 1 says that

$$\mathrm{Dolb}^{p,q}(K) = 0 \quad \text{for} \quad p \geq 0, q \geq 1, \quad \text{and } K \text{ a compact product set in } \mathbb{C}^m.$$

Thus the following is an immediate consequence of Theorem 3:

Theorem 4. *If K is a compact product set in $\mathbb{C}^m$, then*

$$H^q(K, \Omega^p) = 0 \quad \text{for} \quad p \geq 0, q \geq 1.$$

In particular,

$$H^q(K, \mathcal{O}) = 0 \quad \text{for} \quad q \geq 1.$$

We will make decided use of this last statement in the case of compact blocks (Chapter III.3.2).

The Dolbeault cohomology groups $\mathrm{Dolb}^{p,q}(X)$ vanish as soon as $p + q > m = \dim X$. Thus Theorem 3 implies that, for *every* m-dimensional complex manifold X,

$$H^q(X, \Omega^p) = 0 \quad \text{for all} \quad p, q \quad \text{with} \quad p + q > m.$$

In particular

$$H^q(X, \mathcal{O}) = 0 \quad \text{for all} \quad q > m.$$

3. The Analytic de Rham Theory. For every differentiable manifold the differentiable de Rham theory yield the acyclic resolution

$$0 \longrightarrow \mathbb{C} \xrightarrow{i} \mathscr{E} \xrightarrow{d} \mathscr{A}^1 \xrightarrow{d} \mathscr{A}^2 \longrightarrow \cdots$$

of the constant sheaf $\mathbb{C}$ by the complex of differential forms. For complex manifolds X, it is appropriate to consider the sequence

$$(\circ) \qquad 0 \longrightarrow \mathbb{C} \xrightarrow{i} \mathcal{O} \xrightarrow{d} \Omega^1 \xrightarrow{d} \Omega^2 \longrightarrow \cdots$$

The Poincaré Lemma again implies that $(\circ)$ is a resolution of $\mathbb{C}$ over X. This resolution is in general not acyclic (the sheaves Ω^p are not soft, rather they are coherent!). However we do want to note the following:

Theorem 5. *Let X be a complex manifold such that*

$$H^q(X, \Omega^p) = 0 \quad \text{for all} \quad p \geq 0, q \geq 1.$$

Then there exists natural $\mathbb{C}$-isomorphisms

$$H^0(X, \mathbb{C}) \cong \mathrm{Ker}(d\colon \mathcal{O}(X) \to \Omega^1(X))$$
$$H^q(X, \mathbb{C}) \cong \mathrm{Ker}(d\colon \Omega^q(X) \to \Omega^{q+1}(X))/\mathrm{Im}(d\colon \Omega^{q-1}(X) \to \Omega^q(X)).$$

In particular,

$$H^q(X, \mathbb{C}) = 0 \quad \textit{for} \quad q > m = \dim X.$$

Proof: The existence of the isomorphisms follows from the formal de Rham Theorem. The statement about the complex cohomology of X is then trivial since $\Omega^q = 0$ for all $q > m$. □

The vanishing of the complex cohomology of a real $2m$-dimensional manifold X from dimension $m + 1$ on is a very restrictive topological condition on X. For example, since $H^{2m}(X, \mathbb{C}) \cong \mathbb{C}$, this condition is never satisfied by a compact complex manifold X. Thus for compact complex manifolds X, there are always integers $p \geq 0$, $q \geq 1$ so that $H^q(X, \Omega^p) \neq 0$.

The assumptions of Theorem 5 are always satisfied by Stein manifolds. It is not known if there exist non-Stein manifolds of this type (i.e. $H^q(X, \Omega^p) = 0$, $p \geq 0$, $q \geq 1$).

Supplement to §4.1. A Theorem of Hartogs

Let $\Delta_s(0) := \{(z_1, \ldots, z_m) \in \mathbb{C}^m \,|\, |z_\mu| < s, \mu = 1, \ldots, m\}$ denote the polycylinder of radius $s > 0$ about $0 \in \mathbb{C}^m$. Using elementary techniques (Laurent developments, see [SCV], p. 31, Theorem 1.3) one can convince himself of the validity of the following:

Theorem 1. *Let $m \geq 2, s > r > 0$. Then every function holomorphic on the "annular region" $\Delta_s(0) \backslash \overline{\Delta_r(0)}$ can be continued to a function holomorphic on the entire polycylinder $\Delta_s(0)$.*

This and Theorem 4.1 imply the following:

Theorem 2. *Let $m \geq 2$ and $\varphi = \sum_{\mu=1}^{m} a_\mu \, d\bar{z}_\mu \in \mathscr{A}^{0,1}(\mathbb{C}^m)$ be a closed (0, 1)-form with compact support* (i.e. $\operatorname{supp} \varphi = \bigcup_{\mu=1}^{m} \operatorname{supp} a_\mu$ *is compact*). *Then there exists a function $g \in \mathscr{E}(\mathbb{C}^m)$ with compact support so that $\bar{\partial} g = \varphi$.*

Proof: Let $\Delta_r(0)$ be a polycylinder which contains $\operatorname{supp} \varphi$. Take $s > r$ and let $v \in \mathscr{E}(\Delta_s(0))$ be the function guaranteed by Theorem 4.1 so that $\bar{\partial} v = \varphi \,|\, \Delta_s(0)$. In particular

$$\bar{\partial} v = 0 \quad \text{on} \quad \overline{\Delta_s(0) \backslash \operatorname{supp} \varphi}.$$

Thus $v \in \mathcal{O}(\Delta_s(0) \backslash \overline{\Delta_r(0)})$. By Theorem 1, $v \,|\, \Delta_s(0) \backslash \overline{\Delta_r(0)}$ is continuable to a function $h \in \mathcal{O}(\Delta_s(0))$. Define $w := v - h \in \mathscr{E}(\Delta_s(0))$. Then likewise $\bar{\partial} w = \varphi \,|\, \Delta_s(0)$. Since v and h agree on $\Delta_s(0) \backslash \overline{\Delta_r(0)}$, the trivial extension, g, of w to all of $\mathbb{C}^m$ is the desired function. □

From this we are able to easily derive the following fundamental fact:

Theorem 3. (Hartogs' Theorem [SCV], p. 33). *Let G be a domain in $\mathbb{C}^m$, $m \geq 2$, and K a compact set in G such that $G \backslash K$ is connected. Then the restriction homomorphism $\mathcal{O}(G) \to \mathcal{O}(G \backslash K)$ is surjective.*

Proof: We choose a compact set $L \subset G$ with $K \subset \mathring{L}$. Thus if $f \in \mathcal{O}(G \backslash K)$ is arbitrarily given, $f \,|\, G \backslash L$ is always continuable to a differentiable function $v \in \mathscr{E}(G)$. This is accomplished by choosing some $r \in \mathscr{E}(G)$ with $r \,|\, G \backslash L \equiv 1$ and $r \,|\, U(K) \equiv 0$, where $U(K) \subset \mathring{L}$ is a neighborhood of K (see Theorem A.4.4). Then one extends f

by the trivial extension of $r \cdot f$ to all of G. Define $\psi := \bar{\partial} v \in \mathscr{A}^{0,1}(G)$. Since $\psi = \bar{\partial} f = 0$ on $G \backslash L$, one can extend ψ trivially to a $\bar{\partial}$-closed (0, 1)-form $\varphi \in \mathscr{A}^{0,1}(\mathbb{C}^m)$ with compact support. By Theorem 2 there exists a $g \in \mathscr{E}(\mathbb{C}^m)$ with compact support such that $\bar{\partial} g = \varphi$. Since φ vanishes on $\mathbb{C}^m \backslash L$, $g \mid \mathbb{C}^m \backslash L$ is holomorphic. Since φ vanishes identically outside of its compact support, $g \mid W = 0$, where W is *the* unbounded component of $\mathbb{C}^m \backslash L$. Now since every connectivity component of $\mathbb{C}^m \backslash L$ has points of L in its boundary, $U := W \cap (G \backslash L)$ is non-empty. Let $h := v - g$. Then $h \in \mathcal{O}(G)$ and $h \mid U = v \mid U = f \mid U$. Thus by the identity principle (using the connectivity of $G \backslash K$),

$$h \mid G \backslash K = f \mid G \backslash K.$$

In other words, h is the desired holomorphic continuation of f to G. □

Chapter III. Theorems A and B for Compact Blocks in $\mathbb{C}^m$

In this chapter the main results of the theory of coherent analytic sheaves for compact blocks Q in $\mathbb{C}^m$ are proved (see Paragraph 3.2). The standard techniques for coherent sheaves and cohomology theory are used, in particular the fact that $H^q(Q, \mathscr{S}) = 0$ for large q (see Chapter B.2.5 and 3.4). Moreover we will bring into play the fact that $H^q(Q, \mathcal{O}) = 0$ for $q \geq 1$. The basic tool which is derived in this chapter is an attaching lemma for analytic sheaf epimorphisms (Theorem 2.3). The proof of this lemma is based on an attaching lemma of H. Cartan for matrices near the identity (Theorem 1.4) and the Runge approximation theorem (Theorem 2.1).

§ 1. The Attaching Lemmas of Cousin and Cartan

Unless the reader knows the origins of the problems, reference to the fundamental lemmas of Cousin and Cartan as‘attaching" lemmas carries little meaning. Thus we wish to begin this section by remarking that the existence of these lemmas *allows* us to solve attaching problems. For example, suppose that on a cover $U = \{U_i\}$ of a complex space X one has prescribed meromorphic functions m_i on U_i so that $m_j - m_i =: f_{ij} \in \mathcal{O}(U_i \cap U_j)$ whenever $U_i \cap U_j \neq \emptyset$. In other words, one has prescriptions of "principal parts" of meromorphic functions! If $f_i \in \mathcal{O}(U_i)$ can be found so that $f_{ij} = f_i - f_j$ on $U_i \cap U_j$, then the meromorphic function m, which is defined by $m_i - f_i$ on U_i, has the prescribed principal parts. Hence, if one can find such f_i's, one can "attach" the m_i's to each other. Solving this *additive* attaching problem (i.e., given the f_{ij}'s, find the f_i's) in a very special case is the essence of Cousin's attaching lemma.

Cartan's attaching lemma solves the analogous *multiplicative* problem for holomorphic matrices near the identity. This will allow us to attach sheaf epimorphisms.

1. The Lemma of Cousin. We always work in $\mathbb{C}^m$ with the variable $z = (z_1, \ldots, z_m)$. We set $x := \operatorname{Re} z_1$, $y := \operatorname{Im} z_1$ and let

$$E = \{z_1 \in \mathbb{C} \mid a < x < b, c < y < d\}.$$

Thus E is an open, non-empty rectangle in the z_1-plane. Furthermore let $a', b' \in \mathbb{R}$ with $a < a' < b' < b$. Then,

$$E = E' \cap E'', \quad \text{where} \quad E' := \{z \in E \,|\, x < b'\} \quad \text{and} \quad E'' := \{z \in E \,|\, a' < x\}.$$

Let $U \neq \emptyset$ open in $\mathbb{C}^{m-1}$ (using the variables $z_2, \ldots, z_m$) and $B := E \times U$, $B' := E' \times U$, $B'' := E'' \times U$, $D := (E' \cap E'') \times U$. Then $B = B' \cup B''$ and $D = B' \cap B''$.

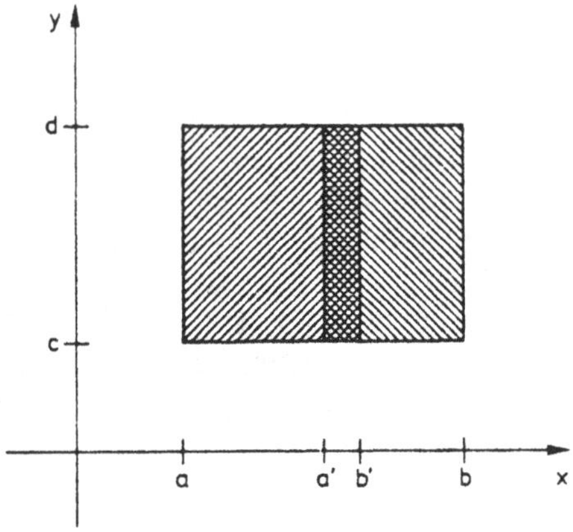

Theorem 1 (Cousin Attaching Lemma). *There exists a real constant K so that for every bounded analytic function $f \in \mathcal{O}(D)$ one can find bounded analytic functions $f' \in \mathcal{O}(B')$ and $f'' \in \mathcal{O}(B'')$ with*

1) $f = f'|D + f''|D$
2) $|f'|_{B'} \leq K|f|_D$, $|f''|_{B''} \leq K|f|_D$.

Proof: We choose a positive real number $\delta < \frac{1}{2}(b' - a')$ and a smooth real valued function $r(x)$ with values between 0 and 1 so that

$$r(x) = 0 \quad \text{for} \quad x < a' + \delta \quad \text{and} \quad r(x) = 1 \quad \text{for} \quad x > b' - \delta.$$

We then set

$$p(z) := \begin{cases} r(x) f(z) & \text{for } z \in D \\ 0 & \text{for } z \in B' \setminus D, \end{cases} \qquad q(z) := \begin{cases} (1 - r(x)) f(z) & \text{for } z \in D \\ 0 & \text{for } z \in B'' \setminus D. \end{cases}$$

Certainly $p \in \mathscr{E}(B')$ and $q \in \mathscr{E}(B'')$ and both functions are holomorphic in $z_2, \ldots, z_m$. Moreover

$$f = p|D + q|D, \qquad |p|_{B'} \leq |f|_D, \qquad |q|_{B''} \leq |f|_D.$$

The fact that f is holomorphic implies that

$$\frac{\partial p}{\partial \bar{z}_1}(z) = \frac{1}{2}\frac{dr}{dx}(x) \cdot f(z), \quad z \in B' \quad \text{and} \quad \frac{\partial q}{\partial \bar{z}_1}(z) = -\frac{1}{2}\frac{dr}{dx}(x) \cdot f(z), \quad z \in B''.$$

The function

$$h(z) := \frac{1}{2}\frac{dr}{dx}(x) \cdot f(z), \qquad z \in B = B' \cup B''$$

is defined and smooth on B and holomorphic in $z_2, \ldots, z_m$.

Furthermore

$$|h|_B \leq M|f|_D, \quad \text{where} \quad M := \sup_{x \in \mathbb{R}} \left| \frac{1}{2}\frac{dr}{dx}(x) \right|.$$

By the corollary to Theorem 2.3.5, there exists a function $\hat{h} \in E(B)$ which is holomorphic in $z_2, \ldots, z_m$ so that

$$\frac{\partial \hat{h}}{\partial \bar{z}} = h, \qquad |\hat{h}|_B \leq 2\rho|h|_B, \qquad \text{where } \rho := \text{diameter of } E.$$

We now define our desired functions.

$$f' := p - \hat{h}|B' \quad \text{and} \quad f'' := q + \hat{h}|B''.$$

Thus $f = f'|D + f''|D$. Since $\partial f'/\partial \bar{z}_1 = 0$ and $\partial f''/\partial \bar{z}_1 = 0$, $f' \in \mathcal{O}(B')$ and $f'' \in \mathcal{O}(B'')$. Estimating in the obvious way,

$$|f'|_B \leq |p|_{B'} + |h|_B \leq |f|_D + 2\rho M|f|_D = K|f|_D,$$

where $K := 1 + 2\rho M$. In the same way $|f''|_{B''} \leq K|f|_D$ and thus the theorem is proved. □

2. Bounded Holomorphic Matrices. Let $p, q \geq 1$ be fixed natural numbers and V a non-empty open subset of $\mathbb{C}^m$. If

$$a = a(z) = (a_{ik}(z))_{1 \leq i \leq p, 1 \leq k \leq q}, \qquad z \in V,$$

is a (p, q)-matrix valued function on V, then we define the norm of a by

$$|a| := \max_{i,k} |a_{ik}|_V.$$

Clearly $|a| < \infty$ if and only if each function $a_{ik}(z)$ is bounded on V.

If a (resp. b) is a bounded (p, q)-matrix (resp. (q, r)-matrix) on V, then $a \cdot b$ is a bounded (p, r)matrix and

$$|a \cdot b| \leq q|a| \cdot |b|.$$

In particular matrix multiplication is continuous with respect to the norm $|\ \ |$.

We let $e = (\delta_{ik})$ denote the $q \times q$ identity matrix and we note the following:

Let $s := q^{-1} - (2q^2)^{-1} > 0$. *Then, for every* (q, q)*-matrix* a *on* V *with* $|a - e| \leq s$, *the inverse matrix* a^{-1} *exists and is bounded on* V *with* $|a^{-1}| \leq 2$.

Proof: Let $h := e - a$. Then $|h^i| \leq q^{i-1}|h|^i \leq q^{-1}(sq)^i$ for all $i \geq 0$. Since $sq < 1$, the sum $a^{-1} := \sum_{i=0}^{\infty} h^i$ converges on V and

$$|a^{-1}| \leq \sum_{i=0}^{\infty} |h^i| \leq q^{-1} \cdot \frac{1}{1 - sq} = 2$$

by the choice of s. □

We note further that,

for every sequence (a_ν) *of* (q, q)*-matrices on* V *such that* $2 \sum_{\nu=0}^{\infty} |a_\nu - e| \leq q^{-1}$,

$$(\#) \qquad |a_0 a_1 \cdots a_n - e| \leq 2 \sum_{\nu=0}^{n} |a_\nu - e| \quad \text{for all} \quad n \geq 0.$$

The proof follows immediately by induction on n from the fact that

$$\begin{aligned} a_0 a_1 \cdots a_n a_{n+1} - e = {} & (a_0 a_1 \cdots a_n - e)(a_{n+1} - e) \\ & + (a_0 a_1 \cdots a_n - e) + (a_{n+1} - e). \end{aligned}$$ □

A (p, q)-matrix function $a(z) = (a_{ik}(z))$ is said to be *holomorphic* on V if each component $a_{ik}(z)$ is a holomorphic function on V. The $\mathbb{C}$-vector space of all (p, q)-matrices which are holomorphic and *bounded* on V is a Banach space with respect to the norm $| \ \ |$.

In the following p is always equal to q. The bounded holomorphic matrix functions on V form a $\mathbb{C}$-algebra. The reader should note that $| \ \ |$ is *not* an algebra norm. We let $B^*(V)$ denote the invertible elements of $B(V)$. Since the inverse of a holomorphic matrix function is holomorphic, the above remarks show that

$$\{a \in B(V) \mid |a - e| < s\} \subset B^*(V)^1$$

For a later construction (in the proof of Cartan's Lemma) we need the following:

Lemma 2. *Let* $g_\nu, h_\nu \in B(V)$ *be sequences with*

$$2 \sum_{0}^{\infty} |g_\nu| < s \quad \textit{and} \quad 2 \sum_{0}^{\infty} |h_\nu| \leq s.$$

---

[1] An $a \in B(V)$ can have a holomorphic inverse a^{-1} which is not bounded (i.e. $B^*(V)$ is a prope[r] subset of $GL(q, \mathcal{O}(V))$.).

Then the products

$$u_n := (e + g_0)(e + g_1) \cdots (e + g_{n-1})(e + g_n) \in B(V)$$
$$v_n := (e + h_n)(e + h_{n-1}) \cdots (e + h_1)(e + h_0) \in B(V)$$

converge uniformly on V *to the matrices* $u, v \in B^*(V)$ *with*

$$|u - e| \leq 2 \sum_{\nu=0}^{\infty} |g_\nu| \quad \text{and} \quad |v - e| \leq 2 \sum_{\nu=0}^{\infty} |h_\nu|.$$

Proof: It suffices to verify the statement for the sequence u_n. By (#) (noting that $s \leq q^{-1}$), it follows that

$$|(e + g_i) \cdots (e + g_j) - e| \leq 2 \sum_{\rho=i}^{j} |g_\rho|, \quad \text{for all} \quad j \geq i \geq 0.$$

In particular $|u_n - e| \leq 2 \sum_{0}^{n} |g_\rho| \leq s$. Thus the sequence $|u_n|$ is bounded by $1 + s$. Hence

$$|u_\mu - u_\nu| = |u_\nu\{(e + g_{\nu+1}) \cdots (e + g_\mu) - e\}| \leq 2q(1 + s) \sum_{\rho=\nu+1}^{\mu} |g_\rho|$$

for all $\mu \geq \nu \geq 0$. Consequently (u_ν) is a Cauchy sequence in $B(V)$, converging to a holomorphic matrix function u. Since $|u_n - e| \leq s$, it is also the case that $|u - e| < s$. Thus, by the choice of s, $u \in B^*(V)$. □

3. The Lemma of Cartan. In this paragraph we use the notation $B = E \times U$, B', B'' and D which was introduced in Paragraph 1. We will use K to denote the constant of the Cousin Lemma. All matrices which enter into our discussions are (q, q)-matrices and we again let $s = q^{-1} - (2q^2)^{-1}$. The matrix norms over B', B'', and D are specified by $|\ |_{B'}$, $|\ |_{B''}$, and $|\ |_D$. The results of the following lemma are basic for the proof of Cartan's Lemma.

Lemma 3. *Let* $\varepsilon \in \mathbb{R}$ *be such that* $0 < \varepsilon < sK^{-1}$ *and* $t = 4q^3K^2\varepsilon < 1$. *Then, for every holomorphic matrix* $a = e + b \in B(D)$ *with* $|b|_D \leq \varepsilon$, *there exist holomorphic matrices*

$$a' = e + b' \in B(B'), \quad a'' = e + b'' \in B(B''), \quad \text{and} \quad \tilde{a} = e + \tilde{b} \in B(D)$$

having the following properties:

1) $|b'|_{B''} \leq K|b|_D$, $|b''|_{B''} \leq K|b|_D$, $|\tilde{b}|_D \leq t|b|_D$,
2) $a = a'|D \cdot \tilde{a} \cdot a''|D$.

Proof: Applying Cousin's Lemma to the q^2 components of b, there exist matrices $b' \in B(B')$, $b'' \in B(B'')$ and $b = b'|D + b''|D$ with $|b'|_{B'} \leq K|b|_D$ and $|b''|_{B''} \leq K|b|_D$. Defining $a' := e + b'$ and $a'' := e + b''$, it follows that

$$(*) \qquad a'|D \cdot a''|D = a + (b'|D \cdot b''|D).$$

Since $K|b|_D \leq s$ (recall how ε was chosen), a' and a'' are *invertible* on B' and B'' respectively (see Lemma 2) and

$$(**) \qquad |a'^{-1}|_{B'} \leq 2, \qquad |a''^{-1}|_{B''} \leq 2.$$

For $\tilde{a} := (a'|D)^{-1} \cdot a \cdot (a''|D)^{-1} \in B(D)$, we therefore have the equation in 2). Note that

$$a = a'|D \cdot (e + \tilde{b}) \cdot a''|D = a'|D \cdot a''|D + a'|D \cdot \tilde{b} \cdot a''|D.$$

Thus, by $(*)$, $\tilde{b} = -(a'|D)^{-1} \cdot b'|D \cdot b''|D \cdot (a''|D)^{-1}$. From this, using the estimates in 1) and $(**)$, it follows that

$$|\tilde{b}|_D \leq 4q^3 \cdot K^2 \cdot |b|_D^2.$$

The definition of t and the fact that $|b|_D \leq \varepsilon$ now imply that $|\tilde{b}|_D \leq t|b|_D$. □

We are now in a position to prove Cartan's Lemma.

Theorem 4 (Cartan's Attaching Lemma). *For every $q \geq 1$, there exists a real constant $\varepsilon > 0$ with the following property: Given a matrix, $a \in B(D)$ with $|a - e|_D < \varepsilon$, there exist invertible matrices $c' \in B^*(B')$, $c'' \in B^*(B'')$ so that*

$$a = c'|D \cdot c''|D$$

and

$$|c' - e|_D \leq 4K|a - e|_D, \qquad |c'' - e|_D \leq 4K|a - e|_D.$$

Proof: In addition to the restriction placed on ε in Lemma 3, we further require that $2t \leq 1$ and $4K\varepsilon \leq s$. We set $a = e + b$ and $L := |b|_D$. We now inductively define three sequences

$$a_\nu = e + b_\nu \in B(D), \qquad a'_\nu = e + b'_\nu \in B(D), \qquad \text{and} \qquad a''_\nu = e + b''_\nu \in B(D)$$

satisfying

$$(\circ) \qquad |b_\nu|_D \leq Lt^\nu, \qquad |b'_{\nu-1}|_{B'} \leq KLt^{\nu-1}, \qquad |b''_{\nu-1}|_{B''} \leq KLt^{\nu-1}, \qquad \text{and}$$

$$(\circ\circ) \qquad a_{\nu-1} = a'_{\nu-1}|D \cdot a_\nu \cdot a''_{\nu-1}|D.$$

For this let $a_0 := a$. If $a'_{\nu-1}$, $a''_{\nu-1}$, and a_ν are already constructed, then $|b_\nu|_D \leq Lt^\nu \leq \varepsilon$ and, applying Lemma 3 to a_ν, we obtain matrices a'_ν, a''_ν and $a_{\nu+1} := \tilde{a}_\nu$. Thus, ($\circ\circ$) is valid with all indices increased by 1. Further (by 1) of Lemma 3),

$$|b_{\nu+1}|_D \leq t|b_\nu|_D, \qquad |b'_\nu|_{B'} \leq K|b_\nu|_D, \qquad \text{and} \qquad |b''_\nu|_{B''} \leq K|b_\nu|_D.$$

Hence, using these inequalities as well as the induction assumption, ($\circ$) is valid for $b_{\nu+1}$, b'_ν and b''_ν.

Now for all $n > 0$, we define

$$u_n := (e + b'_0) \cdot (e + b'_1) \cdots (e + b'_{n-1}) \cdot (e + b'_n) \in B(B')$$
$$v_n := (e + b''_n) \cdot (e + b''_{n-1}) \cdots (e + b''_1) \cdot (e + b''_0) \in B(B'')$$

From ($\circ\circ$) it follows that $a = u_n|D \cdot a_{n+1} \cdot v_n|D$ for all $n \geq 0$. From the estimates in ($\circ$) and the fact that $t \leq \frac{1}{2}$, it follows that

$$(*) \qquad 2\sum_{\nu=0}^{\infty} |b'_\nu|_{B'} = 2KL \sum_{\nu=0}^{\infty} t^\nu = 4K\varepsilon \leq s.$$

Thus, by Lemma 2, the sequence u_n converges on B' to an invertible matrix $c' \in B^*(B')$ such that

$$|c' - e|_{B'} \leq \sum_{\nu=0}^{\infty} |b'_\nu|_{B'} \leq 4K|a - e|_D.$$

For the analogous reason, the sequence v_n converges on B'' to a matrix $c'' \in B^*(B'')$ with $|c'' - e|_{B''} \leq 4K|a - e|_D$. Furthermore, $a_{n+1} = e + b_{n+1}$ converges to e. Thus,

$$a = \lim u_n|D \cdot \lim a_{n+1} \cdot \lim v_n|D = c'|D \cdot c''|D. \qquad \square$$

The above lemma only applies to matrices "near the identity." However the same statement holds for arbitrary *invertible* matrices. Since we don't need this more general version for the proof of Theorems A and B we do not go any further into this matter here.

§ 2. Attaching Sheaf Epimorphisms

We begin by proving an approximation theorem for holomorphic functions in a very special geometric situation. This is done by going back to the definition of the Cauchy integral as a Riemann sum. By means of this approximation theorem and the Cartan Attaching Lemma, we are able to attach epimorphisms of sheaves. Later we will again make decided use of the approximation theorem (see IV.4.4).

In this section we will often write $z = x + iy$ for z_1 and z' for $(z_2, \ldots, z_m)$. We will always use

$$R := \{z \in \mathbb{C} \mid a \le x \le b,\ c \le y \le d\} \subset \mathbb{C}$$

to denote a compact, non-empty rectangle with possibly empty interior and K' will always be an arbitrary non-empty compact subset of the z'-space $\mathbb{C}^{m-1}$. We set

$$K := R \times K'.$$

1. An Approximation Theorem of Runge. By carrying over the well-known methods from the theory of one complex variable, we show the following:

Theorem 1 (Runge). *Let $\delta > 0$ be arbitrary. Then for a given $f \in \mathcal{O}(K)$ there exists a polynomial $\hat{f}$ in z with coefficients holomorphic on K' such that*

$$|f - \hat{f}|_K \le \delta$$

Proof: We may assume that $R \ne \emptyset$. There exists an open rectangle $E \supset R$ and $\tilde{f} \in \mathcal{O}(\bar{E} \times K')$ with $\tilde{f} \mid K = f$. By the Cauchy integral formula, letting ∂E be the oriented boundary of E and setting $w = (z, z') \in K$,

$$f(w) = \frac{1}{2\pi i} \int_{\partial E} \frac{f(\zeta, z')}{\zeta - z} \, d\zeta$$

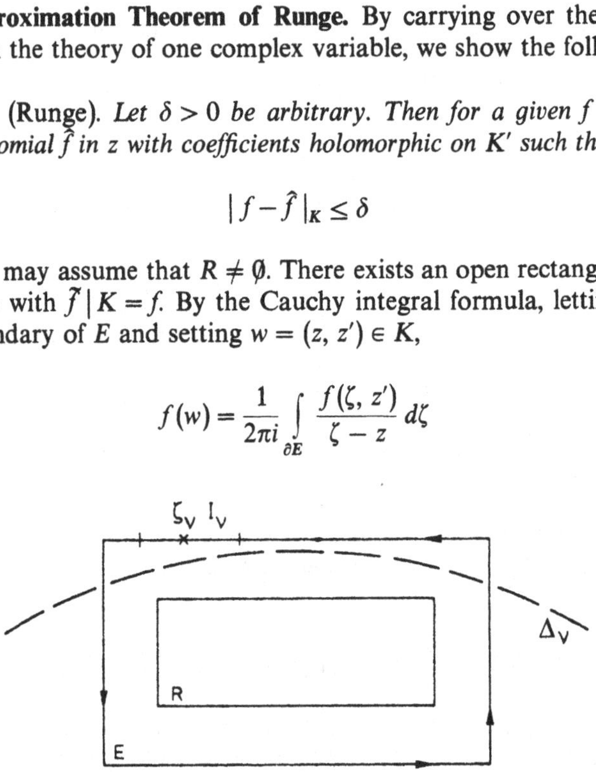

The integrand $k(\zeta, w) := (2\pi i(\zeta - z))^{-1} \cdot f(\zeta, z')$ is uniformly continuous on the compact set $\partial E \times K$. Thus there exists $\rho > 0$, so that, for all $\zeta, \zeta' \in \partial E$ with $|\zeta - \zeta'| \le \rho$ and all $w \in K$,

$$|k(\zeta, w) - k(\zeta', w)| \le \frac{\delta}{2L}, \qquad \text{where } L \text{ is the circumference of } E.$$

We now decompose ∂E into intervals I_ν, $1 \le \nu \le n$, of length $\rho_\nu < \rho$ and fix points $\zeta_\nu \in I_\nu$. Then $k(\zeta_\nu, w)$ is holomorphic on $(\mathbb{C} \setminus \{\zeta_\nu\}) \times K'$ and

$$g(w) := \sum_{\nu=1}^{n} k(\zeta_\nu, w) \rho_\nu$$

is a Riemann sum for the above Cauchy integral. In fact

$$f(w) - g(w) = \sum_{\nu=1}^{n} \int_{I_\nu} (k(\zeta, w) - k(\zeta_\nu, w))\, d\zeta, \qquad w \in K,$$

and thus

$$|f - g|_K \le \sum_{\nu=1}^{n} \frac{\delta}{2L} \rho_\nu = \frac{\delta}{2}.$$

For every point ζ_ν we find an open disk $\Delta_\nu \subset \mathbb{C}$ with $R \subset \Delta_\nu$ and $\zeta_\nu \notin \bar{\Delta}_\nu$. We choose a Taylor polynomial $t_\nu \in \mathbb{C}[z]$ from the Taylor development of $\rho_\nu(2\pi i(\zeta_\nu - z)^{-1})$ about the center of Δ_ν (i.e. the first part of the Taylor series) so that

$$|\rho_\nu(2\pi i(\zeta_\nu - z))^{-1} - t_\nu(z)|_R \le \frac{\delta}{2nLM_\nu}, \quad \text{where} \quad M_\nu := |f(\zeta_\nu, z')|_{K'}.$$

Then

$$\hat{f}(w) := \sum_{\nu=1}^{n} \hat{f}(\zeta_\nu, z') t_\nu(z), \qquad z \in \mathbb{C}, \quad z' \in K',$$

is a polynomial in z with coefficients which are holomorphic on K'. Moreover, for $w \in K$,

$$g(w) - \hat{f}(w) = \sum_{\nu=1}^{n} \hat{f}(\zeta_\nu, z') \cdot [\rho_\nu(2\pi i(\zeta_\nu - z))^{-1} - t_\nu(z)].$$

Therefore

$$|g - \hat{f}|_K \le \sum_{\nu=1}^{h} M_\nu \frac{\delta}{2nLM_\nu} = \frac{\delta}{2}.$$

Combining this with the estimate above,

$$|f - \hat{f}|_K \le \delta. \qquad \square$$

For later applications we note the following corollary. It is necessary to introduce the notion of a (compact) block in the $(z_1, \ldots, z_m)$-space $\mathbb{C}^m$: It is just a compact block in the underlying real space $\mathbb{R}^{2m}$ with respect to the variables $\operatorname{Re} z_1, \ldots, \operatorname{Re} z_m, \operatorname{Im} z_1, \ldots, \operatorname{Im} z_m$.

Corollary (Approximation Theorem for Blocks). *Let $\varepsilon > 0$ and suppose $Q \subset \mathbb{C}^m$ is a compact block. Then for every $f \in \mathcal{O}(Q)$ there is a polynomial $\hat{f} \in \mathbb{C}[z_1, \ldots, z_m]$ so that*

$$|f - \hat{f}|_Q \le \varepsilon$$

Proof: (by induction on m). Thecase of $m = 1$ is clear by Theorem 1. For $m > 1$ let $Q = R \times Q'$, where R is a compact rectangle and Q' is a compact block in

$\mathbb{C}^{m-1}$. By Theorem 1 there exists a polynomial $\hat{f} = \sum_{i=0}^{n} f_i z^i$, with $f_i \in \mathcal{O}(Q')$, such that $|f - \hat{f}|_Q \leq \varepsilon/2$. By the induction hypothesis there exist polynomials $\check{f}_i \in \mathbb{C}[z_2, \ldots, z_m]$ with

$$|f_i - \check{f}_i|_{Q'} \leq \frac{\varepsilon}{2(n+1)T}, \quad \text{where} \quad T := \max_{0 \leq i \leq n} |z_1^i|_R.$$

Defining $\check{f} := \sum_{i=0}^{n} \check{f}_i z_1^i \in \mathbb{C}[z_1, z_2, \ldots, z_m]$, it follows that

$$|\hat{f} - \check{f}|_Q \leq \sum_{i=0} |f_i - \check{f}_i|_{Q'} \cdot |z_1^i|_R \leq \frac{\varepsilon}{2}.$$

Thus $|f - \check{f}|_Q \leq |f - \hat{f}|_Q + |\hat{f} - \check{f}|_Q \leq \varepsilon$. □

Remark: The rectangle R in Theorem 1 can be replaced by an arbitrary compact set (The construction of the disk Δ_ν which contains R and does not contain ζ_ν still works!). Correspondingly the corollary above is valid for product sets in $\mathbb{C}^m$ whose factors are compact and convex.

2. The Attaching Lemma for Epimorphisms of Sheaves. As in the previous paragraph, $K = R \times K'$ is the product of a compact set K' in the z'-space $\mathbb{C}^{m-1}$ with a compact rectangle $R = \{z \in \mathbb{C} \mid a \leq x \leq b, c \leq y \leq d\}$. Given $e \in [a, b]$, we define the subrectangles

$$R^- := \{z \in R \mid x \leq e\} \quad \text{and} \quad R^+ := \{z \in R \mid x \geq e\}.$$

We then set

$$K^- := R^- \times K', \qquad K^+ := R^+ \times K'$$

and

$$P := K^- \cap K^+ = (R^- \cap R^+) \times K'.$$

We consider analytic sheaves on K. Recall that a sheaf $\mathcal{S}$ on a set M in $\mathbb{C}^m$ or more generally in a complex space X is said to be *analytic* if $\mathcal{S}$ is defined on an open neighborhood of M and is analytic there. Correspondingly, sections, homomorphisms, exact sequences, etc. are always defined on open neighborhoods of M.

In the proof of the following theorem we apply both the approximation theorem and Cartan's Attaching Lemma.

Theorem 2 (Attaching Sections). *Let $\mathcal{S}$ be an analytic sheaf on K. Suppose that $t_1^-, \ldots, t_p^- \in \mathcal{S}(K^-)$ and $t_1^+, \ldots, t_q^+ \in \mathcal{S}(K^+)$ are such that their restrictions to P generate the same $\mathcal{O}(P)$-submodule of $\mathcal{S}(P)$:*

$$\sum_{i=1}^{p} O(P) t_i^- \mid P = \sum_{j=1}^{q} O(P) t_j^+ \mid P.$$

Then there exists an invertible, holomorphic (p, p)-matrix on K^-, $a^- \in GL(p, \mathcal{O}(K^-))$, *and sections* $t_1, \ldots, t_p \in \mathcal{S}(K)$ *so that*

$$(t_1|K^-, \ldots, t_p|K^-) = (t_1^-, \ldots, t_p^-)a^-.$$

Proof: By assumption we have equations

$$t_i^- \,|P = \sum_{\alpha=1}^{q} t_\alpha^+ \,|P \cdot u_{\alpha i}, \qquad t_j^+ \,|P = \sum_{\beta=1}^{p} t_\beta^- \,|P \cdot v_{\beta j}, \qquad i, j = 1, \ldots, p, q,$$

with coefficients $u_{\alpha i}, v_{\beta j} \in \mathcal{O}(P)$. We write the sections t_i^- (resp. t_j^+) as row vectors $t^- \in \mathcal{S}^p(K^-)$ (resp. $t^+ \in \mathcal{S}^q(K^+)$). Thus, writing the coefficients $u_{\alpha i}$ and $v_{\beta j}$ in matrix form u and v respectively, we have

(1) $$t^- \,|P = t^+ \,|P \cdot u, \qquad t^+ \,|P = t^- \,|P \cdot v.$$

For $\rho > 0$ we set

$$E := \{z \in \mathbb{C} \,|\, a - \rho < x < b + \rho,\ c - \rho < y < d + \rho\}.$$

$$E' := \{z \in E \,|\, x < e + \rho\} \quad \text{and} \quad E'' := \{z \in E \,|\, x > e - \rho\}.$$

Consequently if U is a bounded open neighborhood of K' in $\mathbb{C}^{m-1}$, then

$$D := (E' \cap E'') \times U$$

is a bounded open neighborhood of P. We choose U and ρ so small that all of the functions $u_{\alpha i}$, $v_{\beta j}$ and a fortiori the matrices u and v are homomorphic on some open neighborhood in $\mathbb{C}^m$ of the compact set $\bar{D} = \overline{(E' \cap E'')} \times \bar{U}$.

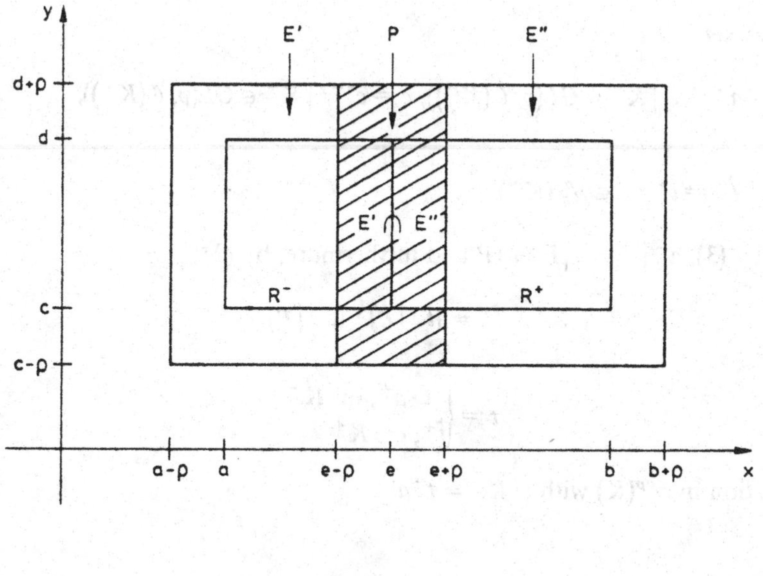

Since $\overline{E' \cap E''}$ is a compact rectangle, the approximation theorem implies that, given $\delta > 0$, there exist holomorphic functions $\hat{u}_{\alpha i} \in \mathcal{O}(\mathbb{C} \times \bar{U})$ which are polynomials in z so that

$$|u_{\alpha i} - \hat{u}_{\alpha i}|_D \leq \delta, \qquad \alpha = 1, \ldots, q, \qquad i = 1, \ldots, p.$$

Using these functions we form the (q, p)-matrix function $\hat{u} := (\hat{u}_{\alpha i})$ which is holomorphic on $\overline{E'' \times U}$ and, since $K^+ \subset E'' \times U$, define the section $\hat{t}^+ \in \mathcal{S}^p(K^+)$ by $\hat{t}^+ := t^+ \cdot \hat{u} | K^+$.

By equation (1) above,

$$\hat{t}^+ | P - t^- | P = t^+ | P \cdot (\hat{u} - u) | P = t^- | P \cdot v | P \cdot (\hat{u} - u) | P.$$

If one introduces the (p, p)-matrix

$$a := e + v(\hat{u} - u)$$

which is holomorphic on $\bar{D}$, then

$$\hat{t}^+ | P = t^- | P \cdot a | P. \tag{2}$$

Since $\hat{u}$ is near u, a is near the identity e. More precisely (see Paragraph 1.2),

$$|a - e|_D \leq q |v|_D |\hat{u} - u|_D \leq q |v|_D \cdot \delta.$$

If δ is taken small enough, then $|a - e|_D$ will be smaller than the ε of the Cartan Attaching Lemma. Thus, for $\delta > 0$ small enough, there exist two holomorphic invertible matrices $c' \in GL(p, \mathcal{O}(E' \times U))$ and $c'' \in GL(p, \mathcal{O}(E'' \times U))$ such that

$$a | D = c' | D \cdot c'' | D. \tag{3}$$

We now set

$$a^- := c' | K^- \in GL(p, \mathcal{O}(K^-)), \; c := c''^{-1} | K^+ \in GL(p, \mathcal{O}(K^+)),$$

and

$$\tilde{t}^+ := t^+ \cdot c \in \mathcal{S}^p(K^+).$$

Then, by (3), $a^- | P = a | P \cdot c | P$ and furthermore, by (2),

$$\tilde{t}^+ = (t^- | P) \cdot (a^- | P)$$

Hence

$$t := \begin{cases} t^- a^-, & \text{on } K^- \\ \tilde{t}^+, & \text{on } K^+ \end{cases}$$

is a section in $\mathcal{S}^p(K)$ with $t | K^- = t^- a^-$. □

Remark: Since the assumptions of Theorem 2 are invariant under exchanging $+$ and $-$, there also exists a (q, q)-matrix $a^+ \in GL(q, \mathcal{O}(K^+))$ so that $(t_1^+, \ldots, t_q^+)a^+$ is continuable to an element of $\mathcal{S}^q(K)$.

The following theorem, which is essentially a straight forward application of Theorem 2, is the main tool for proving Theorems A and B on blocks.

Theorem 3 (Extension Lemma for Sheaf Epimorphisms). *Let $\mathcal{S}$ be an analytic sheaf on K and suppose that*

$$h^-: \mathcal{O}^p|K^- \to \mathcal{S}|K^- \quad \text{and} \quad h^+: \mathcal{O}^q|K^+ \to \mathcal{S}|K^+$$

are analytic epimorphisms having the property that, on $P = K^- \cap K^+$

$$\operatorname{Im}\{(h^-|P)_P: \mathcal{O}^p(P) \to \mathcal{S}(P)\} = \operatorname{Im}\{(h^+|P)_P: \mathcal{O}^q(P) \to \mathcal{S}(P)\}.$$

Then there exist matrices $a^- \in GL(p, \mathcal{O}(K^-))$ and $a^+ \in GL(q, \mathcal{O}(K^+))$ such that the maps $h^- \cdot a^-$ and $h^+ \cdot a^+$ are extendible to sheaf homomorphisms $\varphi: \mathcal{O}^p|K \to \mathcal{S}$ and $\psi: \mathcal{O}^p|K \to \mathcal{S}$. In particular, the sum $\varphi + \psi: \mathcal{O}^{p+q}|K \to S$ is surjective.

Proof: It is enough to prove the existence of a^-. For this, let $t_1^-, \ldots, t_p^- \in \mathcal{S}(K^-)$ (resp. $t_1^+, \ldots, t_q^+ \in \mathcal{S}(K^+)$) be the images under h_p^- (resp. h_p^+) of the canonical generators of $\mathcal{O}^p(K^-)$ (resp. $\mathcal{O}^q(K^+)$). By assumption the restriction of these sections generate the same $\mathcal{O}(P)$-submodule of $\mathcal{S}(P)$. Hence, by Theorem 2, there exists $a^- \in GL(p, \mathcal{O}(K^-))$ and there are sections $t_1, \ldots, t_p \in \mathcal{O}(K)$ so that

$$(t_1|K^-, \ldots, t_p|K^-) = (t_1^-, \ldots, t_p^-)a^-.$$

Thus the map $\varphi: \mathcal{O}^p|K \to \mathcal{S}$, defined by $(f_{1x}, \ldots, f_{px}) \mapsto \sum_{i=1}^{p} f_{ix} t_{ix}$, is an extension to K of $h^- \cdot a^-$. Since a^- induces an invertible germ $a_x^- \in GL(p, \mathcal{O}_x)$ and since h^- is a sheaf epimorphism, $\varphi|K^-$ is likewise a sheaf epimorphism. Thus the sum $\varphi + \psi: \mathcal{O}^{p+q}|K \to S$ is a sheaf epimorphism. □

§ 3. Theorems A and B

As in the last section (see 2.1), Q denotes a non-empty compact block in $\mathbb{C}^m$. Thus

$$Q = R \times Q',$$

where $R = \{z \in \mathbb{C} \mid a \leq x \leq b,\ c \leq y \leq d\}$ is a compact rectangle in the $z' = z_1$-plane and Q' is a compact block in the $z' = (z_2, \ldots, z_m)$-space $\mathbb{C}^{m-1}$. The *dimension* $d(Q)$ is defined inductively by $d(Q) := d(R) + \dim Q'$, where, in the case of $m = 1$,

$$d(R) := \begin{cases} 0, & \text{if} \quad a = b \quad \text{and} \quad c = d \\ 2, & \text{if} \quad a < b \quad \text{and} \quad c < d \\ 1, & \text{otherwise} \end{cases}$$

It follows that

$$0 \le d(Q) \le 2m,$$

and in fact $d(Q)$ is just the topological dimension of Q.

Whenever $a < b$ we define, for every $e \in [a, b]$, the set

$$Q(e) := R(e) \times Q', \quad \text{where} \quad R(e) := \{z \in R \mid \operatorname{Re}\{z\} = e\}.$$

Then $Q(e)$ is a compact block in $\mathbb{C}^m$ having dimension exactly 1 less than $d(Q)$. Sets of this type are used in the following to make induction proofs on $d(Q)$ possible.

1. Coherent Analytic Sheaves on Compact Blocks. An analytic sheaf $\mathcal{S}$ on Q is called *coherent on* Q if there exists an open neighborhood U of Q in $\mathbb{C}^m$ and a coherent analytic sheaf $\hat{\mathcal{S}}$ on U such that $\hat{\mathcal{S}} \mid Q = \mathcal{S}$.

If $\mathcal{S}$ *and* $\mathcal{T}$ *are coherent* $\mathcal{O}$*-sheaves on* Q *and* $h: \mathcal{S} \to \mathcal{T}$ *is an* $\mathcal{O}$*-homomorphism, then the sheaves* $\mathcal{K}er\, h$, $\mathcal{I}m\, h$, *and* $\mathcal{C}oker\, h$ *are also analytic and coherent on* Q.

Proof: There exists an open neighborhood U of Q and coherent analytic sheaves $\hat{\mathcal{S}}, \hat{\mathcal{T}}$ on U such that $\hat{\mathcal{S}} \mid Q = \mathcal{S}$ and $\hat{\mathcal{T}} \mid Q = \mathcal{T}$. Further $h: \mathcal{S} \to \mathcal{T}$ is a section $h \in \Gamma(Q, \mathcal{H}om(S, T))$. By Theorem A.4.1 there exists an open neighborhood $\hat{U} \subset U$ of B and a section $\hat{h} \in \Gamma(\hat{U}, \mathcal{H}om(\hat{\mathcal{S}}, \hat{\mathcal{T}}))$ with $\hat{h} \mid Q = h$. Thus $h: \mathcal{S} \to \mathcal{T}$ has been continued to an $\mathcal{O}$-homomorphism $\hat{h}: \hat{\mathcal{S}} \mid \hat{U} \to \hat{\mathcal{T}} \mid \hat{U}$. It was shown in Chapter A.2.3 that $\mathcal{K}er\, \hat{h}$, $\mathcal{I}m\, \hat{h}$, and $\mathcal{C}oker\, \hat{h}$ are all analytic and coherent on U. Therefore the desired result follows, since $\mathcal{K}er\, h = \mathcal{K}er\, \hat{h} \mid Q$, etc. □

Remark: If M is a closed subset of a complex space X, then one has the more general notion of a coherent analytic sheaf on M. The above statements also hold in this more general situation, the proofs carrying over word for word.

2. The Formulations of Theorems A and B and the Reduction of Theorem B to Theorem A. The main results for coherent analytic sheaves on compact blocks were summarized by Cartan and Serre in the form of two theorems:

Theorem A. *For every coherent* $\mathcal{O}$*-sheaf* $\mathcal{S}$ *on a compact block* $Q \subset \mathbb{C}^m$, *there exists a natural number* p *and an exact* $\mathcal{O}$*-sequence*

$$\mathcal{O}^p \mid Q \to \mathcal{S} \to 0.$$

Theorem B. *For every coherent $\mathcal{O}$-sheaf $\mathcal{S}$ on a compact block $Q \subset \mathbb{C}^m$,*

$$H^q(Q, \mathcal{S}) = 0 \quad \text{for all} \quad q \geq 1.$$

One can also formulate Theorem A as follows:

There exist p sections in $\mathcal{S}(Q)$ whose germs at each point $z \in Q$ generate the stalk $\mathcal{S}_z$ as an $\mathcal{O}_z$-module.

Before proceeding, we note an important immediate consequence of Theorem B:

Every exact $\mathcal{O}$-sequence on Q,

$$\mathcal{S} \xrightarrow{h} \mathcal{T} \longrightarrow 0,$$

between coherent analytic sheaves induces an exact $\mathcal{O}(Q)$-sequence

$$\mathcal{S}(Q) \xrightarrow{h_Q} \mathcal{T}(Q) \longrightarrow 0.$$

Proof: Associated to the exact sequence $0 \longrightarrow \mathcal{K}er\, h \longrightarrow \mathcal{S} \longrightarrow \mathcal{T} \longrightarrow 0$, we have the cohomology sequence

$$\cdots \longrightarrow \mathcal{S}(Q) \xrightarrow{h_Q} \mathcal{T}(Q) \longrightarrow H^1(Q, \mathcal{K}er\, h) \qquad \cdots$$

We proved in Paragraph 1 that $\mathcal{K}er\, h$ was a coherent analytic sheaf on Q. Thus by Theorem B $H^1(Q, \mathcal{K}er\, h) = 0$. □

Since blocks are in particular product sets in $\mathbb{C}^m$, we already have Theorem B at our disposal in the case of compact blocks and the structure sheaf $\mathcal{O}$ (Theorem 2.4.4). Since in general $H^q(X, \mathcal{S}^p)$ is isomorphic to the p-fold direct sum of $H^q(X, \mathcal{S})$ with itself, we thus have

$$H^q(Q, \mathcal{O}^p) = 0, \qquad p, q \geq 1.$$

This simple case of Theorem B makes it possible to reduce Theorem B to Theorem A in the case of compact blocks and arbitrary coherent sheaves.

Theorem 1. *In the case of compact blocks, Theorem A implies Theorem B.*

Proof: Let $\mathcal{S}$ be a coherent $\mathcal{O}$-sheaf on Q. We will show that, given a natural number $k \geq 1$, there exists a coherent analytic sheaf $\mathcal{S}_k$ on Q and isomorphisms

$$H^q(Q, \mathcal{S}) \simeq H^{q+k}(Q, \mathcal{S}_k) \quad \text{for all} \quad q \geq 1.$$

The claim follows immediately from this, since the vanishing theorem for compact blocks implies that all of the groups $H^{q+k}(Q, \mathcal{S}_k)$ vanish when $k \geq 3^m$.

It is enough to consider the case of $k = 1$, as the general case follows from this one by repetition. Assuming Theorem A, there exists $p \geq 1$ and an epimorphism $h\colon \mathcal{O}^p | Q \to \mathcal{S}$. Thus $\mathcal{S}_1 := \mathcal{K}er\, h$ is a coherent analytic sheaf on Q and the short exact sequence

$$0 \to \mathcal{S}_1 \to \mathcal{O}^p | Q \to \mathcal{S} \to 0$$

gives rise to the exact cohomology sequence

$$\cdots \longrightarrow H^q(Q, \mathcal{O}^p) \longrightarrow H^q(Q, \mathcal{S}) \xrightarrow{\delta_q} H^{q+1}(Q, \mathcal{S}_1) \longrightarrow H^{q+1}(Q, \mathcal{O}^p) \longrightarrow \cdots$$

Since for $q \geq 1$ all of the groups $H^q(Q, \mathcal{O}^p)$ vanish, the maps δ_q are bijective. □

3. The Proof of Theorem A for Compact Blocks. We give the proof of Theorem A for compact blocks by induction on $d := d(Q)$. In the case of $d = 0$ the claim is trivial, because Q is a point $z \in \mathbb{C}^m$ and $\mathcal{S} = \mathcal{S}_z$ is a finite $\mathcal{O}_z$-module. Thus we consider the case of $d \geq 1$.

We let A_d and B_d denote the corresponding statements of Theorems A and B for all compact blocks of dimension $d(Q) \leq d$. Then A_d implies B_d (Theorem 1). Furthermore, as a consequence of B_d, it follows that every exact sequence $\mathcal{S} \to \mathcal{T} \to 0$ of coherent $\mathcal{O}$-sheaves on Q with $d(Q) \leq d$ induces an exact sequence of sections $\mathcal{S}(Q) \to \mathcal{T}(Q) \to 0$. If we denote this consequence by F_d, then it is enough to show that

$$A_{d-1} \quad \textit{and} \quad F_{d-1} \quad \textit{imply} \quad A_d.$$

Without loss of generality we may assume that $a < b$. Then $Q(e)$ is a $(d-1)$-dimensional compact block and, using the induction assumption A_{d-1}, there exists a natural number $p(e)$ and an exact $\mathcal{O}$-sequence

$$\mathcal{O}^{p(e)} | Q(e) \xrightarrow{\varphi_e} \mathcal{S} | Q(e) \to 0.$$

There exists an open neighborhood $U_e \subset \mathbb{C}^m$ of $Q(e)$, a coherent analytic sheaf $\hat{\mathcal{S}}$ on U_e with $\hat{\mathcal{S}} | U_e \cap Q = \mathcal{S} | U_e \cap Q$ and a section $\hat{h}_e \in \Gamma(U_e, \mathcal{H}om(\mathcal{O}^{p(e)}, \mathcal{S}))$ with $\hat{h}_e | Q(e) = \varphi_e$. Since the sheaf $\mathcal{C}oker\, \hat{h}_e$ is coherent on U_e and its support does not intersect $Q(e)$, one can find $\varepsilon > 0$, so that the d-dimensional block

$$Q(e)_\varepsilon := R(e)_\varepsilon \times Q', \quad \text{where} \quad R(e)_\varepsilon := \{z \in R \,|\, e - \varepsilon \leq x \leq e + \varepsilon\}$$

is contained in U_e and the homomorphism induced by $\hat{h}_e$,

$$h(e)\colon \mathcal{O}^{p(e)} | Q(e)_\varepsilon \to \mathcal{S} | Q(e)_\varepsilon,$$

is surjective. Since the interval $[a, b]$ is compact, the block Q can be covered by finitely many of these $Q(e)_\varepsilon$'s, say $Q_1, \ldots, Q_l$. We may assume that these blocks come from a decomposition $a = e_0 < e_1 \cdots < e_l = b$ of the interval $[a, b]$ with $Q_j = \{(z, z') \in Q \,|\, e_{j-1} \leq \operatorname{Re} z \leq e_j\}$. For each Q_j we have an integer $p_j > 0$ and an $\mathcal{O}$-epimorphism $h_j\colon \mathcal{O}^j | Q_j \to \mathcal{S} | Q_j$.

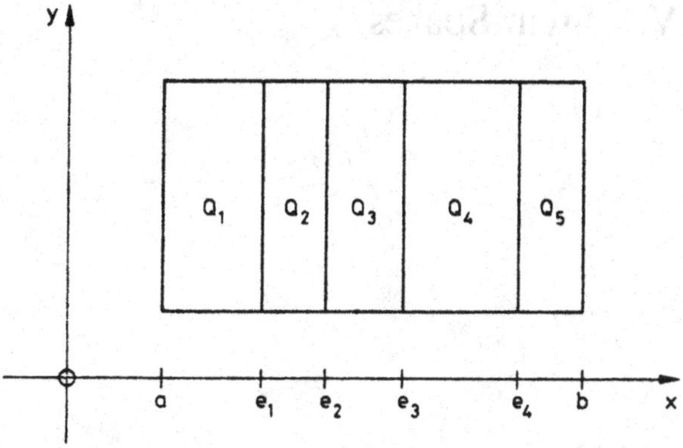

We first consider h_1 and h_2. Since $Q_1 \cap Q_2 = Q(e_1)$ is a $(d-1)$-dimensional block, F_{d-1} implies that the induced homomorphisms of sections,

$$\mathcal{O}^{p_1}(Q(e_1)) \to \mathcal{S}(Q(e_1)) \quad \text{and} \quad \mathcal{O}^{p_2}(Q(e_1)) \to \mathcal{S}(Q(e_1)),$$

are surjective. Thus, by Theorem 2.3, there exists an analytic epimorphism

$$h_{1,2}: \mathcal{O}^{p_1+p_2}|Q_1 \cup Q_2 \to \mathcal{S}|Q_1 \cup Q_2$$

Repeating this procedure for $h_{1,2}$ and h_3, we obtain in the same way an epimorphism

$$h_{1,2,3}: \mathcal{O}^{p_1+p_2+p_3}|Q_1 \cup Q_2 \cup Q_3 \to \mathcal{S}|Q_1 \cup Q_2 \cup Q_3.$$

Continuing on in the obvious way, after $(l-1)$ steps we have the desired analytic epimorphism

$$h_{1,2,\dots,l}: \mathcal{O}^{p_1+\cdots p_l}|Q \to \mathcal{S}|Q.$$

Chapter IV. Stein Spaces

Stein spaces are complex spaces for which Theorem B is valid. Theorem A is a consequence of Theorem B and thus is automatically true for such spaces. A complex space is Stein if it possesses a Stein exhaustion. Particular Stein exhaustions are the exhaustions by blocks. Every weakly holomorphically convex space in which every compact analytic subset is finite can be exhausted by blocks and consequently is a Stein space.

§ 1. The Vanishing Theorem $H^q(X, \mathscr{S}) = 0$

In this section the central notion *of a Stein* set is introduced. Compact Stein sets are constructed from compact blocks in $\mathbb{C}^n$ by means of a lifting process. The main tools for this are the coherence theorem for finite maps and Theorem B for blocks.

It is shown that complex spaces X which are exhausted by compact Stein sets have the property that $H^q(X, \mathscr{S}) = 0$ for all $q \geq 2$, where $\mathscr{S}$ is an arbitrary coherent analytic sheaf. Moreover, whenever such an X possesses a so-called Stein exhaustion, the group $H^1(X, \mathscr{S})$ vanishes.

1. Stein Sets and Consequences of Theorem B. The following language is convenient:

Definition 1 (Stein Sets). *A closed subset P of a complex space X is called a Stein set (in X) if Theorem B is valid on P (i.e. for every coherent analytic sheaf $\mathscr{S}$, $H^q(P, \mathscr{S}) = 0$ for all $q \geq 1$). A complex space which is itself a Stein set is called a Stein space.*

It follows that compact blocks in $\mathbb{C}^m$ are Stein sets. Applying the vanishing of the *first* cohomology groups, one obtains the following theorem in exactly the same way it was proved for blocks (p. 97):

Theorem 1. *Let P be a Stein set in X and suppose that h: $\mathscr{S} \to \mathscr{T}$ is an analytic epimorphism between coherent analytic sheaves over P. Then the induced homomorphism of sections, h_P: $\mathscr{S}(P) \to \mathscr{T}(P)$, is surjective.*

We say that the module of sections $\mathscr{S}(P)$ generates the stalk $\mathscr{S}_x$, $x \in P$, if the image of $\mathscr{S}(P)$ in $\mathscr{S}_x$ via restriction $\mathscr{S}(P) \to \mathscr{S}_x$, $s \to s(x)$, generates $\mathscr{S}_x$ as an $\mathcal{O}_x$-module.

Theorem 2 (Theorem A for Stein Sets). *Let P be a Stein set in X and $\mathscr{S}$ a coherent analytic sheaf on P. Then $\mathscr{S}(P)$ generates every stalk $\mathscr{S}_x$, $x \in P$.*

Proof: Let $x \in P$ be fixed. We denote by $\mathscr{M}$ the coherent sheaf of ideals of all germs of holomorphic functions which vanish at x. In other words $\mathscr{M}_p = \mathcal{O}_p$ for $p \neq x$ and $\mathscr{M}_x = \mathfrak{m}(\mathcal{O}_x) =$ the maximal ideal of $\mathcal{O}_x$. Defining $\mathscr{N} := \mathscr{M} | P$, it also follows that $\mathscr{N}\mathscr{S}$ is coherent over P. By Theorem 1 the sheaf epimorphism, $\mathscr{S} \to \mathscr{S}/\mathscr{N}\mathscr{S}$, induces an epimorphism of sections, $\mathscr{S}(P) \xrightarrow{\varepsilon} \mathscr{S}/\mathscr{N}\mathscr{S}(P)$. Now

$$(\mathscr{S}/\mathscr{N}\mathscr{S})_p = 0 \quad \text{for} \quad p \neq x \quad \text{and} \quad (\mathscr{S}/\mathscr{N}\mathscr{S})_x \xrightarrow{\sim} \mathscr{S}_x/\mathfrak{m}(\mathcal{O}_x)\mathscr{S}_x.$$

Thus ε is just the restriction map $\mathscr{S}(P) \to \mathscr{S}_x$ followed by the quotient epimorphism $\mathscr{S}_x \to \mathscr{S}_x/\mathfrak{m}(\mathcal{O}_x)\mathscr{S}_x$.

Now let $e_1, \ldots, e_m$ be a generating system of the finite dimensional $\mathbb{C}$-vector space $\mathscr{S}_x | \mathfrak{m}(\mathcal{O}_x)\mathscr{S}_x$ and let $s_1, \ldots, s_m \in \mathscr{S}(P)$ be sections with $\varepsilon(s_\mu) = e_\mu$, $1 \leq \mu \leq m$. By a well-known theorem in the theory of local rings,[1] the germs $s_{1x}, \ldots, s_{mx} \in \mathscr{S}_x$ generate the $\mathcal{O}_x$-module $\mathscr{S}_x$. □

Corollary. *Let P be a compact Stein set in X and $\mathscr{S}$ a coherent analytic sheaf on X. Then there exists an integer $p \geq 1$ and an $\mathcal{O} | P$-sheaf epimorphism,*

$$\mathcal{O}^p | P \xrightarrow{h} \mathscr{S}$$

such that the associated homomorphism of sections, $\mathcal{O}^p(P) \xrightarrow{h_P} \mathscr{S}(P)$, is likewise surjective.

Proof: Theorem 2 implies that $\mathscr{S}(P)$ generates $\mathscr{S}_x$ for every $x \in P$. Thus the compactness of P allows us to choose a *finite* open cover $\{U_\mu\}_{1 \leq \mu \leq m}$ of P and sheaf homomorphisms h_μ: $\mathcal{O}^{p_\mu} | P \to \mathscr{S}$ which are surjective on U_μ. Setting $p := \sum_{\mu=1}^{m} p_\mu$ and $h := \sum_{\mu=1}^{m} h_\mu$, it follows that h: $\mathcal{O}^p | P \to \mathscr{S}$ is surjective on P.

2. Construction of Compact Stein Sets Using the Coherence Theorem for Finite Maps. From compact blocks in $\mathbb{C}^m$ we obtain other compact Stein sets by a lifting process. These are very important for the further development of the theory.

---

[1] What we need for the above proof is a simple consequence of Nakayama's Lemma (see [AS], p. 213):

Let R be a noetherian ring with maximal ideal $\mathfrak{m}$ and suppose that M is a finitely generated R-module. Then the elements $x_1, \ldots, x_p \in M$ generate the R-module M if and only if their equivalence classes $\bar{x}_1, \ldots, \bar{x}_p \in M/\mathfrak{m}M$ generate the $R/\mathfrak{m}$-vector space $M/\mathfrak{m}M$.

Theorem 3. *Let X be a complex space and $P \subset X$ a set with the following properties:*

1) *There is an open neighborhood U of P in X, a domain V in $\mathbb{C}^m$ and a finite holomorphic map $\tau: U \to V$.*
2) *There exists a compact block Q in $\mathbb{C}^m$ with*

$$Q \subset V \quad \textit{and} \quad \tau^{-1}(Q) = P.$$

Then P is a compact Stein set in X.

Proof: We note first that P is likewise compact. Let $\mathscr{S}$ be a coherent analytic sheaf on P. There exists an open neighborhood U' of P which is contained in U such that $\mathscr{S}$ is analytic and coherent on U'. Since $\tau: U \to V$ is finite, there exists a domain $V' \subset V$ with $V' \supset Q$ and $\tau^{-1}(V') \subset U'$. The restriction of τ to the corresponding map $\tau^{-1}(V') \to V'$ is again finite. Thus $\mathscr{T} := \tau_*(\mathscr{S} | \tau^{-1}(V'))$ is a coherent sheaf over V'. Since $P = \tau^{-1}(Q)$, the map $\tau | P: P \to Q$ is also finite. Hence by Theorem I.1.5 there exist isomorphisms $H^q(P, \mathscr{S}) \cong H^q(B, (\tau | P)_*(\mathscr{S}))$, $q \geq 0$. But $(\tau | P)_*(\mathscr{S}) = \tau_*(\mathscr{S} | P) = \mathscr{T} | Q$ is coherent. Consequently, applying Theorem B for compact blocks in $\mathbb{C}^m$, $H^q(Q, \mathscr{T}) = 0$ for all $q \geq 1$. Thus the groups $H^q(P, \mathscr{S})$ vanish for $q \geq 1$. □

3. Exhaustions of Complex Spaces by Compact Stein Sets. If X is a topological space, then a sequence $\{K_\nu\}_{\nu \geq 1}$ of compact subsets of X is called *an exhaustion of* X whenever the following hold:

1) Every K_ν is contained in the interior of $K_{\nu+1}$: $K_\nu \subset \mathring{K}_{\nu+1}$.
2) The space X is the union of all the K_ν's: $X = \bigcup_{\nu=1}^{\infty} K_\nu$.

If X has such an exhaustion then *every* compact set $K \subset X$ is contained in some K_ν and X itself is locally compact.

Using an exhaustion, sections in sheaves are frequently constructed by the following simple principle:

Let $\{K_\nu\}_{\nu \geq 1}$ be an exhaustion of X by compact sets. Let $\mathscr{S}$ be a sheaf on X and $s_\nu \in \mathscr{S}(K_\nu)$ a sequence of sections having the property that $s_{\nu+1} | K_\nu = s_\nu$, $\nu \geq 1$. Then there exists a unique section $s \in \mathscr{S}(X)$ such that

$$s | K_\nu = s_\nu, \qquad \nu \geq 1.$$

The proof is trivial. □

A simple result in general topology states that *every locally-compact, second countable topological space possesses an exhaustion.* For complex spaces we have the following obvious remark:

A complex space X (with countable topology) is exhaustable by a sequence of compact Stein sets if and only if every compact set $K \subset X$ is contained in a compact Stein set $P \subset X$.

Example: Every *open* block in $\mathbb{C}^m$ as well as $\mathbb{C}^m$ itself is exhaustable by compact Stein sets (namely compact blocks).

4. The Equations $H^q(X, \mathcal{S}) = 0$ for $q \geq 2$. If X is a paracompact space and $\mathcal{S}$ is a sheaf of abelian groups on X, then the cohomology $H^*(X, \mathcal{S})$ can be computed via a flabby resolution

$$(*) \qquad 0 \to \mathcal{S} \to \mathcal{S}^0 \to \cdots \mathcal{S}^{q-2} \to \mathcal{S}^{q-1} \to \mathcal{S}^q \to \cdots$$

as the cohomology of the associated complex of sections (see Chapter B.1).

If $K \subset X$ is compact, then the restriction of $(*)$ is a flabby resolution of $\mathcal{S} | K$. One has a commutative diagram,

$$\left(\substack{*\\ *}\right) \quad \begin{array}{ccccccccccccc}
0 \to \mathcal{S}(X) & \xrightarrow{i} & \mathcal{S}^0(X) & \xrightarrow{d_0} & \cdots & \longrightarrow & \mathcal{S}^{q-2}(X) & \xrightarrow{d_{q-2}} & \mathcal{S}^{q-1}(X) & \xrightarrow{d_{q-1}} & \mathcal{S}^q(X) & \xrightarrow{d_q} & \cdots \\
\downarrow & & \downarrow & & & & \downarrow & & \downarrow & & \downarrow & & \\
0 \to \mathcal{S}(K) & \longrightarrow & \mathcal{S}^0(K) & \longrightarrow & \cdots & \longrightarrow & \mathcal{S}^{q-2}(K) & \longrightarrow & \mathcal{S}^{q-1}(K) & \longrightarrow & \mathcal{S}^q(K) & \longrightarrow & \cdots,
\end{array}$$

where the vertical maps are the restrictions. It follows that

$$0 = H^q(K, \mathcal{S}) = \operatorname{Ker}(d_q | K)/\operatorname{Im}(d_{q-1} | K)$$

if and only if the bottom row is exact at the place $\mathcal{S}^q(K)$. One would expect that if this exactness were the case for all sets of some exhaustion $\{K_\nu\}_{\nu \geq 1}$ for X, then the top row would also be exact at $\mathcal{S}^q(X)$. In this direction we prove the following:

Theorem 4. *Let X be paracompact and $\mathcal{S}$ a sheaf of abelian groups on X. Let $q \geq 2$ and suppose that $\{K_\nu\}_{\nu \geq 1}$ is an exhaustion of X by compact sets K_ν having the property that*

$$H^{q-1}(K_\nu, \mathcal{S}) = H^q(K_\nu, \mathcal{S}) = 0, \qquad \text{all } \nu.$$

Then $H^q(X, \mathcal{S}) = 0$.

Proof: For $K := K_\nu$ we consider the diagram $\left(\substack{*\\ *}\right)$ for each ν. Since $H^q(X, \mathcal{S}) = \operatorname{Ker} d_q/\operatorname{Im} d_{q-1}$, in order to prove the vanishing, given a section $\alpha \in \operatorname{Ker} d_q$, we must find $\beta \in \mathcal{S}^{q-1}(X)$ with $d_{q-1}(\beta) = \alpha$. Thus it suffices to inductively construct a sequence $\beta_\nu \in \mathcal{S}^{q-1}(K_\nu)$ with the following properties:

1) $(d_{q-1} | K_\nu)\beta_\nu = \alpha | K_\nu$,
2) $\beta_{\nu+1} | K_\nu = \beta_\nu$.

Then by a remark in Paragraph 3 there exists a unique section $\beta \in \mathcal{S}^{q-1}(X)$ with $\beta | K_\nu = \beta_\nu$. Since $\left(\substack{*\\ *}\right)$ is commutative, it will then follow that

$$(d_{q-1}\beta) | K_\nu = (d_{q-1} | K_\nu)(\beta | K_\nu) = \alpha | K_\nu$$

for all ν (i.e. $d_{q-1}\beta = \alpha$).

We now construct the sequence β_ν. Since

$$\alpha|K_\nu \in \operatorname{Ker}(d_q|K_\nu) \quad \text{and} \quad H^q(K_\nu, \mathscr{S}) = \operatorname{Ker}(d_q|K_\nu)/\operatorname{Im}(d_{q-1}|K_\nu) = 0,$$

there exists a sequence $\beta'_\nu \in \mathscr{S}^{q-1}(K_\nu)$ which satisfies (1):

$$(d_{q-1}|K_\nu)\beta'_\nu = \alpha|K_\nu \quad \text{for all } \nu.$$

We define $\beta_1 := \beta'_1$. Let $\beta_1, \ldots, \beta_\nu$ be already constructed satisfying (1) and (2). Then $(d_{q-1}|K_\nu)\beta_\nu = \alpha|K_\nu$ and hence $(d_{q-1}|K_\nu)(\beta'_{\nu+1}|K_\nu - \beta_\nu) = 0$. In other words $\beta'_{\nu+1}|K_\nu - \beta_\nu \in \operatorname{Ker}(d_{q-1}|K_\nu)$. But

$$H^{q-1}(K_\nu, \mathscr{S}) = \operatorname{Ker}(d_{q-1}|K_\nu)/\operatorname{Im}(d_{q-2}|K_\nu) = 0.$$

Therefore there exists $\gamma'_\nu \in \mathscr{S}^{q-2}(K_\nu)$ with

$$d_{q-2}(\gamma'_\nu) = \beta'_{\nu+1}|K_\nu - \beta_\nu.$$

Now, due to the fact that $q \geq 2$, $\mathscr{S}^{q-2}$ is flabby on X. So there exists $\gamma_\nu \in \mathscr{S}^{q-2}(X)$ which is a continuation to X of γ'_ν. We correct $\beta'_{\nu+1}$ as follows:

$$\beta_{\nu+1} := \beta'_{\nu+1} - (d_{q-2}\gamma_\nu)|K_{\nu+1} \in \mathscr{S}^{q-1}(K_{\nu+1}).$$

Since $d_{q-1} \circ d_{q-2} = 0$,

$$(d_{q-1}|K_{\nu+1})\beta_{\nu+1} = (d_{q-1}|K_{\nu+1})\beta'_{\nu+1} = \alpha|K_{\nu+1}.$$

Furthermore $(d_{q-2}\gamma_\nu)|K_\nu = (d_{q-2}|K_\nu)\gamma'_\nu = \beta'_{\nu+1}|K_\nu - \beta_\nu$. Hence

$$\beta_{\nu+1}|K_\nu = \beta'_{\nu+1}|K_\nu - (d_{q-2}\gamma_\nu)|K_\nu = \beta_\nu. \qquad \square$$

The following is now immediate:

Theorem 5. *Let X be a complex space which is exhaustable by a sequence $\{P_\nu\}_{\nu \geq 1}$ of compact Stein sets. Let $\mathscr{S}$ be a coherent analytic sheaf on X. Then*

$$H^q(X, \mathscr{S}) = 0 \quad \text{for all} \quad q \geq 2.$$

5. Stein Exhaustions and the Equation $H^1(X, \mathscr{S}) = 0$. We want to analyze under which additional assumptions on the exhaustion used in the proof of Theorem 4 the first cohomology group can be shown to vanish. For this we begin with a complex space X and exhaustion $\{P_\nu\}_{\nu \geq 1}$ by compact Stein subsets of X. The commutative diagram

$$\begin{array}{ccccccccc}
0 & \longrightarrow & \mathscr{S}(X) & \xrightarrow{\;i\;} & \mathscr{S}^0(X) & \xrightarrow{\;d_0\;} & \mathscr{S}^1(X) & \xrightarrow{\;d_1\;} & \cdots \\
 & & \downarrow & & \downarrow & & \downarrow & & \\
0 & \longrightarrow & \mathscr{S}(P_\nu) & \xrightarrow{\;i\;} & \mathscr{S}^0(P_\nu) & \xrightarrow{\;d_0|P_\nu\;} & \mathscr{S}^1(P_\nu) & \xrightarrow{\;d_1|P_\nu\;} & \cdots
\end{array}$$

is exact at $\mathscr{S}^0(X)$ and $\mathscr{S}^0(P_\nu)$. Thus, using the injection i, we interpret $\mathscr{S}(X)$ and $\mathscr{S}(P_\nu)$ as subgroups of $\mathscr{S}^0(X)$ and $\mathscr{S}^0(P_\nu)$ respectively with

$$\mathscr{S}(X) = \operatorname{Ker} d_0 \quad \text{and} \quad \mathscr{S}(P_\nu) = \operatorname{Ker} d_0 | P_\nu.$$

In order to prove that $H^1(X, \mathscr{S}) = 0$ one must show that for every $\alpha \in \operatorname{Ker} d_1$ there exists $\beta \in \mathscr{S}^0(X)$ with $d_0 \beta = \alpha$. The choice of a sequence $\beta'_\nu \in \mathscr{S}^0(P_\nu)$ with $(d_0 | P_\nu)\beta'_\nu = \alpha | P_\nu$ can be carried out as before, because $H^1(P_\nu, \mathscr{S}) = \operatorname{Ker}(d_1 | P_\nu)/\operatorname{Im}(d_0 | P_\nu) = 0$. However the construction of a sequence $\beta_\nu \in \mathscr{S}^0(P_\nu)$ which, along with the property $(d_0 | P_\nu)\beta_\nu = \alpha | P_\nu$, additionally satisfies $\beta_{\nu+1} | P_\nu = \beta_\nu$ is no longer possible. This is due to the fact that $\beta'_{\nu+1} | P_\nu - \beta_\nu$ lies in $\operatorname{Ker} d_0 | P_\nu = \mathscr{S}(P_\nu)$ and, since $\mathscr{S}$ is *not a flabby sheaf*, it is not possible to continue it to a section in $\mathscr{S}$ over all of X.

In the previous case the equations $\beta_{\nu+1} | P_\nu = \beta_\nu$ were used in order to obtain by successive continuation a section β with $d_{q-1}\beta = \alpha$. This can be done, however, in another way: Given a sequence $\beta_\nu \in \mathscr{S}^0(P_\nu)$, one can determine a "correction sequence" $\delta_\nu \in \mathscr{S}(P_{\nu-1})$ which, instead of (2), satisfies the following:

$$(\beta_{\nu+1} + \delta_{\nu+1}) | P_{\nu-1} = (\beta_\nu + \delta_\nu) | P_{\nu-1}, \qquad \nu \geq 2.$$

Then, by the remark in Paragraph 3, the sequence of pairs β_ν, δ_ν determine a section $\beta \in \mathscr{S}^0(X)$ with $\beta | P_\nu = (\beta_{\nu+1} + \delta_{\nu+1}) | P_\nu$. If in addition $(d_0 | P_\nu)\beta_\nu = \alpha | P_\nu$ and $\delta_{\nu+1} \in \mathscr{S}(P_\nu) = \operatorname{Ker}(d_0 | P_\nu)$, then

$$d_0 \beta | P_\nu = (d_0 | P_\nu)(\beta_{\nu+1} | P_\nu) + (d_0 | P_\nu)(\delta_{\nu+1}) = \alpha | P_\nu,$$

for all ν. In other words $d_0 \beta = \alpha$.

In the following, instead of *continuing sections*, an *approximation* of $\beta'_{\nu+1} | P_\nu - \beta_\nu \in \mathscr{S}(P_\nu)$ by sections in $\mathscr{S}(X)$ enters. However for this one needs a good topology on the $\mathbb{C}$-vector space $\mathscr{S}(P_\nu)$. This is one of the main reasons that the case $q = 1$ is significantly more difficult than that of $q \geq 2$. In the following definition we list the key properties that are needed in order to prove that $H^1(X, \mathscr{S}) = 0$.

Definition 6 (Stein Exhaustion). *Let X be a complex space and $\mathscr{S}$ a coherent sheaf on X. An exhaustion $\{P_\nu\}_{\nu \geq 1}$ of X by compact Stein sets is called a Stein exhaustion of X relative to $\mathscr{S}$ if the following are satisfied:*

a) *Every $\mathbb{C}$-vector space $\mathscr{S}(P_\nu)$ possesses a semi-norm $| \ |_\nu$ such that the subspace $\mathscr{S}(X) | P_\nu \subset \mathscr{S}(P_\nu)$ is dense in $\mathscr{S}(P_\nu)$.*[2]
b) *Every restriction map $\mathscr{S}(P_{\nu+1}) \subset \mathscr{S}(P_\nu)$ is bounded. In other words, there exists a positive real number M_ν so that $|s | P_\nu|_\nu \leq M_\nu |s|_{\nu+1}$ for all $s \in \mathscr{S}(P_{\nu+1})$, $\nu \geq 1$.*

---

[2] A semi-norm has all of the properties of a norm with the one exception that $|x| = 0$ no longer implies $x = 0$. Semi-normed vector spaces are topological spaces which are not necessarily Hausdorff. Thus sequences can have more than one limit.

c) *If* $(s_j)_{j\in\mathbb{N}}$ *is a Cauchy sequence in* $\mathcal{S}(P_\nu)$, *then the restricted sequence* $(s_j|P_{\nu-1})_{j\in\mathbb{N}}$ *has a limit in* $\mathcal{S}(P_{\nu-1})$, $\nu \geq 2$.

d) *If* $s \in \mathcal{S}(P_\nu)$ *and* $|s|_\nu = 0$, *then* $s|P_{\nu-1} = 0$, $\nu \geq 2$.

Maintaining our earlier notation, we now show the following:

Theorem 7. *Let* X *be a complex space and* $\mathcal{S}$ *a coherent analytic sheaf on* X. *Further suppose that there exists a Stein exhaustion* $\{P_\nu\}$ *for* X *relative to* $\mathcal{S}$. *Then given a section* $\alpha \in \operatorname{Ker} d_1$ *there exist two sequences* $\beta_\nu \in \mathcal{S}^0(P_\nu)$ *and* $\delta_\nu \in \mathcal{S}(P_{\nu-1})$, $\nu \geq 3$, *with the following properties:*

1) $$(d_0|P_\nu)\beta_\nu = \alpha|P_\nu$$

2) $$(\beta_{\nu+1} + \delta_{\nu+1})|P_{\nu-1} = (\beta_\nu + \delta_\nu)|P_{\nu-1}.$$

The sequences β_ν, δ_ν *define a section* $\beta \in \mathcal{S}^0(X)$ *with* $\beta|P_{\nu-1} = (\beta_\nu + \delta_\nu)|P_{\nu-1}$ *such that* $d_0\beta = \alpha$. *In particular* $H^1(X, \mathcal{S}) = 0$.

Proof: We have just observed that the existence of two sequences β_ν, δ_ν having properties (1) and (2) results in the existence of a section β with $d_0\beta = \alpha$. This obviously implies that $H^1(X, \mathcal{S}) = 0$.

The construction of sequences β_ν, δ_ν is carried out in three steps. We may assume that the restriction $\mathcal{S}(P_{\nu+1}) \to \mathcal{S}(P_\nu)$ are *contractions* (i.e. $M_\nu \leq 1$ for all ν).

1) We first construct the sequence β_ν. As in the proof of Theorem 4 one chooses a sequence $\beta'_\nu \in \mathcal{S}^0(P_\nu)$ with $(d_0|P_\nu)\beta'_\nu = \alpha|P_\nu$. For this one uses the vanishing of $H^1(P_\nu, S) = \operatorname{Ker}(d_1|P_\nu)/\operatorname{Im}(d_0|P_\nu)$. We obtain the β_ν's inductively from the sequence β'_ν. One begins with $\beta_1 := \beta'_1$. Let $\beta_1, \ldots, \beta_\nu$ be already constructed satisfying (1). Define

$$\gamma'_\nu := \beta'_{\nu+1}|P_\nu - \beta_\nu.$$

Then

$$(d_0|P_\nu)\gamma'_\nu = \alpha|P_\nu - \alpha|P_\nu = 0 \qquad (\text{i.e. } \gamma'_\nu \in \mathcal{S}(P_\nu)).$$

By a) of Definition 6, γ'_ν is approximable by sections in $\mathcal{S}(X)$. We thus may choose $\gamma_\nu \in \mathcal{S}(X)$ such that

$$|\gamma'_\nu - \gamma_\nu|P_\nu|_\nu \leq q^\nu, \text{ where } 0 < q < 1 \left(\text{say } q := \frac{1}{2}\right).$$

We now define

$$\beta_{\nu+1} := \beta'_{\nu+1} - \gamma_\nu|P_{\nu+1} \in \mathcal{S}^0(P_{\nu+1})$$

Then, as it should be,

$$(d_0|P_{\nu+1})\beta_{\nu+1} = (d_0|P_{\nu+1})\beta'_{\nu+1} - (d_0\gamma_\nu)|P_{\nu+1} = \alpha|P_{\nu+1} - 0.$$

2) We now construct the sequence δ_ν. The differences

$$\beta_{\nu+1}|P_\nu - \beta_\nu = \gamma'_\nu - \gamma_\nu|P_\nu \in \mathscr{S}(P_\nu)$$

no longer vanish (as in Theorem 4). Nevertheless they are "small":

$$(*) \qquad |\beta_{\nu+1}|P_\nu - \beta_\nu|_\nu \le q^\nu.$$

For every $\nu \ge 1$ we consider the sequence

$$s_j^{(\nu)} := \beta_{\nu+j}|P_\nu - \beta_\nu \in S(P_\nu), \qquad j = 0, 1, \dots.$$

As one easily verifies by direct substitution,

$$(0) \qquad s_j^{(\nu)} - s_{j-1}^{(\nu+1)}|P_\nu = \beta_{\nu+1}|P_\nu - \beta_\nu \quad \text{for all} \quad \nu \text{ and } j.$$

In the third step below we will show that, for all $\nu \ge 1$, the sequence $s_j^{(\nu)}|P_{\nu-1} \in \mathscr{S}(P_{\nu-1})$ has a limit in $\mathscr{S}(P_{\nu-1})$. Let δ_ν be such a limit, $\nu \ge 1$. Now, all of the restrictions are continuous (by b) of Definition 6).

Thus

$$\lim_{j\to\infty} s_j^{(\nu+1)}|P_{\nu-1} = \delta_{\nu+1}|P_{\nu-1},$$

and consequently

$$\lim_{j\to\infty} (s_j^{(\nu)} - s_{j-1}^{(\nu+1)})|P_{\nu-1} = \delta_\nu - \delta_{\nu+1}|P_{\nu-1}.$$

Together with (0), this implies that

$$|((\delta_\nu - \delta_{\nu+1})|P_{\nu-1}) - ((\beta_{\nu+1} - \beta_\nu)|P_{\nu-1})|_{\nu-1} = 0.$$

Thus (2) follows directly from d) of Definition 6.

3) We verify here the fact that $s_j^{(\nu)}|P_{\nu-1}$ has a limit in $\mathscr{S}(P_{\nu-1})$. By c) of Definition 6 it is enough to show that for all ν,

the sequence $s_j^{(\nu)} \in \mathscr{S}(P_\nu)$ is a Cauchy sequence in $\mathscr{S}(P_\nu)$.

Since all maps $\mathscr{S}(P_{\nu+\mu}) \to \mathscr{S}(P_\nu)$ are *contractions*, the estimate $(*)$ implies that

$$|\beta_{\nu+\mu}|P_\nu - \beta_{\nu+\mu-1}|P_\nu|_\nu \le |\beta_{\nu+\mu}|P_{\nu+\mu-1} - \beta_{\nu+\mu-1}|_{\nu+\mu-1} \le q^{\nu+\mu-1}$$

for all $\mu \ge 1$. As a consequence, for all i and j with $j > i$,

$$|s_j^{(\nu)} - s_i^{(\nu)}|_\nu \le \sum_{\mu=i+1}^{j} q^{\nu+\mu-1} \le \frac{q^\nu}{1-q} \cdot q^i.$$

This clearly implies that $s_j^{(\nu)}$ is a Cauchy sequence in $\mathscr{S}(P_\nu)$. □

An exhaustion $\{P_\nu\}_{\nu \geq 1}$ of X by compact Stein sets P_ν is called a *Stein exhaustion of* X whenever it is a Stein exhaustion relative to *every* coherent sheaf $\mathscr{S}$. Thus in closing this section we have the following clean formulation:

Theorem 8 (Exhaustion Theorem). *Every complex space X which has a Stein exhaustion is a Stein space.*

§ 2. Weak Holomorphic Convexity and Stones

In this section (analytic) stones are defined. They are used in Section 4 for the construction of Stein exhaustions. One is naturally led to the notion of a "stone" while carrying out a careful study of the fundamental idea of holomorphically convex complex spaces.

1. The Holomorphically Convex Hull. Let X be a complex space with structure sheaf $\mathcal{O} = \mathcal{O}_X$. We denote by red: $\mathcal{O} \to \mathcal{O}_{\text{red}\,X} := \mathcal{O}/\mathfrak{n}(\mathcal{O})$ the reduction map. Since $\mathcal{O}_{\text{red}\,X} \subset \mathscr{C}_X$, every section $h \in \mathcal{O}(X)$ determines a *complex-valued continuous* function red $h \in \mathscr{C}(X)$. Thus for every point $x \in X$ the "value" $h(x) := (\text{red } h)(x) \in \mathbb{C}$ as well as the absolute value

$$|h(x)| := |(\text{red } h)(x)| \geq 0$$

is well-defined (see Chapter A.3.5). As a consequence, for every subset $M \subset X$, we have

$$|h|_M := \sup_{x \in M} |h(x)|, \qquad 0 \leq |h|_M \leq \infty.$$

Since red h is continuous on X, it follows that $|h|_M < \infty$ whenever M is compact.

Definition 1 (Holomorphically Convex Hull). *For any set M in X the holomorphically convex hull of M in X is the set*

$$\hat{M} := \bigcap_{h \in \mathcal{O}(X)} \{x \in X \mid |h(x)| \leq |h|_M\}.$$

We often use the more precise notation $\hat{M}_X$ instead of $\hat{M}$. The elementary properties of the hull operator "$\hat{\ }$" are listed in the following:

Theorem 2. For any subset M in X, the following hold:

1) $\hat{M}$ is closed in X. In particular, for every $p \in X \setminus \hat{M}$, there exists a function $h \in \mathcal{O}(X)$ with $|h|_{\hat{M}} < 1 < |h(p)|$.
2) $M \subset \hat{M}$ and $\hat{\hat{M}} = \hat{M}$.

3) If $M \subset M'$, then $\hat{M} \subset \hat{M}'$.
4) If $\varphi: Y \to X$ is a holomorphic map, then

$$\widehat{\varphi^{-1}(M)}_Y \subset \varphi^{-1}(\hat{M}_X).$$

The proof follows directly from the definition of $\hat{}$.

We note that 1) above immediately implies (by taking powers of h) that $|h(p)|$ and $|h|_M$ can be chosen to be arbitrarily small and large respectively.

A particular consequence of 3) is that,

for arbitrary sets $M, M' \subset X$, $\widehat{M \cap M'} \subset \hat{M} \cap \hat{M}'$. Thus

$$\widehat{M \cap M'} = M \cap M', \quad \textit{whenever} \quad \hat{M} = M \quad \textit{and} \quad \hat{M}' = M'.$$

The statement 4) contains the following:

If $U \subset X$ is an open subspace of X which contains M, then $\hat{M}_U \subset \hat{M}_X$.

The hull operator satisfies a certain product rule:

If $X \times X'$ is the product of the complex spaces X and X', then

$$\widehat{M \times M'} \subset \hat{M} \times \hat{M}' \quad \textit{for all sets} \quad M \subset X \quad \textit{and} \quad M' \subset X'.$$

Proof: Obviously $\widehat{M \times X'} \subset \hat{M} \times X'$ and analogously $\widehat{X \times M'} \subset X \times \hat{M}'$. Since $M \times M' \subset (M \times X') \cap (X \times M')$, the claim follows. □

In general $\hat{M}$ is bigger than M. However, in many important cases $\hat{M} = M$. For example,

for every compact block Q in $\mathbb{C}^m$, $\hat{Q} = Q$.

Proof: By the above product rule it is enough to consider the case of $m = 1$. In this situation, given $p \in \mathbb{C} \setminus Q$, there exists a disk

$$T := \{z \in \mathbb{C} \mid |z - a| \leq r\}, \qquad a \in \mathbb{C}, \qquad r > 0,$$

such that $Q \subset T$ and $p \notin T$. Defining $f := z - a \in \mathcal{O}(\mathbb{C})$, it follows that $|f(p)| > |f|_Q$ (i.e. $p \notin \hat{Q}$). □

2. Holomorphically Convex Spaces. The holomorphically convex hull $\hat{K}$ of a compact set $K \subset X$ is by no means always compact. Spaces having this property are given special prominence by the following definition.

Definition 3 (Holomorphically Convex). *A complex space X is called holomorphically convex if the holomorphically convex hull $\hat{K}$ of any compact $K \subset X$ is itself compact in X.*

Compact complex spaces are always holomorphically convex. If X is holomorphically convex, then so is its reduction red $X = (X, \mathcal{O}_{\text{red}\,X})$. Furthermore, *if X and X' are holomorphically convex, then the product $X \times X'$ is also holomorphically convex.*

This follows immediately from the product rule for the hull operator $\hat{}$, since every compact set in $X \times X'$ is contained in a product $\hat{K} \times \hat{K}'$ of compact sets.

There exists a simple sufficient condition for holomorphic convexity:

Theorem 4. *Let X be a complex space and suppose that, given an infinite discrete set $D \subset X$, there exists a holomorphic function $h \in \mathcal{O}(X)$ which is unbounded on D (i.e. $|h|_D = \infty$). Then X is holomorphically convex.*

Proof: If K is compact in X, then $|h|_K < \infty$ for all $h \in \mathcal{O}(X)$. By assumption the hull $\hat{K} = \bigcap_h \{x \in X \mid |h(x)| \leq |h|_K\}$ doesn't contain a sequence of points which is discrete in X. Since $\hat{K}$ is always closed, this implies its compactness. □

At the end of this section we will see that the condition in Theorem 4 is also necessary for the holomorphic convexity of X. The proof of this fact is substantially more demanding.

Corollary to Theorem 4. *Let G be a domain in $\mathbb{C}^m$ such that, for every boundary point $p \in \partial G$, there exists a holomorphic function f on an open neighborhood U of $\bar{G}$ with*

$$G \cap \{x \in U \mid f(x) = 0\} = \emptyset \quad \textit{and} \quad f(p) = 0.$$

Then G is holomorphically convex.

Proof: We will show that for $X := G$ the condition of Theorem 4 is fulfilled. For this let D be an infinite discrete set in G. If D is unbounded in $\mathbb{C}^m$, then some coordinate function of $\mathbb{C}^m$ is unbounded on D. If D is bounded, then it has an accumulation point $p \in \partial G$. Let f be the function guaranteed by the assumption for the point p. Then $h := f^{-1} | G \in \mathcal{O}(G)$ is unbounded on G. □

It now follows immediately that *every domain G* in the plane $\mathbb{C}^1$ is holomorphically convex. Consequently every product domain

$$G_1 \times G_2 \times \cdots \times G_m \subset \mathbb{C}^m, \qquad G_\mu := \text{domain in } \mathbb{C},$$

is holomorphically convex. In particular

every open block (i.e. the interior of a block) as well as every open polycylinder in $\mathbb{C}^m$ is holomorphically convex.

Remark: The corollary further implies that every *linearly convex* domain G in $\mathbb{C}^m (\cong \mathbb{R}^{2m})$ is holomorphically convex. This follows since through every boundary point $p \in \partial G$ of such a domain there exists a "supporting hyperplane" E with $G \cap E = \emptyset$. The hyperplane E can be described by an equation $l(z) + \overline{l(z)} = 0$, where l is a *linear*, holomorphic function. Certainly $l(p) = 0$, but $G \cap \{z \in \mathbb{C}^m \mid l(z) = 0\} = \emptyset$.

3. Stones. It is of interest in Stein theory to introduce a weakening of the notion of holomorphic convexity. We will need a condition like the following: For every compact set K in X there exists a neighborhood W of K in X such that $\hat{K} \cap W$ is compact. In order to better understand this condition, we begin by introducing a simplified language.

Definition 5 (Stone). *A pair (P, π) is called an (analytic) stone in X whenever the following conditions are satisfied:*

1) *P is a non-empty, compact set in X and $\pi: X \to \mathbb{C}^m$ is a holomorphic map of X into $\mathbb{C}^m$.*
2) *There exists a (compact) block Q in $\mathbb{C}^m$ and an open set W in X such that $P = \pi^{-1}(Q) \cap W$.*

Since $P \subset \pi^{-1}(Q)$ implies that $\hat{P} \subset \pi^{-1}(Q)$ and since $\hat{Q} = Q$, it follows that $\hat{P} \cap W \subset P \cap W = P$. Thus $\hat{P} \cap W = P$.

The interior $\mathring{Q}$ of Q has, with respect to $\pi | W$, the open set $P^0 := \pi^{-1}(Q) \cap W$ as preimage. We call P^0 the *analytic interior* of the stone (P, π). It is clear that all open sets W with $\pi^{-1}(Q) \cap W = P$ lead to the same set P^0.

Since P^0 is open and contained in P, P^0 is contained in the interior of P, $\mathring{P}$. However it is not always the case that $P^0 = \mathring{P}$. For example, let X be compact and $\pi: X \to \mathbb{C}^m$ be the constant map $x \mapsto 0$. Then the pair (X, π) is a stone no matter what block $Q \subset \{0\}$ is used. In the case $\dim Q < 2m$, however, $\mathring{Q} = \emptyset$ and consequently $X^0 = \emptyset$.

Theorem 6. *Let K be a compact set in X. Then the following are equivalent:*

i) *There exists an open neighborhood W of K in X so that $\hat{K} \cap W$ is compact.*
ii) *There exists an open relatively compact neighborhood W of K in X such that the boundary ∂W does not intersect $\hat{K}$.*
iii) *There exists a stone (P, π) in X with $K \subset P^0$.*

Proof: i)$\Rightarrow$ii): This is trivial since one can take W itself to be relatively compact in X.

ii)$\Rightarrow$iii): Since $\partial W \cap \hat{K} = \emptyset$, given a point $p \in \partial W$ there exists a holomorphic function $h \in \mathcal{O}(X)$ such that $|h|_K < 1 < |h(p)|$ (see Theorem 2.2). Thus $\max\{|\text{Re } h|_K, |\text{Im } h|_K\} < 1$ and, raising h to a power if necessary, $\max\{|\text{Re } h(p)|, |\text{Im } h(p)|\} > 1$. Since ∂W is compact, continuity implies that there exist finitely many sections $h_1, \ldots, h_m \in \mathcal{O}(X)$ so that

$$(*) \qquad \begin{aligned} &\max_{1 \le \mu \le m} \{|\text{Re } h_\mu|_K, |\text{Im } h_\mu|_K\} < 1 \\ &\max_{1 \le \mu \le m} \{|\text{Re } h_\mu(p)|, |\text{Im } h_\mu(p)|\} > 1 \end{aligned}$$

for all $p \in \partial W$.

Let $Q := \{z_1, \ldots, z_m) \in \mathbb{C}^m \mid |\text{Re } z_\mu| \le 1, |\text{Im } z_\mu| \le 1\}$ denote the "unit block" in $\mathbb{C}^m$. Then the sections $h_1, \ldots, h_m \in \mathcal{O}(X)$ determine a holomorphic map

$$\pi: X \to \mathbb{C}^m, \qquad x \to (\text{red } h_1(x), \ldots, \text{red } h_m(x)).$$

From $(*)$ we see that

$$\pi(\partial W) \cap Q = \emptyset, \qquad K \subset \pi^{-1}(\mathring{Q}) \cap W.$$

Thus $P := \pi^{-1}(Q) \cap W$ is compact and $K \subset P^0$. Hence (P, π) is the desired stone.

iii) $\Rightarrow$ i): Let $W \subset X$ be the open set associated to the stone (P, π). Since $\hat{P} \cap W = P$ and $K \subset P$, it follows that $\hat{K} \cap W \subset P \cap W$. But P is compact. Thus $\hat{K} \cap W$ is compact and W is the desired neighborhood of K. □

The following theorem concerning stones is quite important:

Theorem 7. *Let (P, π) be a stone in X and Q an associated compact block in $\mathbb{C}^m$. Then there exist open neighborhoods U and V of P and Q in X and $\mathbb{C}^m$ respectively with $\pi(U) \subset V$ and $P = \pi^{-1}(Q) \cap U$ such that the induced map $\pi | U: U \to V$ is proper.*[2]

Proof: Let $W \subset X$ be the associated (via Definition 5) open set to the stone (P, π). We may assume that W is relatively compact. Thus ∂W and $\pi(\partial W)$ are also compact. Since $\partial W \cap \pi^{-1}(Q)$ is empty, $V := \mathbb{C}^m \backslash \pi(\partial W)$ is an open neighborhood of Q. The set $U := W \cap \pi^{-1}(V) = W \backslash \pi^{-1}(\pi(\partial W))$ is open in X, $\pi(U) \subset V$, and $\pi | U: U \to V$ is *proper.*[3] Furthermore

$$\pi^{-1}(Q) \cap U = \pi^{-1}(Q) \cap W \backslash \pi^{-1}(Q) \cap \pi^{-1}(\pi(\partial W)).$$

The set $\pi^{-1}(Q) \cap \pi^{-1}(\pi(\partial W))$ is empty, because $Q \cap \pi(\partial W)$ is empty. Thus $\pi^{-1}(Q) \cap U = P$ and, in particular U is a neighborhood of P. □

4. Exhaustions by Stones and Weakly Holomorphically Convex Spaces. Let (P, π) and $('P, '\pi)$ be stones in X with associated maps $\pi: X \to \mathbb{C}^m$, $'\pi: X \to \mathbb{C}^{'m}$ and blocks Q, $'Q$.

Definition 8 (Inclusion of Stones). *The stone (P, π) is said to be contained in the stone $('P, '\pi)$, in symbols $(P, \pi) \subset ('P, '\pi)$, if the following are satisfied:*

1) *The set P lies in the analytic interior of $'P$: $P \subset 'P^0$.*
2) *The space $\mathbb{C}^{'m}$ is a direct product $\mathbb{C}^m \times \mathbb{C}^n$ and there exists a point $q \in \mathbb{C}^n$ so that $Q \times \{q\} \subset Q'$.*
3) *There exists a holomorphic map $\varphi: X \to \mathbb{C}^n$ so that*

$$'\pi = (\pi, \varphi) \qquad (\text{i.e. } \; '\pi(x) = (\pi(x), \; \varphi(x)), \quad x \in X).$$

This inclusion relation is *transitive* on the set of analytic stones in X. If $(P, \pi) \subset ('P, '\pi)$, then

$$P \subset 'P, \qquad P^0 \subset 'P^0, \qquad \text{and} \qquad \dim Q \leq \dim 'Q.$$

---

[3] A continuous map f between locally compact topological spaces is called *proper* if the f-preimages of every compact set is compact. The proof of the following is trivial.

If $f: X \to Y$ is a continuous mapping between topological spaces and W is an open relatively compact subset of X, then the induced map $W - f^{-1}(f(\partial W)) \to Y - f(\partial W)$ is proper.

Definition 9 (Exhaustion by Stones). *A sequence* $\{(P_\nu, \pi_\nu)\}_{\nu \geq 1}$ *of stones in* X *is called an exhaustion of* X *by stones if the following hold:*

1) $$(P_\nu, \pi_\nu) \subset (P_{\nu+1}, \pi_{\nu+1}) \text{ for all } \nu \geq 1.$$

2) $$\bigcup_{\nu=1}^{\infty} P_\nu^0 = X.$$

Since $P_\nu \subset P_{\nu+1}^0$, every compact set $K \subset X$ is contained in some P_j^0.

Theorem 10. *The following statements about the complex space* X *are equivalent:*

i) *There exists an exhaustion of* X, $\{(P_\nu, \pi_\nu)\}_{\nu \geq 1}$, *by analytic stones.*
ii) *For every compact set* $K \subset X$ *there exists an open set* $W \subset X$ *so that* $\hat{K} \subset W$ *is compact.*

Proof: i)$\Rightarrow$ii): Since $K \subset P_j^0$ for some j, this is contained in Theorem 6.

ii)$\Rightarrow$i): Let $\{(K_\nu)\}_{\nu \geq 1}$ be an exhaustion of X by compact sets. Using this exhaustion we inductively construct an exhaustion of X by stones. Let (P_{j-1}, π_{j-1}) with $\pi_{j-1}: X \to \mathbb{C}^{m_{j-1}}$ be already constructed so that $K_{j-1} \subset P_{j-1}^0$. Let $Q_{j-1} \subset \mathbb{C}^{m_{j-1}}$ be an associated block. Let $K_j \cup P_{j-1}$ be the compact set in ii). Then, by Theorem 6, there exists a stone (P_j, π_j^*) with $K_j \cup P_{j-1} \subset P_j^0$. Let $\pi_j^*: X \to \mathbb{C}^n$, $Q_j^* \subset \mathbb{C}^n$ be an associated block, and $W \subset X$ be an open set such that $P_j = \pi_j^{*-1}(Q_j^*) \cap W$.

We choose a block $Q_j' \subset \mathbb{C}^{m_{j-1}}$ with $Q_{j-1} \subset \mathring{Q}_j'$ so that the compact set $\pi_{j-1}(P_j) \subset \mathbb{C}^{m_{j-1}}$ is contained in $\mathring{Q}_j'$. We now set

$$\pi_j := (\pi_{j-1}, \pi_j^*): X \to \mathbb{C}^{m_{j-1}} \times \mathbb{C}^n \quad \text{and} \quad Q_j := Q_j' \times Q_j^*.$$

Certainly

$$\pi_j^{-1}(Q_j) \cap W = \pi_{j-1}^{-1}(Q_j') \cap (\pi_j^{*-1}(Q_j^*) \cap W) = \pi_{j-1}^{-1}(Q_j') \cap P_j.$$

Since $\pi_{j-1}(P_j) \subset \mathring{Q}_j'$, it follows that $\pi_{j-1}^{-1}(Q_j') \subset P_j$. Thus $\pi_j^{-1}(Q_j) \cap W = P_j$ and we have shown that (P_j, π_j) is a stone in X with Q_j as an associated block.

From the above construction it is obvious that (P_{j-1}, π_{j-1}) is contained in (P_j, π_j). Since $\bigcup_1^\infty P_\nu^0 \supset \bigcup_1^\infty K_\nu = X$, it follows that $\{(P_\nu, \pi_\nu)\}_{\nu \geq 1}$ is an exhaustion of X by stones.

Definition 11. (Weak Holomorphic Convexity). *A complex space* X *is called weakly holomorphically convex if the equivalent conditions of Theorem 10 are fulfilled.*

5. Holomorphic Convexity and Unbounded Holomorphic Functions. In this paragraph we show that the converse to Theorem 4 is also true:

Theorem 12. *Let X be holomorphically convex and D be an infinite discrete set in X. Suppose that for every $p \in D$ there is given a real number $r_p > 0$. Then there exists a holomorphic function $h \in \mathcal{O}(X)$ so that*

$$|h(p)| \geq r_p \quad \textit{for all} \quad p \in D.$$

This theorem is proved here only for complex manifolds. In the case of arbitrary complex spaces one is confronted with a convergence problem which can be handled but which requires further considerations (Chapter V.6.7).

In order to prepare for the proof of Theorem 12, we choose an exhaustion $(K_n)_{n \geq 1}$ of X by compact sets with $\hat{K}_n = K_n$. Since D is discrete, the sets

$$D_0 := D \cap K_1, \qquad D_\nu := D \cap (K_{\nu+1} \backslash K_\nu), \qquad \nu > 0,$$

are finite and have the property that $D_\mu \cap D_\nu = \emptyset$ when $\mu \neq \nu$. Obviously $D = \bigcup_0^\infty D_\nu$.

Let q be an arbitrary point in D. Then $q \in D_t$ for some t and consequently $q \notin K_t$. Thus there exists $g_q \in \mathcal{O}(X)$ with $|g_q|_{K_t} < 1$ and $|g_q(q)| > 1$. Let

$$D_t'(q) := \{p \in D_t \mid |g_q(p)| \geq |g_q(q)|\} \quad \text{and} \quad D_t''(q) := D_t \backslash D_t'(q).$$

Certainly $q \in D_t'(q)$. Furthermore, by multiplying g_q by a constant if necessary, we may assume that

$$|g_q(p)| < 1 \quad \text{for all} \quad p \in D_t''(q).$$

Now we list the different points in D_t as $x_{t1}, x_{t2}, \ldots, x_{tn_t}$ and write D as a sequence $(x_\nu)_{\nu \geq 0}$ enumerated as follows:

$$x_{11}, \ldots, x_{1n_1}, x_{21}, \ldots, x_{2n_2}, \ldots, x_{j1}, \ldots, x_{jn_j}, \ldots$$

For every ν let $t := t(\nu)$ be the index with $x_\nu \in D_t$. We claim that

there exists a sequence $(h_\nu)_{\nu \geq 0}$, $h_\nu \in \mathcal{O}(X)$, so that for all ν

1) $$|h_\nu(p)| \geq r_p + 2 + \sum_{i=1}^{\nu-1} |h_i(p)| \quad \textit{for all} \quad D_t'(x_\nu)$$

and

2) $$|h_\nu|_{K_t} \leq n_t^{-1} 2^{-t}, \qquad |h_\nu(p)| \leq n_t^{-1} 2^{-t} \quad \textit{for all} \quad p \in D_t''(x_\nu).$$

Proof (by induction): Suppose $h_0, \ldots, h_{\nu-1}$ have been constructed. Since the modulus of g_{x_ν} is larger than 1 on $D_t'(x_\nu)$ and less than 1 on $K_t \cup D_t''(x_\nu)$, $h_\nu := g_{x_\nu}^s$ does the job if s is chosen to be sufficiently large. □

We set $f_i = \sum_{j=1}^{n_i} h_{ij}$, where h_{ij} is the function h_k which is associated to $x_k = x_{ij} \in D_i$. By 2) above we have $|h_{ij}|_{K_i} \leq n_i^{-1} 2^{-i}$. Thus

3) $$|f_i|_{K_i} \leq 2^{-i}, \qquad i = 1, 2, \ldots.$$

From these preparations it is clear that the desired function h should be the limit of the infinite series $\sum_{\nu=0}^{\infty} h_\nu$ which formally is the same as $\sum_{i=1}^{\infty} f_i$ It is at this point where the convergence difficulties arise in the non-reduced case. Thus from now on we assume that X is reduced. Thus $\mathcal{O}(X) \subset \mathscr{C}(X)$ and 3) implies that

$\sum_{i=1}^{\infty} f_i$ *converges uniformly and absolutely on compact subsets to a continuous function h with*

$$h = \sum_{\nu=0}^{\infty} h_\nu = \sum_{i=1}^{\infty} f_i \quad \text{and} \quad \left|\sum_{i>t}^{\infty} f_i\right|_{K_{t+1}} \le 2^{-t} < 1, \qquad t = 0, 1, \ldots.$$

We want to estimate the value of $h(p)$ for $p \in D$, say $p = x_{tk} \in D_t'(x_{tk})$. For such a point, $p \in K_{t+1}$ and thus

$$(*) \qquad |h(p)| \ge \left|\sum_{i=1}^{t} f_i(p)\right| - \left|\sum_{i>t} f_i(p)\right| \ge \left|\sum_{i=1}^{t} f_i(p)\right| - 1.$$

There exists a largest index l with $p \in D_t'(x_{tl})$, where $1 \le k \le l \le n_t$. For all m satisfying $l < m \le n_t$, $p \in D_t''(x_{tm})$ and consequently $|h_{tm}(p)| \le n_t^{-1} 2^{-t}$. Hence

$$\left|\sum_{m>l}^{n_t} h_{tm}(p)\right| \le \sum_{m>l}^{n_t} n_t^{-1} 2^{-t} < 2^{-t} < 1.$$

Thus, since $f_t = (h_{t1} + \cdots + h_{tl}) + \sum_{m>l}^{n_t} h_{tm}$,

$$(**) \qquad \left|\sum_{i=1}^{t} f_i(p)\right| \ge \left|\sum_{i=1}^{t-1} f_i(p) + \sum_{j=1}^{l} h_{tj}(p)\right| - 1.$$

Since $p \in D_t'(x_{tl})$, we can apply 1) to show that

$$|h_{tl}(p)| \ge r_p + 2 + \sum_{i=1}^{t-1} \sum_{j=1}^{n_i} |h_{ij}(p)| + \sum_{j=1}^{l-1} |h_{tj}(p)|.$$

Hence

$$(*_*{*}) \quad \left|\sum_{i=1}^{t-1} f_i(p) + \sum_{j=1}^{l} h_{tj}(p)\right|$$
$$\ge |h_{tl}(p)| - \left|\sum_{i=1}^{t-1} \sum_{j=1}^{n_i} h_{ij}(p) + \sum_{j=1}^{l-1} h_{tj}(p)\right| \ge r_p + 2.$$

The estimates $(*)$, $(**)$, and $(*_*{*})$ together yield $|h(p)| \ge r_p$. □

It remains to show that the constructed $h \in \mathscr{C}(X)$ is in fact holomorphic on X. If X is a manifold then this follows from the classical theorem that a uniform limit of holomorphic functions is holomorphic. This statement is also valid on reduced complex spaces (Theorem 8, Chapter V.6.6). Thus Theorem 12 above is completely proved when X is a manifold and is proved modulo Theorem 8 in V.6.6 for reduced complex spaces. For the non-reduced case, some additional considerations are needed (see V.6.7). For these we need the following remark:

If $(h_\nu)_{\nu \geq 1}$ is the sequence constructed above and $(m_\nu)_{\nu \geq 0}$ is an arbitrary sequence of positive integers, then the sequence $(h_\nu^{m_\nu})_{\nu \geq 0}$ has properties 1) and 2) and in particular the series $\sum_0^\infty h_\nu^{m_\nu}$ converges (for reduced spaces) uniformly on compact subsets to a function $g \in \mathscr{C}(X)$ with $|g(p)| \geq r_p$ for all $p \in D$.

§ 3. Holomorphically Complete Spaces

In this section the notion of an analytic block is introduced. Complex spaces which possess exhaustions by analytic blocks are called *holomorphically complete*.

1. Analytic Blocks. For every stone (P, π) in X with an associated block $Q \subset \mathbb{C}^m$, there exist neighborhoods U and V of P and Q respectively such that $\pi(U) \subset V$, $P = \pi^{-1}(Q) \cap U$, and the induced map $\pi | U: U \to V$ is proper.

Definition 1 (Analytic Blocks). *A stone (P, π) is called an analytic block in X if U and V can be chosen so that $\pi | U: U \to V$ is finite.*[4]

It is clear that every block $Q \subset \mathbb{C}^m$ has the associated analytic block (Q, id) in $\mathbb{C}^m$.

Theorem 2. *If (P, π) is an analytic block in X, then P is a compact Stein set in X.*

Proof: This follows immediately by applying Theorem 1.3 to the map $\pi | U: U \to V$. □

There exists an important and easily described class of spaces in which every stone is an analytic block. For this we first note the following general fact:

Let X be a complex space in which every compact analytic set is finite. Then every proper holomorphic map $f: U \to Y$ of an open subset U in X into an arbitrary complex space Y is finite.

Proof: Every fiber $f^{-1}(y)$, $y \in Y$, is a compact analytic subset of U and consequently is a compact subset of X. Thus every f-fiber is finite. □

---

[4] By assumption finite maps are always proper.

The following is now obvious:

Theorem 3. *If X is a complex space such that every compact analytic subset is finite, then every stone in X is an analytic block.*

2. Holomorphically Spreadable Spaces. We now introduce a classical notion of Stein theory.

Definition 4 (Holomorphically Spreadable). *A complex space X is called holomorphically spreadable if given $p \in X$ there exist finitely many function $f_1, \ldots, f_r \in \mathcal{O}(X)$ so that p is an isolated point of the set $\{x \in X \mid f_1(x) = \cdots = f_r(x) = 0\}$.*

Every domain in $\mathbb{C}^m$ is obviously holomorphically spreadable. A complex space is called holomorphically separable if, given $x_1, x_2 \in X$ with $x_1 \neq x_2$, there exists $f \in \mathcal{O}(X)$ such that $f(x_1) \neq f(x_2)$. It can be shown that every holomorphically separable space is holomorphically spreadable. The proof of this is elementary, but nevertheless uses dimension theory and will therefore not be given here.

The following is a simple consequence of the maximum principle.

Theorem 5. *In a holomorphic separable or holomorphically spreadable complex space X, every compact analytic subset A is finite.*

Proof: Let B be a connected component of A. Thus every function $f \mid B$, $f \in \mathcal{O}(X)$, is by the maximum principle constant on B. If X is holomorphically separable and p, p' were distinct points in B, then there would exist $h \in \mathcal{O}(X)$ such that $h(p) \neq h(p')$. Since this can't happen, $B = \{p\}$.

If X is holomorphically spreadable, then there exist functions $f_1, \ldots, f_r \in \mathcal{O}(X)$ such that p is an isolated point of $\{x \in X \mid f_1(x) = \cdots = f_r(x) = 0\}$. Since $f_i \mid B = 0$ for all i, $B = \{p\}$. □

Corollary 1. *In a holomorphically spreadable space X, every stone is an analytic block.*

Proof: This is now clear by Theorem 3. □

Corollary 2. *If A is a compact analytic subset of the complex space X and there exists an analytic block (P, π) in X with $A \subset P$, then A is finite.*

Proof: Let $U \subset X$ and $V \subset \mathbb{C}^m$ with $P \subset U$ be such that $\pi \mid U \colon U \to V$ is finite. Then U is holomorphically spreadable. To see this just note that $p \in U$ is an isolated point of $\{x \in U \mid f_1(x) = \cdots = f_m(x) = 0\}$, where $\pi(p) = (c_1, \ldots, c_m)$ and $f_\mu = (z_\mu - c_\mu) \circ \pi \in \mathcal{O}(U)$.[5] Thus by Theorem 5 A is finite in U. □

3. Holomorphically Convex Spaces. An exhaustion $\{(P_\nu, \pi_\nu)\}_{\nu \geq 1}$ of X by stones is called an *exhaustion of X by blocks* whenever every (P_ν, π_ν) is an analytic block.

---

[5] In general if $f \colon X \to Y$ is finite and Y is holomorphically spreadable, then X is holomorphically spreadable.

Theorem 7. *The following statements about X are equivalent.*

i) *There exists an exhaustion $\{(P_\nu, \pi_\nu)\}$ of X by blocks.*
ii) *X is weakly holomorphically convex and every compact analytic subset of X is finite.*

Proof: i)$\Rightarrow$ii): By Theorem 2.10, X is weakly holomorphically convex. Furthermore, every compact analytic subset A of X is contained in some block P_j. Thus, by Corollary 2 to Theorem 5, A is finite.

ii)$\Rightarrow$i): By Theorem 2.10, there exists an exhaustion of X, $\{(P_\nu, \pi_\nu)\}$, by stones. By Theorem 3, every stone in X is an analytic block. □

Definition 8 (Holomorphically Complete). *A complex space X is called holomorphically complete if the equivalent conditions in Theorem 7 are fulfilled.*

Every holomorphically convex domain G in $\mathbb{C}^m$, (in particular, any domain in $\mathbb{C}$) is holomorphically complete.

In the next section it is shown that holomorphically complete spaces are Stein spaces.

§ 4. Exhaustions by Analytic Blocks are Stein Exhaustions

Let $\mathscr{S}$ be a coherent sheaf on the complex space X and suppose that (P, π) is an analytic block in X. In this section a procedure is developed to provide the $\mathbb{C}$-vector space $\mathscr{S}(P)$ with a "good semi-norm." The properties of such semi-norms are explained in detail with the main motivation for the considerations being conditions a), b) and c) of Definition 1.6 (Stein Exhaustion). The basic result is that an exhaustion $\{(P_\nu, \pi_\nu)\}_\nu$ by analytic blocks, where the $\mathscr{S}(P_\nu)$'s are equipped with good semi-norms, is a Stein exhaustion.

1. Good Semi-norms. We need the following remark:

If X is a complex manifold and $\mathscr{J}$ is a coherent subsheaf of $\mathcal{O}^l$ on X, $1 \le l < \infty$, then, with respect to the topology of convergence on compact subsets, the module of sections $\mathscr{J}(X)$ is a closed vector subspace of $\mathcal{O}^l(X)$.

Proof: Let $f_j \in \mathscr{J}(X)$ be a sequence having a limit $f \in \mathcal{O}^l(X)$. Thus at every point $x \in X$ the sequence of germs $f_{jx} \in \mathscr{J}_x$ converges to $f_x \in \mathcal{O}_x^l$. Since every $\mathcal{O}_x$-submodule of $\mathcal{O}_x^l$ is closed (see AS, p. 87), $f_x \in \mathscr{J}_x$ for all $x \in X$. Hence $f \in \mathscr{J}(X)$. □

Now let (P, π), $\pi: X \to \mathbb{C}^m$, be an analytic block in a complex space X having associated block $Q \subset \mathbb{C}^m$ with $Q \neq \emptyset$. Using Theorem 2.7, we choose neighborhoods U and V of P and Q so that $\pi(U) \subset V$, $P = \pi^{-1}(Q) \cap U$, and the induced map $\tau: U \to V$, where $\tau := \pi | U$ is finite. Then $P^0 = \tau^{-1}(\mathring{Q})$ is the analytic interior of P.

By the direct image theorem for finite maps, the image sheaf

$$\mathcal{T} := \tau_*(\mathcal{S} \mid U)$$

is coherent on V for every coherent $\mathcal{S}$ on X. Thus by Theorem A for blocks in $\mathbb{C}^m$ there exists $l \geq 1$ and an *$\mathcal{O}$-epimorphism*

$$\varepsilon: \mathcal{O}^l \mid Q \to \mathcal{T} \mid Q$$

which induces an $\mathcal{O}(Q)$-epimorphism of modules of sections

$$\varepsilon_Q: \mathcal{O}^l(Q) \to \mathcal{T}(Q).$$

Since $P = \tau^{-1}(Q)$, there exists a canonical $\mathbb{C}$-vector space isomorphism $\varphi: \mathcal{S}(P) \to \mathcal{T}(Q)$. For every section $s \in \mathcal{S}(P)$, we define $|s|$ by

$$|s| := \inf\{|f|_Q \mid f \in \mathcal{O}^l(Q) \text{ with } \varepsilon_Q(f) = \varphi(s)\}.$$

Theorem 1. *The map* $| \quad |: \mathcal{S}(P) \to \mathbb{R}$ *is a semi-norm on* $\mathcal{S}(P)$. *For every section* $s \in \mathcal{S}(P)$ *with* $|s| = 0$, *it follows that* $s \mid P^0 = 0$.

Proof: It is clear that $| \quad |$ is a semi-norm on $\mathcal{S}(P)$. In fact it is just the quotient semi-norm on $\mathcal{O}^l(Q)/\text{Ker } \varepsilon_Q = \mathcal{T}(Q)$ (using $| \quad |_Q$) transported to $\mathcal{S}(P)$ by ι. Since Ker ε_Q is in general not a closed subspace of $\mathcal{O}^l(Q)$, the semi-norm $| \quad |$ is not necessarily a norm.

Let $|s| = 0$. Then there exists $h \in \mathcal{O}^l(Q)$ with $\varepsilon_Q(h) = \iota(s)$ and a sequence $h_j \in \text{Ker } \varepsilon_Q$ so that $\lim_j |h - h_j|_Q = 0$. In particular $\lim_j (h_j \mid Q) = h \mid Q$ in the topology of uniform convergence on compact subsets. Since $\mathcal{K}er\ \varepsilon$ is a coherent subsheaf of $\mathcal{O}^l \mid Q$, the introductory remark above implies that $\varepsilon_Q(h \mid \mathring{Q}) = 0$. Thus $\iota(s) \mid \mathring{Q} = \varepsilon_Q(h \mid Q) = 0$, and, since $P^0 = \tau^{-1}(Q)$, it follows that $s \mid P^0 = 0$. □

In the following we call any semi-norm on $\mathcal{S}(P)$ which is obtained as above by a sheaf epimorphism $\varepsilon: \mathcal{O}^l \mid Q \to \pi_*(S \mid U) \mid Q$ a good semi-norm.

2. The Compatibility Theorem. Suppose that along with (P, π) we have another analytic block $('P, '\pi)$ on X, where $'\pi: X \to \mathbb{C}^{'m}$ and $'Q \subset \mathbb{C}^{'m}$ is the associated euclidean block. We fix good semi-norms $| \quad |, '| \quad |$ and $\mathcal{S}(P), \mathcal{S}('P)$ as well as the associated $\mathbb{C}$-vector space epimorphisms

$$\alpha: \mathcal{O}^l(Q) \to \mathcal{S}(P), \qquad '\alpha: \mathcal{O}^{'l}('Q) \to \mathcal{S}('P).$$

We will keep this notation for the remainder of this section.

In the case $P \subset P'$ it is important to know if the restriction map,

$$\rho: \mathcal{S}('P) \to \mathcal{S}(P), \qquad s \to s \mid P,$$

is bounded (i.e. with respect to the semi-norms). When $\mathbb{C}^{'m} = \mathbb{C}^m \times \mathbb{C}^n$ and $Q \cong Q \times \{q\} \subset {'Q}$, then the $\mathbb{C}$-linear restriction

$$\mathcal{O}^{'l}('Q) \to \mathcal{O}^{'l}(Q), \qquad h \to h\,|\,Q,$$

is obviously a contraction. This fact implies the following basic lemma.

Lemma. *Let $P \subset {'P}$, $\mathbb{C}^{'m} = \mathbb{C}^m \times \mathbb{C}^n$, and $Q \times \{q\} \subset {'Q}$ with $q \in \mathbb{C}^n$. Then there exists a bounded $\mathbb{C}$-linear map $\eta\colon \mathcal{O}^{'l}('Q) \to \mathcal{O}^{l}(Q)$ so that*

$$(*) \qquad \begin{array}{ccc} \mathcal{O}^{'l}('Q) & \xrightarrow{\ '\alpha\ } & \mathcal{S}('P) \\ \downarrow{\scriptstyle \eta} & & \downarrow{\scriptstyle \rho} \\ \mathcal{O}^{l}(Q) & \xrightarrow{\ \alpha\ } & \mathcal{S}(P) \end{array}$$

is commutative: $\alpha \circ \eta = \rho \circ {'\alpha}$.

Proof: Let $(e_1, \ldots, e_{'l})$ be the natural basis of $\mathcal{O}^{'l}('Q)$. We choose sections $g_\mu \in \mathcal{O}^{l}(Q)$, $\mu = 1, \ldots, {'l}$, with

$$\alpha(g_\mu) = \rho \circ {'\alpha}(e_\mu).$$

Obviously the map

$$\eta\colon \mathcal{O}^{'l}('Q) \to \mathcal{O}^{l}(Q), \qquad (f_1, \ldots, f_{'l}) \to \sum_{\mu=1}^{'l} (f_\mu\,|\,Q) g_\mu,$$

does the job. □

The following is now immediate:

Theorem 2 (Compatibility Theorem). *If $P \subset {'P}$, $\mathbb{C}^{'m} = \mathbb{C}^m \times \mathbb{C}^n$ and $Q \times \{q\} \subset {'Q}$ with $q \in \mathbb{C}^n$, then the restriction map $\rho\colon \mathcal{S}('P) \to \mathcal{S}(P)$ is bounded.*

Proof: Since $'\alpha$, α determine the semi-norms $'|\ \ |$, $|\ \ |$, the commutativity of $(*)$ implies that whatever bounds exist for η can be used as bounds for ρ as well.

3. The Convergence Theorem. The main result of this section is based on the following fact:

Let Q, Q^* be blocks in $\mathbb{C}^m$ with $Q \subset \mathring{Q}^*$. Suppose that (h_j) is a Cauchy sequence in $\mathcal{O}^{l}(Q^*)$. Then the restricted sequence $(h_j\,|\,Q)$ has a limit in $\mathcal{O}^{l}(Q)$.

Proof: The sequence $h_j\,|\,Q^*$ converges to some $h \in \mathcal{O}^{l}(Q^*)$ in the topology of uniform convergence on compact subsets.[6] Since $Q \subset \mathring{Q}^*$, it follows that $\lim_j |h - h_j|_Q = 0$. □

[6] The spaces $\mathcal{O}^{l}(Q)$ are normed, but not complete. On the other hand, the spaces $\mathcal{O}^{l}(\mathring{Q})$ are complete (with respect to the topology of uniform convergence on compact subsets), but they are not normable (see Chapter V.6.1). This discrepancy makes the introduction of intermediate blocks unavoidable.

Now again let (P, π) and $('P, \pi)$ be analytic blocks in X. We maintain the notation of the last section, assuming further that the conditions 1) and 2) of the inclusion definition (Definition 2.8) are satisfied:

$$P \subset 'P^0, \qquad \mathbb{C}'^m = \mathbb{C}^m \times \mathbb{C}^n, \qquad \text{and} \qquad Q \times \{q\} \subset 'Q \quad \text{with} \quad q \in \mathbb{C}^n.$$

Theorem 3 (Convergence Theorem). *If (s_j) is a Cauchy sequence in $\mathscr{S}('P)$, then the restricted sequence $(s_j|P)$ has a limit $s \in \mathscr{S}(P)$. The section $s|P^0$ is uniquely determined.*

Proof: Since $Q \times \{q\} \subset '\mathring{Q}$, there exists a block $Q^* \subset \mathbb{C}^m$ with $Q \subset \mathring{Q}^*$ and $Q^* \times \{q\} \subset 'Q$. In a way completely analogous to the Lemma above, one now construct a map $\eta^*: \mathcal{O}'^l('Q) \to \mathcal{O}^l(Q^*)$ so that

$$\begin{array}{ccc} \mathcal{O}'^l('Q) & & \\ \downarrow{\scriptstyle \eta^*} & \searrow{\scriptstyle \eta} & \\ \mathcal{O}^l(Q^*) & \xrightarrow[\omega]{} & \mathcal{O}^l(Q) \end{array}$$

is commutative. In this case ω denotes the obvious restriction map.

Since $'\alpha$ determines the norm on $\mathscr{S}('P)$, to every Cauchy sequence (s_j) one can find a Cauchy sequence (h_j) in $\mathcal{O}'^l('Q)$ with $'\alpha(h_j) = s_j$. Then $(\eta^*(h_j))$ is a Cauchy sequence in $\mathcal{O}^l(Q^*)$. Hence, by the fact mentioned at the first of this paragraph, the sequence $\eta(h_j) = \eta^*(h_j)|Q$ has a limit in $\mathcal{O}^l(Q)$. Consequently $s := \alpha(h) \in \mathscr{S}(P)$ is a limit of $\alpha\eta(h_j) = \rho(s_j) = s_j|P$.

If $\hat{s} \in \mathscr{S}(P)$ is another such limit, then $|\hat{s} - s| = 0$ and therefore, by Theorem 1, $\hat{s}|P^0 = s|P^0$. □

4. The Approximation Theorem. Again let (P, π) and $('P, '\pi)$ be analytic blocks in X with $\mathbb{C}'^m = \mathbb{C}^m \times \mathbb{C}^n$ and $Q \times \{q\} \subset 'Q$. We choose neighborhoods $'U$ and $'V$ of $'P$ and $'Q$ respectively so that $'\pi$ induces a finite map $'\pi|'U: 'U \to 'V$ with $'P = '\pi^{-1}('Q) \subset 'U$. We set

$$Q_1 := (Q \times \mathbb{C}^n) \cap 'Q \quad \text{and} \quad P_1 := '\pi^{-1}(Q_1) \cap 'U.$$

Obviously P_1 is compact and $'V$ is also a neighborhood of Q_1. Moreover, $(P_1, '\pi)$ is an analytic block in X with associated block $Q_1 \subset \mathbb{C}'^m$ and $P_1 \subset 'P$.

Now let $\mathscr{S}$ be a coherent sheaf on X and $'\varepsilon: \mathcal{O}'^l|'Q \to '\pi_*(\mathscr{S}|'U)|'Q$ a sheaf epimorphism which induces the map $'\alpha: \mathcal{O}'^l('Q) \to \mathscr{S}('P)$ and the good semi-norm $'|\ |$.

By restricting to Q_1 one gets a sheaf epimorphism $\varepsilon_1: \mathcal{O}'^l|Q_1 \to '\pi_*(\mathscr{S}|'U)|Q_1$. Since $'\pi_*(\mathscr{S}|'U)(Q_1) \cong \mathscr{S}(P_1)$, the map ε_1 determines a $\mathbb{C}$-linear surjection $\alpha_1: \mathcal{O}'^l(Q_1) \to \mathscr{S}(P_1)$ and the associated semi-norm $|\ \ |_1$ on $\mathscr{S}(P_1)$. By construction the diagram

$$\begin{array}{ccc} \mathcal{O}'^l('Q) & \xrightarrow{'\alpha} & \mathcal{S}('P) \\ \downarrow & & \downarrow \rho_1 \\ \mathcal{O}'^l(Q_1) & \xrightarrow{\alpha_1} & \mathcal{S}(P_1), \end{array}$$

where the vertical maps are natural restrictions, is commutative.

Since these maps are all continuous, and since, by the Runge approximation theorem for blocks (see Chapter III.2.1), $\mathcal{O}'^l('Q)$ is dense in $\mathcal{O}'^l(Q_1)$,

the space $\mathcal{S}('P)|P_1$ is dense in $\mathcal{S}(P_1)$.

We now assume that (P, π) is contained in $('P, '\pi)$. Then $P \subset 'P^0$, and there exists a holomorphic map $\varphi: X \to \mathbb{C}^n$ so that

$$'\pi(x) = (\pi(x), \varphi(x)) \in \mathbb{C}^m \times \mathbb{C}^n = \mathbb{C}^{'m}, \quad \text{for all} \quad x \in X.$$

In this situation the support P_1 of the analytic block $(P_1, '\pi)$ can be decomposed in this following way:

There exists a compact set $\tilde{P}$ in X such that $\tilde{P} \cap P = \emptyset$, and

$$P_1 = P \cup \tilde{P}.$$

Proof: Since $'\pi = (\pi, \varphi)$ and $Q_1 = (Q \times \mathbb{C}^n) \cap 'Q$, one can easily verify that

$$P_1 = \pi^{-1}(Q) \cap 'P.$$

Since $P \subset 'P$ and $P \subset \pi^{-1}(Q)$, it follows that $P \subset P_1$. But there exists a neighborhood U of P in X so that $P = \pi^{-1}(Q) \cap U$. Hence $\tilde{P} := P_1 \setminus P$ is compact. □

The above decomposition of P_1 has as a simple (but important) consequence that whenever $(P, \pi) \subset ('P, '\pi)$ the restriction map $\sigma: \mathcal{S}(P_1) \to \mathcal{S}(P)$ is *surjective.* Since $Q \times \{q\} \subset Q_1$, the assumptions of Theorem 2 are satisfied and σ is *continuous.* These observations allow us to now prove a Runge Theorem for coherent sheaves on analytic blocks.

Theorem 4 (Runge Approximation Theorem). *If (P, π) and $('P, '\pi)$ are analytic blocks in X with $(P, \pi) \subset ('P, '\pi)$, then for every coherent sheaf $\mathcal{S}$ on X, the space $\mathcal{S}('P)|P$ is dense in $\mathcal{S}(P)$.*

Proof: The restriction map $\mathcal{S}('P) \to \mathcal{S}(P)$ is factored into the two other restrictions

$$\mathcal{S}('P) \xrightarrow{\rho_1} \mathcal{S}(P_1) \quad \text{and} \quad \mathcal{S}(P_1) \xrightarrow{\sigma} \mathcal{S}(P).$$

We have already shown that $\rho_1(\mathcal{S}('P)) = \mathcal{S}('P)|P_1$ is dense in $\mathcal{S}(P_1)$. Since σ is both surjective and continuous, it therefore follows that $\sigma\rho_1(\mathcal{S}('P)) = \mathcal{S}('P)|P$ is dense in $\mathcal{S}(P)$. □

5. Exhaustions by Analytic Blocks are Stein Exhaustions. It is now relatively easy to prove the following essential result:

Theorem 5. *Every exhaustion $\{(P_\nu, \pi_\nu)\}_{\nu \geq 1}$ of a complex space X by analytic blocks is a Stein exhaustion of X.*

Proof: First, by Theorem 3.2, every set P_ν is a compact Stein set. On each module of sections $\mathcal{S}(P_\nu)$ we fix a good semi-norm $|\quad|_\nu$. Then conditions b) and c) of Definition 1.6 are satisfied. Further we may assume that the restrictions $\mathcal{S}(P_{\nu+1}) \to \mathcal{S}(P_\nu)$ do not increase the semi-norms.

It remains to show that condition a) is also fulfilled (i.e. for every ν, $\mathcal{S}(X)|P_\nu$ is dense in $\mathcal{S}(P_\nu)$) and it is enough to verify this for $\nu := 1$. Thus let $s \in \mathcal{S}(P_1)$ and $\delta \in \mathbb{R}$ with $\delta > 0$ be given. We choose a sequence $\delta_i \in \mathbb{R}$, $\delta_i > 0$, with $\sum_{i=1}^{\infty} \delta_i < \delta$ and inductively determine by the Runge Theorem (Theorem 4) a sequence $s_i \in \mathcal{S}(P_i)$ with

$$s_1 := s \quad \text{and} \quad |s_{i+1}|P_i - s_i|_i < \delta_i, \qquad i = 1, 2, \ldots.$$

Then $(s_j|P_{i+1})_{j>i}$ is a Cauchy sequence in $\mathcal{S}(P_{i+1})$. By the Convergence Theorem (Theorem 3), the restricted sequence $(s_j|P_i)$ has a limit $t_i \in \mathcal{S}(P_i)$. Since all of the restriction maps $\mathcal{S}(P_{i+1}) \to \mathcal{S}(P_i)$ are bounded, $t_{i+1}|P_i$ is also the limit of the sequence $(s_j|P_i)$. The uniqueness part of Theorem 3 implies that $t_{i+1}|P_i^0 = t_i|P_i^0$. But the sets $\{P_\nu^0\}$ exhaust X. Thus the t_i's determine a global section $t \in \mathcal{S}(X)$ with $t|P_i = t_i$, $i \geq 1$. Since $|\quad|_1 \leq |\quad|_i$, the equation

$$t|P_1 - s = t_1 - s_j|P_1 + \sum_{i=1}^{j-1} (s_{i+1}|P_1 - s_i|P_1)$$

yields the estimate

$$|t|P_1 - s|_1 \leq |t_1 - s_j|P_1|_1 + \sum_{i=1}^{j-1} \delta_i.$$

Letting $j \to \infty$, $|t|P_1 - s|_1 < \delta$, and thus every section $s \in \mathcal{S}(P_1)$ can be approximated by global sections $t \in \mathcal{S}(X)$. □

One can now combine Theorem 5 with Theorem 1.8 and Definition 3.8 and prove the main theorem of Stein theory:

Fundamental Theorem. *Every holomorphically complete space $(X, \mathcal{O})$ is a Stein space. For every coherent analytic sheaf $\mathcal{S}$ on X, then holomorphic completeness implies the following:*

A) *The module of sections $\mathcal{S}(X)$ generates every stalk $\mathcal{S}_x$, $x \in X$, as an $\mathcal{O}_x$-module.*
B) *For all $q \geq 1$, $H^q(X, \mathcal{S}) = 0$.*

Furthermore, the original axioms stated in Section 4 imply both A) *and* B).

Chapter V. Applications of Theorems A and B

In this chapter we give some of the significant classical applications of Theorems A and B. Along with the Cousin problems and the Poincaré problem, we carry out a detailed discussion of Stein algebras. We also become quite involved with the problem of topologizing the module of sections $\mathcal{S}(X)$ of a coherent sheaf.

It is necessary for us to use facts from complex analysis which cannot be proved here. For example we need general results from dimension theory, the Riemann continuation theorems for functions holomorphic on normal spaces, and the normalization theorem.

§ 1. Examples of Stein Spaces

In this section the idea of a Stein space is explained and clarified through recipes for construction, examples, and counterexamples. In doing this, important theorems can, to a certain extent, only be reported.

1. Standard Constructions. If X_α, $\alpha \in A$, are the connected components of a Stein space X, then for every nonempty $A' \subset AX' = \bigcup_{\alpha \in A'} X_\alpha$ is a Stein space. We summarize in the following several other simple recipes for making Stein spaces.

Theorem 1. *Let $(X, \mathcal{O}_X)$ be Stein. Then the following hold:*

a) *Every holomorphically convex open subspace $(U, \mathcal{O}_U)$ of X is Stein.*
b) *Every closed complex subspace $(Y, \mathcal{O}_Y)$ of X is Stein.*
c) *If $h \in \mathcal{O}_X(X)$, $h \not\equiv 0$, and $H := \{x \in X \mid |h(x)| = 0\}$, then the difference $X \backslash H$ is Stein.*
d) *If $f: Z \to X$ is a finite holomorphic map of a complex space Z into X, then Z is Stein.*
e) *If X' is Stein, then the cartesian product $X \times X'$ is also Stein.*

Proof: a) This is clear, because X being holomorphically spreadable implies U is as well. b) Since the injection $\imath: Y \to X$ is finite, this is a special case of d). c) If X has no singularities, then $h^{-1} \in \mathcal{O}(X \backslash H)$ is a function which is unbounded on any sequence of points approaching H. If X is arbitrary, then the proof is too difficult to reproduce here. d) Since f is proper, Z is also holomorphically convex. The finiteness of f implies Z is holomorphically spreadable as well. e) Since X and X' are holomorphically spreadable, $X \times X'$ is likewise holomorphically spreadable. The product rule $K_1 \times K_2 \subset \hat{K}_1 \times \hat{K}_2$ implies that $X \times X'$ is holomorphically convex. □

Remarks: 1) Every $\mathbb{C}^m$, $1 \leq m < \infty$, is Stein. Thus, by 1,b) above, every closed submanifold of some $\mathbb{C}^m$ is likewise Stein. One can in fact prove the converse statement (see [CA], p. 122 ff.):

Embedding Theorem. *Every n-dimensional Stein manifold can be biholomorphically mapped onto a closed complex submanifold of* $\mathbb{C}^{2n+1}$.

The unit disk is even realizable as a closed complex curve without signularities in $\mathbb{C}^2$.

One can also prove the following generalization of the embedding theorem for Stein spaces:

Let X be a finite dimensional Stein space and $p \in X$ an arbitrary point. Then there are finitely many functions $f_1, \ldots, f_n \in \mathcal{O}(X)$ with the following properties:

1) *The holomorphic map $f\colon X \to \mathbb{C}^n, x \to (f_1(x), \ldots, f_n(x))$, is injective and proper. In particular in the topology of $\mathbb{C}^n$, the subspace $f(X)$ is homeomorphic to X.*
2) *There exists a neighborhood U of p in X which is mapped biholomorphically by $f \mid U$ onto a closed complex subspace Y of a polydisk $\Delta \subset \mathbb{C}^n$.*

The property 2) above is always fulfilled as long as $f_{1p}, \ldots, f_{np} \in \mathcal{O}_p$ generate the ideal $\mathfrak{m}(\mathcal{O}_p)$ (compare Section 4.2).

2) In 1956 K. Stein [38] showed that every unramified covering space of a Stein space is Stein. An important generalization of this was recently proved by P. LeBarz [2]. By using the solvability of the Levi problem, he showed the following:

Theorem. *Let $\pi\colon Z \to X$ be a holomorphic map between complex spaces having the following property: Every point $x \in X$ has an open neighborhood U so that $\pi^{-1}(U)$ is the disjoint union of complex spaces $W_1, W_2, \ldots$ such that every induced map $\pi_\nu \mid W_\nu\colon W_\nu \to U$ is finite. Then, if X is a Stein space, Z is likewise Stein.*

In particular it follows that every local-analytically ramified cover of a Stein space is Stein.

3) From the beginning of the subject, the following question has held great interest in Stein theory: Is a complex space X which is exhaustable by a sequence $X_1 \Subset X_2 \Subset X_3 \Subset \cdots$ of open Stein subspaces itself a Stein space? Every compact analytic subspace of such a space X is certainly finite. Nevertheless, if a Runge condition is not imposed on the X_ν's, the space X is not necessarily holomorphically convex. In fact E. Fornaess [11] described a 3-*dimensional, non-holomorphically convex, complex manifold M having an exhaustion*

$M_1 \Subset M_2 \Subset M_3 \Subset \cdots$ by open submanifolds M_i each of which is biholomorphically equivalent to the unit ball B (or the unit polydisk) in $\mathbb{C}^3$.

The maps $B \simeq M_i$ are explicitly defined with the deciding fact being that for all $\varepsilon > 0$ the polynomial map

$$\tau: \mathbb{C}^3 \to \mathbb{C}^3, \qquad (z, w, t) \mapsto (z, zw + \varepsilon t, (zw - 1)w + 2\varepsilon wt),$$

is *injective* on the set $\{(z, w, t) \mid |t| < 1/2\varepsilon\}$. By a refinement of the construction [13] one can even obtain a limit manifold M on which all holomorphic functions are constant. Fornaess recently found 2-dimensional examples of the same phenomena (see [12]).

The Fornaess manifold M is certainly not isomorphic to a domain in $\mathbb{C}^3$. On the contrary, in 1938 Behnke and Stein [4] proved the following:

Exhaustion Theorem. *Every domain B in $\mathbb{C}^m$ which is exhaustable by a sequence of Stein domains $B_1 \Subset B_2 \Subset \cdots$ is itself Stein.*

This theorem holds more generally for any *unramified Riemann domain B over* $\mathbb{C}^m$ (for this idea see Section 5 as well as [19], in particular Theorem D″, p. 161).

In 1956 Stein [38] showed the following:

Let X be a reduced complex space and $X_1 \Subset X_2 \Subset \cdots$ be an exhaustion of X by Stein domains. If every pair $(X_\nu, X_{\nu+1})$ is Runge (i.e. $\mathcal{O}(X_{\nu+1})$ is dense in $\mathcal{O}(X_\nu)$ in the topology of compact convergence), then $X = \bigcup X_\nu$ is Stein.

Recently A. Markoe [25] proved the following:

Let X be reduced and $X_1 \Subset X_2 \Subset \cdots$ be an exhaustion of X by Stein domains. Then X is Stein if and only if $H^1(X, \mathcal{O}) = 0$.

2. Stein Coverings. The following fact is trivial to prove, but powerful:

Let X be a complex space and U_1, U_2 holomorphically convex (resp. Stein) subspaces of X. Then $U_1 \cap U_2$ is holomorphically convex (resp. Stein).

Proof: Suppose U_1 and U_2 are holomorphically convex and let $K \subset U_1 \cap U_2$ be compact. By Theorem IV.2.2

$$\hat{K}_{U_1 \cap U_2} \subset \hat{K}_{U_1} \cap \hat{K}_{U_2}.$$

Since $\hat{K}_{U_1 \cap U_2}$ is closed in any case and $\hat{K}_{U_1}$, $\hat{K}_{U_2}$ are compact by assumption, $\hat{K}_{U_1 \cap U_2}$ is compact. If U_1 and U_2 are Stein, then, being a holomorphically convex subspace of a Stein space, $U_1 \cap U_2$ is Stein. □

The language introduced in the following is now appropriate.

Definition 2 (Stein Covering). *An open cover $\mathfrak{U} = \{U_i\}$, $i \in I$, of X is called Stein if $\mathfrak{U}$ is locally finite and each U_i is Stein.*

The introductory remark above shows that if $\mathfrak{U}$ is a Stein cover, then all of the intersections $U(i_0, \ldots, i_q) = U_{i0} \cap \cdots \cap U_{iq}$ are also Stein. Hence Theorem B implies that such a $\mathfrak{U}$ is acyclic for every coherent $\mathcal{O}$-sheaf $\mathcal{S}$ over X. Thus the following is a consequence of the Leray Theorem.

Theorem 3. *If* $\mathfrak{U}$ *is a Stein cover of an arbitrary complex space* X *and* $\mathcal{S}$ *is a coherent sheaf of* $\mathcal{O}$*-modules on* X, *then*

$$H^q_a(\mathfrak{U}, \mathcal{S}) = H^q(\mathfrak{U}, \mathcal{S}) = H^q(X, \mathcal{S})$$

for all $q \geq 0$.

Corollary. *If* X *is a compact complex space, then there exists a natural number* $a = a(X)$ *so that for every coherent sheaf* $\mathcal{S}$ *on* X

$$H^q(X, \mathcal{S}) = 0$$

for all $q \geq a$.

Proof: There exists a finite Stein cover of X! □

Since every point $x \in X$ has a neighborhood basis of Stein spaces, there exist arbitrarily fine Stein covers. More precisely,

every open cover $\mathfrak{W}$ *of* X *has a refinement* $\mathfrak{U}$ *which is Stein.*

Proof: Without loss of generality we may assume that $\mathfrak{W} = \{W_j\}$, $j \in J$, is already locally finite and that $\bar{W}_j$ is always compact in X. By the Shrinking Theorem we may choose a cover $V = \{V_j\}, j \in J$, of X so that $V_j \Subset W_j$ for all $j \in J$. Now each compact set $\bar{V}_j$ has a finite cover $\{U_{j1}, \ldots, U_{jn}\}$ by Stein open sets where $U_{jk} \subset W_j$. The collection $\{U_{jk}\}$ is therefore a Stein cover of X. □

3. Differences of Complex Spaces. If A is an analytic set which is *everywhere at least 2-codimensional* in a *normal* complex space X, then the second Riemann Continuation Theorem says that every function f which is holomorphic on $X \backslash A$ is continuable to all of X. If $A \neq \emptyset$, then the difference $X \backslash A$ is *not* holomorphically convex. This follows because, given a point $p \in A$ it is impossible to find a function $f \in \mathcal{O}(X \backslash A)$ which is unbounded on sequences approaching p. The following is a more general version of this remark.

Theorem 4. *Let* $(X, \mathcal{O}_X)$ *be an arbitrary complex space and* A *an analytic set in* X *which is at least 2-codimensional at some* $p \in A$. *Then the difference space* $(Y, \mathcal{O}_Y)$, *where* $Y = X \backslash A$ *and* $\mathcal{O}_Y := \mathcal{O}_X | Y$, *is not holomorphically convex and therefore not Stein.*

Proof: It is enough to show that the reduction $(Y, \mathcal{O}_{\text{red } Y})$ is not holomorphically convex. Let $(\tilde{X}, \mathcal{O}_{\tilde{X}})$ be a normalization of $(X, \mathcal{O}_{\text{red } X})$ and $\xi: \tilde{X} \to X$ the *finite* normalization map (see Chapter A.3.8). Then $\tilde{A} := \xi^{-1}(A)$ is at least 2-codimensional in $\tilde{X}$ at every point in $\xi^{-1}(p)$. Thus, given a section $f \in \mathcal{O}_{\text{red } Y}(Y)$, the function $\tilde{f} := f \circ \xi \in \mathcal{O}_{\tilde{X}}(\tilde{X} \backslash \tilde{A})$ is holomorphically continuable across all points of $\xi^{-1}(p)$. Since ξ is finite and $\tilde{f}$ is continuous at all points of $\xi^{-1}(p)$, the finitely many numbers in the set $\tilde{f}(\xi^{-1}(p)) \subset \mathbb{C}$ are the only possible accumulation points for f on sequences approaching p. Thus every $f \in \mathcal{O}_{\text{red } Y}(Y)$ is bounded at p and, by Theorem IV.2.12, Y is therefore not holomorphically convex. □

The above theorem implies that a difference space $Y := X \backslash A$, where $A \neq \emptyset$, can be Stein only if A is everywhere 1-dimensional. Such analytic sets are called *analytic hypersurfaces*. In a reduced space X the zero set of a coherent principal ideal sheaf $\mathcal{J} \neq 0$, where $\mathcal{J}_x \neq \{0\}$ for all $x \in X$, is an analytic hypersurface. (Recall that $\mathcal{J}$ is a principal ideal sheaf when $\mathcal{J}_x$ is a principal ideal in $\mathcal{O}_x$ for all $x \in X$.) In the case of complex manifolds the converse is true. That is, the sheaf of germs of holomorphic functions which vanish on an analytic hypersurface H is a principal ideal sheaf (see Chapter A.3.5). The following is a nice application of Theorem A.

Theorem 5. *Let X be a reduced complex space and H an analytic hypersurface in X so that the sheaf of germs of holomorphic functions which vanish on H is a principal ideal sheaf $\mathcal{J}$. Then, if X is a Stein space, it follows that $Y := X \backslash H$ is also Stein.*

Proof: It suffices, given $p \in H$, to find $f \in \mathcal{O}(Y)$ which is unbounded on sequences approaching p. For this we construct a *meromorphic* function $h \in \mathcal{M}(X)$ which is holomorphic on Y and whose germ $h_p \in \mathcal{M}_p$ satisfies $h_p \cdot a_p = 1$ for some $a_p \in \mathfrak{m}(\mathcal{O}_p)$. It is clear that $f := h | Y$ will do the job.

We begin by defining an $\mathcal{O}$-subsheaf of $\mathcal{M}$: $\mathcal{S} := \bigcup_{x \in X} \mathcal{S}_x$ where $\mathcal{S}_x := \{f_x \in \mathcal{M}_x | f_x \mathcal{J}_x \subset \mathcal{O}_x\}$. For all $x \in Y$ we have $\mathcal{J}_x = \mathcal{O}_x$. Thus $\mathcal{S}_x = \mathcal{O}_x$ for all $x \in Y$. By assumption $\mathcal{J}$ is a principal ideal sheaf. Hence, given a point $p \in H$, there exists a neighborhood $U = U(p)$ so that $\mathcal{J}_U = g\mathcal{O}_U$ for some $g \in \mathcal{O}(U)$. Since $\mathcal{J}_p \neq \mathcal{O}_p$, it follows that $g_p \in \mathfrak{m}(\mathcal{O}_p)$. Now X is reduced and H is a 1-codimensional analytic set which is nowhere dense in X. Thus no germ g_x, $x \in U$, is a zero divisor in $\mathcal{O}_x$. Consequently $g^{-1} \in \mathcal{M}(U)$ and $\mathcal{S}_U = \mathcal{O}_U g^{-1} \cong \mathcal{O}_U$. Hence the sheaf $\mathcal{S} \subset \mathcal{M}$ is locally-free and therefore is coherent.

Since X is Stein, given a point $p \in H$, Theorem A guarantees the existence of a section $h \in \mathcal{S}(X)$ which generates the stalk $\mathcal{S}_p$. In other words, there exists $v_p \in \mathcal{O}_p$ so that $g_p^{-1} = v_p \cdot h_p$. But $\mathcal{S}$ is the same as $\mathcal{O}$ on Y. So $h | Y \in \mathcal{O}(Y)$ and, since $g_p \in \mathfrak{m}(\mathcal{O}_p)$, it follows that $1 = a_p h_p$ where $a_p := v_p g_p \in \mathfrak{m}(\mathcal{O}_p)$. □

It should be remarked that Theorem 5 is also valid for arbitrary complex spaces. However the above proof must be modified. The main point is that, even though the germs g_x can be zero divisors in $\mathcal{O}_x$, they are still "active" elements [CAG].

In the case that X is not a complex manifold, the assumptions on H in

Theorem 5 are quite restrictive. For example, even in a normal complex space X, a hypersurface may not be locally defined as the zero set of a single function. An instructive example for this and other phenomena is the *affine* cone Q^3 defined by the polynomial $q := z_1^2 + z_2^2 + z_3^2 + z_4^2$. The pair

$$(Q^3, \mathcal{O}_{Q^3}), \quad \text{where} \quad Q^3 := \{z \in \mathbb{C}^4 \mid q(z) = 0\} \quad \text{and} \quad \mathcal{O}_{Q^3} := (\mathcal{O}_{\mathbb{C}^4}/q\mathcal{O}_{\mathbb{C}^4}) \mid Q^3,$$

is a 3-dimensional, normal Stein space which has only an isolated singularity at 0. The two equations $z_1 = iz_2$, $z_3 = iz_4$ define a hypersurface H in Q^3 which in every neighborhood of 0 is *not* describable by a single function. The difference space $Q^3\backslash H$ is *not Stein*. The reader can find a detailed discussion of this example and the higher dimensional Segre cones in [20].

Without proof we report the following:

If X is a normal Stein space and H is a hypersurface in X, then $X\backslash H$ is Stein in either of the following situations:

1) $\dim X = 2$
2) *At each of its points H is locally the zero set of a single holomorphic function.*

Case 1) was proved by R. R. Simka (On the complement of a curve on a Stein space of dimension two, Math. Z. 82, 63-66(1963)) using the desingularization of X.

4. The Spaces $\mathbb{C}^2\backslash\{0\}$ and $\mathbb{C}^3\backslash\{0\}$. Let $m \geq 2$ and $d \geq 2$ be natural numbers with $d \leq m$. In $\mathbb{C}^m$ with coordinates $z_1, \ldots, z_m$ we consider the $(m-d)$-dimensional analytic plane

$$E^{m-d} := \{(z_1, \ldots, z_m) \in \mathbb{C}^m \mid z_1 = \cdots z_d = 0\}.$$

Since $d \geq 2$ the space $\mathbb{C}^m\backslash E^{m-d}$ is not Stein (Theorem 4). Every set

$$U_i := \{(z_1, \ldots, z_m) \in \mathbb{C}^m \mid z_i \neq 0\}, \qquad 1 \leq i \leq d,$$

is an *open Stein subspace* of $\mathbb{C}^m\backslash E^{m-d}$. Consequently the d sets $U_1, \ldots, U_d$ form an acyclic cover $\mathfrak{U}$ of $\mathbb{C}^m\backslash E^{m-d}$. Thus, by Theorem 3,

$$H^q(\mathbb{C}^m\backslash E^{m-d}, \mathcal{O}) \cong H^q_a(\mathfrak{U}, \mathcal{O}) \quad \text{for all} \quad q = 0, 1, \ldots.$$

In particular

$$H^q(\mathbb{C}^m\backslash E^{m-d}, \mathcal{O}) = 0 \quad \text{for all} \quad q \geq d.$$

These groups cannot vanish for every $q = 1, \ldots, d-1$, since (e.g. use Theorem 5.2) $\mathbb{C}^m\backslash E^{m-d}$ would be Stein. In fact it can be shown that

$$H^{d-1}(\mathbb{C}^m\backslash E^{m-d}, \mathcal{O}) \neq 0 \quad \text{and} \quad H^q(\mathbb{C}^m\backslash E^{m-d}, \mathcal{O}) = 0 \quad \text{for} \quad g = 1, \ldots, d-2.$$

We want to go through the calculations in two special cases. From now on we let $d = m$ and consequently $E^{m-d} = \{0\}$. We consider the first cohomology group

$$H^1(\mathbb{C}^m \backslash \{0\}, \mathcal{O}) = H^1_a(\mathfrak{U}, \mathcal{O}) = Z^1_a(\mathfrak{U}, \mathcal{O})/B^1_a(\mathfrak{U}, \mathcal{O}), \quad \text{where} \quad \mathfrak{U} = \{U_1, \ldots, U_m\}.$$

a) Let $m = 2$. Then $Z^1_a(\mathfrak{U}, \mathcal{O}) \subset C^1_a(\mathfrak{U}, \mathcal{O}) = \mathcal{O}(D)$, where $D = U_1 \cap U_2$. Since $C^2_a(\mathfrak{U}, \mathcal{O}) = 0$, it follows that $Z^1_a(\mathfrak{U}, \mathcal{O}) = \mathcal{O}(D)$. Furthermore $C^0_a(\mathfrak{U}, \mathcal{O}) = \mathcal{O}(U_1) \oplus \mathcal{O}(U_2)$ and the coboundary map $d_a\colon C^0_a(\mathfrak{U}, \mathcal{O}) \to C^1_a(\mathfrak{U}, \mathcal{O})$ is given by $(f_1, f_2) \mapsto f_2|D - f_1|D$. In this case f_1 and f_2 have *convergent Laurent series* on U_1 and U_2 respectively:

$$f_1 = \sum_{\mu,\nu \in \mathbb{Z}} a_{\mu\nu} z_1^\mu z_2^\nu \quad \text{with} \quad a_{\mu\nu} = 0 \quad \text{for all} \quad \nu < 0,$$

$$f_2 = \sum_{\mu,\nu \in \mathbb{Z}} b_{\mu\nu} z_1^\mu z_2^\nu \quad \text{with} \quad b_{\mu\nu} = 0 \quad \text{for all} \quad \mu < 0.$$

The functions in $\mathcal{O}(D)$ are the convergent Laurent series on $D = \mathbb{C}^2 \backslash \{z_1 z_2 = 0\}$,

$$f = \sum_{\mu,\nu \in \mathbb{Z}} c_{\mu\nu} z_1^\mu z_2^\nu,$$

where the $c_{\mu\nu}$'s are not required to satisfy any extra conditions. Since such a series is in $B^1_a(\mathfrak{U}, \mathcal{O})$ if and only if it can be written in the form $f_2|D - f_1|D$,

$$B^1_a(\mathfrak{U}, \mathcal{O}) = \left\{ f = \sum_{\mu,\nu \in \mathbb{Z}} c_{\mu\nu} z_1^\mu z_2^\nu \in \mathcal{O}(D) \,\middle|\, c_{\mu\nu} = 0 \text{ for all } \mu, \nu \text{ with } \mu < 0, \nu < 0 \right\}.$$

Hence the infinite dimensional $\mathbb{C}$-vector space

$$V := \left\{ h = \sum_{\mu,\nu \in \mathbb{Z}} d_{\mu\nu} z_1^\mu z_2^\nu \in \mathcal{O}(D) \,\middle|\, d_{\mu\nu} \in 0 \text{ whenever } \mu \geq 0 \text{ or } \nu \geq 0 \right\}$$

is complementary to $B^1_a(\mathfrak{U}, \mathcal{O})$ in $\mathcal{O}(D)$. Thus

$$H^1_a(\mathfrak{U}, \mathcal{O}) = \mathcal{O}(D)/B^1_a(\mathfrak{U}, \mathcal{O}) \cong V.$$

In particular we see that

the $\mathbb{C}$-vector space $H^1(\mathbb{C}^2 \backslash \{0\}, \mathcal{O})$ is not finite dimensional.

In the next example we show that the first cohomology groups with coefficients n $\mathcal{O}$ for non-Stein domains in $\mathbb{C}^3$ may vanish.

b) Let $m = 3$. As earlier we write $U_{ij} = U_i \cap U_j$ and $U_{ijk} = U_{ij} \cap U_k$. Then

$$C^0_a(\mathfrak{U}, \mathcal{O}) = \bigoplus_{i=1}^{3} \mathcal{O}(U_i), \; C^1_a(\mathfrak{U}, \mathcal{O}) = \mathcal{O}(U_{12}) \oplus \mathcal{O}(U_{23}) \oplus \mathcal{O}(U_{31}),$$

and

$$C_a^2(\mathfrak{U}, \mathcal{O}) = \mathcal{O}(U_{123}).$$

The map $d_a^1: C_a^1(\mathfrak{U}, \mathcal{O}) \to C_a^2(\mathfrak{U}, \mathcal{O})$ is given by

$$(f_{12}, f_{23}, f_{31}) \to f_{12}|U_{123} + f_{23}|U_{123} + f_{31}|U_{123},$$

where $f_{12} \in \mathcal{O}(U_{12})$, $f_{23} \in \mathcal{O}(U_{23})$, and $f_{31} \in \mathcal{O}(U_{31})$. Thus

$$Z_a^1(\mathfrak{U}, \mathcal{O}) = \{(f_{12}, f_{23}, f_{31}) \in C_1^1(\mathfrak{U}, \mathcal{O}) \mid f_{12} + f_{23} + f_{31} = 0 \text{ on } U_{123}\}.$$

The map $d_a^0: C_a^0(\mathfrak{U}, \mathcal{O}) \to C_a^1(\mathfrak{U}, \mathcal{O})$ is given by

$$(f_1, f_2, f_3) \mapsto (f_{12}, f_{23}, f_{31})$$
$$:= (f_2|U_{12} - f_1|U_{12}, f_3|U_{23} - f_2|U_{23}, f_1|U_{31} - f_3|U_{31}),$$

where $f_i \in \mathcal{O}(U_i)$.

We now derive the vanishing of the first cohomology which was originally shown by Cartan in 1937 [8]:

$$H^1(\mathbb{C}^3\backslash\{0\}, \mathcal{O}) = H_a^1(\mathfrak{U}, \mathcal{O}) = Z_a^1(\mathfrak{U}, \mathcal{O})/\mathrm{Im}\, d_a^0 = 0.$$

Proof: Given a triple $(f_{12}, f_{23}, f_{34}) \in C_a^1(\mathfrak{U}, \mathcal{O})$ with $f_{12} + f_{23} + f_{31} = 0$ on U_{123}, we must find functions $f_i \in \mathcal{O}(U_i)$, $1 \leq i \leq 3$, so that $f_{12} = f_2 - f_1$, $f_{23} = f_3 - f_2$, and $f_{31} = f_1 - f_3$.

We begin by considering the Laurent developments of the f_{ij}'s:

$$f_{12} = \sum a_{\mu\nu\rho} z_1^\mu z_2^\nu z_3^\rho,$$
$$f_{23} = \sum b_{\mu\nu\rho} z_1^\mu z_2^\nu z_3^\rho,$$
$$f_{31} = \sum c_{\mu\nu\rho} z_1^\mu z_2^\nu z_3^\rho.$$

Since $f_{12} \in \mathcal{O}(U_{12})$, every $a_{\mu\nu\rho}$ vanishes for $\rho < 0$. The corresponding coefficients of f_{23} and f_{31} vanish:

$$(*) \quad a_{\mu\nu\rho} = 0 \quad \text{for} \quad \rho < 0, \quad b_{\mu\nu\rho} = 0 \quad \text{for} \quad \mu < 0, \quad \text{and} \quad c_{\mu\nu\rho} = 0 \quad \text{for} \quad \nu < 0.$$

Furthermore $f_{12} + f_{23} + f_{31} = 0$. Thus we always have $a_{\mu\nu\rho} + b_{\mu\nu\rho} + c_{\mu\nu\rho} = 0$ and consequently

$$(**) \qquad a_{\mu\nu\rho} = 0, \quad \text{whenever} \quad \mu < 0 \text{ and } \nu < 0,$$
$$(**) \qquad b_{\mu\nu\rho} = 0, \quad \text{whenever} \quad \nu < 0 \text{ and } \rho < 0,$$
$$c_{\mu\nu\rho} = 0, \quad \text{whenever} \quad \rho < 0 \text{ and } \mu < 0.$$

For every Laurent series $g = \sum g_{\mu\nu\rho} z_1^\nu z_2^\mu z_3^\rho$ we let g^{+++}, g^{+-+}, g^{-+-}, etc. denote the respective subseries of g which correspond to $\mu \geq 0$, $\nu \geq 0$, and $\rho \geq 0$, $\mu \geq 0$, $\nu < 0$, and $\rho \geq 0$, $\mu < 0$, $\nu \geq 0$, and $\rho < 0$, etc. Using this notation, we state equations (∗) and (∗∗) in a convenient way:

$$(\circ)\qquad \begin{aligned} f_{12} &= f_{12}^{+++} + f_{12}^{+-+} + f_{12}^{-++}, \\ f_{23} &= f_{23}^{+++} + f_{23}^{+-+} \qquad\quad + f_{23}^{++-}, \\ f_{31} &= f_{31}^{+++} \qquad\quad + f_{31}^{-++} + f_{31}^{++-}. \end{aligned}$$

Since $f_{12} + f_{23} + f_{31} = 0$, we have the following identities:

$$(\circ\circ)\quad f_{12}^{+++} = -(f_{23}^{+++} + f_{31}^{+++}), \qquad -f_{31}^{-++} = f_{12}^{-++}, \qquad -f_{12}^{+-+} = f_{23}^{+-+},$$
$$-f_{23}^{++-} = f_{31}^{++-}.$$

We now set

$$f_1 := f_{31}^{+++} + f_{31}^{-++}, \qquad f_2 := -f_{23}^{+++} + f_{12}^{+-+}, \qquad f_3 := f_{23}^{++-}.$$

Note that $f_{31}^{+++} \in \mathcal{O}(\mathbb{C}^3)$ and $f_{31}^{-++} \in \mathcal{O}(U_1)$. Thus $f_1 \in \mathcal{O}(U_1)$. Analogously $f_2 \in \mathcal{O}(U_2)$ and $f_3 \in \mathcal{O}(U_3)$. Furthermore (∘) and (∘∘) imply the following:

$$f_2 - f_1 = -(f_{23}^{+++} + f_{31}^{+++}) + f_{12}^{+-+} - f_{31}^{-++} = f_{12}^{+++} + f_{12}^{+-+} + f_{12}^{-++} = f_{12},$$
$$f_3 - f_2 = f_{23}^{++-} + f_{23}^{+++} - f_{12}^{+-+} = f_{23}^{++-} + f_{23}^{+++} + f_{23}^{+-+} = f_{23},$$
$$f_1 - f_3 = f_{31}^{+++} + f_{31}^{-++} - f_{23}^{++-} = f_{31}^{+++} + f_{31}^{-++} + f_{31}^{++-} = f_{31}. \qquad \square$$

Since $H^1(\mathbb{C}^3\setminus\{0\}, \mathcal{O}) = 0$ and, as was remarked above, the groups $H^q(\mathbb{C}^3\setminus\{0\}, \mathcal{O})$ vanish for all $q \geq 3$, Theorem 3.7 implies that

$$H^2(\mathbb{C}^3\setminus\{0\}, \mathcal{O}) \neq 0.$$

In fact this space is even infinite dimensional.

The fact that $H^1(\mathbb{C}^3\setminus\{0\}, \mathcal{O}) = 0$ is a special case of the following:

Let X be a Stein manifold, A an analytic subset of X which is at least 3-codimensional, and $Y := X\setminus A$. Then $H^1(Y, \mathcal{O}) = 0$.

This in turn is contained in the following cohomological generalization of the second Riemann Continuation Theorem (see [33]):

Let X be a complex manifold and A an analytic subset of X which is at least r-codimensional. Then for every locally-free analytic sheaf $\mathcal{S}$ the restriction homomorphisms

$$H^q(X\setminus A, \mathcal{S}) \to H^q(X, \mathcal{S}), \qquad q = 0, 1, \ldots, r-2,$$

are bijective.

5. Classical Examples. For years there was an outstanding problem in function theory which withstood numerous attacks by many distinguished mathematicians using quite strong techniques. This so called Caratheodory conjecture asserted the existence of a non-constant analytic function on any given non-compact Riemann surface. In 1948 some work of H. Behnke and K. Stein which had been completed in 1943, appeared in the Mathematische Annalen. Among other things this contains the following generalization of the classical Runge approximation theorem ([5], p. 445):

Let X be a non-compact Riemann surface and B a domain in X so that $X \setminus B$ has no compact connectivity components. Then every holomorphic function on B is uniformly approximable on compact subsets by functions holomorphic on X.

From this it follows immediately that

every non-compact Riemann surface is Stein.

The proof of the Behnke-Stein approximation theorem rests on a generalized Cauchy integral formula

$$f(z) = \frac{1}{2\pi i} \int f(\zeta) A(\zeta, z)\, d\zeta,$$

where the meromorphic differential form $A(\zeta, z)\, d\zeta$ replaces the classical Cauchy kernel $d\zeta/\zeta - z$. The difficulty of constructing this kernel is overcome by using techniques from the theory of compact Riemann surfaces (e.g. Schottky Verdopplung). In the meantime proofs using methods from real analysis have been found. For example in [ARC] p. 239 there is a proof which is due to Malgrange.

Using the methods and results of their work [5], Behnke and Stein showed in 1948 [6] that the Mittag-Leffler Partial Fraction Theorem and the Weierstrass Product Theorem (i.e. the Cousin Theorems) are valid on non-compact Riemann surfaces. The following lemma appears at the end of their paper:

Hilfssatz C: *Let D be a discrete set in a non-compact Riemann surface X. For every $p \in D$ let z_p be a local coordinate at p. Suppose that at all $p \in D$ there is prescribed a finite Laurent-Taylor series $h_p = \sum_{\nu=-m_p}^{n_p} a_\nu z_p^\nu$, $0 \leq m_p, n_p < \infty$. Then there exists a function H which is meromorphic on X, holomorphic on $X \setminus D$, and whose Laurent development at p with respect to z_p agrees with h_p up to the n_p-th term.*

The reader can find detailed expositions on questions in this general area in the lecture notes of A. Huckleberry (Riemann Surfaces: A View Toward Several Complex Variables, Math. Inst. Münster, WS 1974/1975) as well as the Heidelberger Taschenbuch 184 of O. Forster (Riemann Surfaces, Springer-Verlag, Heidelberg 1977).

A domain B in $\mathbb{C}^m$ is called a *domain of holomorphy* if there is a function $f \in \mathcal{O}(B)$ which is *singular* at every boundary point $p \in \partial B$ (see [EFV], p. 38ff). The classical theorem of Cartan-Thullen says the following:

A domain B in $\mathbb{C}^n$ is a domain of holomorphy if and only if it is Stein.

A reduced complex space X together with an *open* holomorphic map $\varphi: X \to \mathbb{C}^m$ is called a *Riemann domain over* $\mathbb{C}^m$ if every fiber $\varphi^{-1}(\varphi(x))$, $x \in X$, is discrete in X. If φ is in addition a local homeomorphism, then X is said to be *unramified*. Every Riemann domain over $\mathbb{C}^m$ is holomorphically spreadable. Moreover the following holds:

Every Stein Riemann domain X over $\mathbb{C}^m$ is a domain of holomorphy (i.e. there exists $f \in \mathcal{O}(X)$ which is not continuable to any "properly larger" Riemann domain).

This result cannot be turned around. In fact the following was shown in [20]:

There is a 2-sheeted, ramified domain of holomorphy X over $\mathbb{C}^3$ which is not Stein (X can be chosen as manifold).

The question of whether or not there exists such a domain over $\mathbb{C}^2$ appears to be still open.

In his ninth work [32] Oka showed in 1953 that

every unramified domain of holomorphy over $\mathbb{C}^m$ is Stein.

On the other hand the following was shown in [18]:

There exists a non-compact 2-dimensional complex manifold Y with the following properties:

1) *Y is a ramified, finite-sheeted domain of meromorphy over the complex projective plane $\mathbb{P}_2$.*
2) *Every holomorphic function on Y is constant and in particular Y is neither a domain of holomorphy nor holomorphically convex.*

In the proofs of all of these statements the notion of *pseudoconvexity* plays a deciding roll. We can't go any further into this matter here. □

The simplest *non-Stein domains* in $\mathbb{C}^2$ are the *non-complete proper Reinhardt domains* (see [BT], p. 52). The notched bicylinder

$$D^* := \{(z_1, z_2) \in \mathbb{C}^2 \mid |z_1| < 1, |z_2| < 1\} \setminus \{(z_1, z_2) \in \mathbb{C}^2 \mid (|z_1| - 1)^2 + (|z_2| - \tfrac{1}{2})^2 \leq \tfrac{1}{16}\}$$

is an example of such. It is a complex manifold which is homeomorphic to a 4-cell and is holomorphically-separable. However it is not Stein. It is easy to construct a complex manifold X which is homeomorphic to a 4-cell, but on which there are no non-constant holomorphic functions. One just takes the complex projective plane $\mathbb{P}^2$ and bends some complex line in it to a *non-complex analytic differentiable surface* F. Thus the domain $X := \mathbb{P}^2 \setminus F \subset \mathbb{P}^2$ is homeomorphic to $\mathbb{R}^4$. Now every non-constant $h \in \mathcal{O}(X)$ must be singular on F. However a theorem of Hartogs states that if the singularity set of an analytic function is a 2-codimensional topological manifold, then that manifold must be an analytic hypersurface. Thus every function analytic on X must be constant.

6. Stein Groups. A *complex Lie group* G is called a *Stein group* if the underlying complex manifold is Stein. Every *abelian* Lie group is complex analytically isomorphic to a product group $\mathbb{C}^m \times \mathbb{C}^{*n} \times T$, where T is a so called *toroid* group (i.e. $\mathcal{O}(T) = \mathbb{C}$). Complex tori are obviously toroid groups. There are however non-compact toroid groups. The following is obvious:

The Lie group $A \cong \mathbb{C}^m \times \mathbb{C}^{*n} \times T$ is holomorphically convex (resp. Stein) if and only if the toroid part T is compact (resp. 0).

Every simply-connected, connected, *solvable* complex Lie group is isomorphic as a complex manifold to some $\mathbb{C}^m$ and is consequently Stein. All of the linear groups $GL(m, \mathbb{C})$ are Stein. Every *semi-simple*, connected, complex Lie group is complex analytically isomorphic to a *closed* complex subgroup of some $GL(m, \mathbb{C})$ and is therefore Stein.

The Stein groups were intensively studied by Y. Matsushima [26, 27]. In particular he gave a Lie algebraic characterization of such groups. For a complex Lie group G we denote by $Z^0(G)$ the connected component of the identity $e \in G$ of the center of G.

Theorem. *The following statements about complex Lie groups are equivalent:*

i) *G is holomorphically convex (resp. Stein).*
ii) *The toroid part of $Z^0(G)$ is compact, equal to a compact torus (resp. 0, i.e. $Z^0(G) \cong \mathbb{C}^m \times \mathbb{C}^{*n}$).*

One sees in particular that G is Stein if and only if it is holomorphically spreadable.

The following can also be shown:

If G is connected and Stein, then the underlying complex manifold of G is affine algebraic.

It should be noted that the realization of G as an affine algebraic variety may not be related to any algebraic group structure on G.

§ 2. The Cousin Problems and the Poincaré Problem

The Cousin problems and the Poincaré problem are classical problems from 19th century complex analysis. They had a tremendous impact on the development of several complex variables. The reader can find the sheaf theoretical formulation and solution of these problems in the work [9], [35] which was published in 1953. These readings are emphatically recommended.

1. The Cousin I Problem. For every complex space X, the *$\mathcal{O}$-sheaf of germs of meromorphic functions on X* is defined as the *sheaf of fractions*

$$\mathcal{M} := \mathcal{O} \quad \text{where} \quad \mathcal{M}_x = (\mathcal{O}_x)_{N_x}, \qquad x \in X,$$

and where N is the multiplicative set of elements of $\mathcal{O}$ which are not zero-divisors (see Chapter A.4.5). It should be noted that $\mathcal{M}$ is *not* a coherent $\mathcal{O}$-sheaf.

The structure sheaf $\mathcal{O}$ is an $\mathcal{O}$-subsheaf of $\mathcal{M}$. The quotient sheaf

$$\mathcal{H} := \mathcal{M}/\mathcal{O}$$

is called the *sheaf of germs of principal parts on X*. An element of $\mathcal{H}(X)$ is called a *principal part distribution* on X. Associated to the short exact sequence $0 \to \mathcal{O} \to \mathcal{M} \to \mathcal{H} \to 0$ we have the long exact cohomology sequence

$$0 \longrightarrow \mathcal{O}(X) \longrightarrow \mathcal{M}(X) \xrightarrow{\varphi} \mathcal{H}(X) \xrightarrow{\zeta} H^1(X, \mathcal{O}) \longrightarrow H^1(X, \mathcal{M}) \quad \cdots.$$

In particular, for any meromorphic function $h \in \mathcal{M}(X)$ we have its principal part distribution $\varphi(h) \in \mathcal{H}(X)$. For any principal part distribution $s \in \mathcal{H}(X)$ there is a cover $\mathfrak{U} = \{U_i\}$ of X and meromorphic functions $h_i \in \mathcal{M}(U_i)$ with $\varphi(h_i) = s \mid U_i$. On U_{ij} we have $g_{ij} := h_j - h_i \in \mathcal{O}(U_{ij})$ and the family (g_{ij}) is an (alternating) 1-cocycle in $Z^1_a(\mathfrak{U}, \mathcal{O})$ which represents the cohomology class $\zeta(s) \in H^1(X, \mathcal{O})$.

Every family $\{U_i, h_i\}$, $h_i \in \mathcal{M}(U_i)$, with $h_j - h_i \in \mathcal{O}(U_{ij})$ determines a principal part distribution $s \in \mathcal{H}(X)$. One calls $\{U_i, h_i\}$ an s-representing *Cousin* I *distribution.* A meromorphic function $h \in \mathcal{M}(X)$ satisfying $\varphi(h) = s$ is one such that $h - h_i$ is holomorphic on U_i for all i (see Section 2 of the introduction to this book).

The classical *Cousin* I *problem* (also called the *additive Cousin problem*) amounts to asking for a characterization of the principal part distributions which belong to meromorphic functions. The long exact cohomology sequence gives an immediate answer:

A principal part distribution $s \in \mathcal{H}(X)$ is the principal part distribution of a meromorphic function $h \in \mathcal{M}(X)$ (i.e. $s = \varphi(h)$) if and only if $\zeta(s) = 0 \in H^1(X, \mathcal{O})$.

One says that the first Cousin problem is *universally solvable* on X whenever φ is surjective. Since Im φ = Ker ζ, this is the case if and only if $\zeta\colon \mathcal{H}(X) \to H^1(X, \mathcal{O})$ is the zero map, and this happens exactly when $H^1(X, \mathcal{O}) \to H^1(X, \mathcal{M})$ is injective. We summarize these facts as follows:

Theorem 1. *The Cousin* I *problem on a given complex space X is universally solvable if and only if the natural homomorphism $H^1(X, \mathcal{O}) \to H^1(X, \mathcal{M})$ is injective. In particular the Cousin* I *problem is universally solvable for all spaces X with $H^1(X, \mathcal{O}) = 0$.*

Theorem B implies that the additive Cousin problem is universally solvable for all Stein spaces. Oka [30] first proved this for domains of holomorphy in $\mathbb{C}^m$. In the case of non-compact Riemann surfaces this is just the Mittag-Leffler Theorem (see [6]).

The sufficient condition $H^1(X, \mathcal{O}) = 0$ for universal solvability of the Cousin I problem is also satisfied on all compact, Kähler manifolds X whose first Betti number is zero. Thus for example the first Cousin problem is universally solvable for all projective rational manifolds, in particular for all complex projective spaces $\mathbb{P}^m$.

In Section 1.4 we showed that $H^1(\mathbb{C}^3\setminus\{0\}, \mathcal{O}) = 0$. Thus in this *non*-Stein case the Cousin I problem is universally solvable. On the other hand there are principal part distributions on $\mathbb{C}^2\setminus\{0\}$, where $H^1(\mathbb{C}^2\setminus\{0\}, \mathcal{O}) \neq 0$, which do not belong to a meromorphic function. More generally in the 2-dimensional case we have the following (compare to Theorem 5.2):

The first Cousin problem is universally solvable for a domain B in $\mathbb{C}^2$ if and only if B is Stein.

With this we see that the sufficient condition $H^1(B, \mathcal{O}) = 0$ (B a domain in $\mathbb{C}^2$) is also necessary for the universal solvability of the Cousin I problem.

2. The Cousin II Problem. We let $\mathcal{O}_x^*$ and $\mathcal{M}_x^*$ denote the groups of units in the rings $\mathcal{O}_x$ and $\mathcal{M}_x$ respectively. The sets

$$\mathcal{O}^* := \bigcup_{x \in X} \mathcal{O}_x^* \quad \text{and} \quad \mathcal{M}^* := \bigcup_{x \in X} \mathcal{M}_x^*$$

are open in $\mathcal{O}$ and $\mathcal{M}$. They are subsheaves of abelian groups with respect to multiplication and $\mathcal{O}^*$ is a subsheaf of $\mathcal{M}^*$. The sections in $\mathcal{O}^*(X)$ are just the nowhere vanishing analytic functions on X. The quotient sheaf

$$\mathcal{D} := \mathcal{M}^*/\mathcal{O}^*$$

is called the *sheaf of germs of divisors on* X, the sections in $\mathcal{D}$ being called *divisors*. We write the operation in the group of divisors $\mathcal{D}(X)$ *additively*.

Associated to the short exact sequence $1 \to \mathcal{O}^* \to \mathcal{M}^* \to \mathcal{D} \to 0$, we have the long exact cohomology sequence:

$$1 \longrightarrow \mathcal{O}^*(X) \longrightarrow \mathcal{M}^*(X) \xrightarrow{\psi} \mathcal{D}(X) \xrightarrow{\eta} H^1(X, \mathcal{O}^*) \longrightarrow H^1(X, \mathcal{M}^*) \longrightarrow \cdots.$$

Thus for any $h \in \mathcal{M}^*(X)$ we have its divisor $\psi(h) \in \mathcal{D}(X)$. Divisors $\psi(h)$ of meromorphic functions are called *principal divisors*. They are denoted in the classical way by (h). It follows that

$$(gh) = (g) + (h) \quad \textit{for all} \quad g, h \in \mathcal{M}^*(X).$$

For every divisor $D \in \mathcal{D}(X)$ there exists a cover $\mathfrak{U} = \{U_i\}$ of X and meromorphic functions $h_i \in \mathcal{M}^*(U_i)$ with $\psi(h_i) = D \mid U_i$. On U_{ij} we have $g_{ij} := h_j h_i^{-1} \in \mathcal{O}(U_{ij})$, where the family (g_{ij}) is an (alternating) 1-cocycle in $Z_a^1(\mathfrak{U}, \mathcal{O}^*)$ which represents the cohomology class $\eta(D) \in H^1(X, \mathcal{O}^*)$.

Every family $\{U_i, h_i\}$, $h_i \in \mathcal{M}^*(U_i)$, with $h_j h_i^{-1} \in \mathcal{O}^*(U_{ij})$ determines a divisor

$D \in \mathscr{D}(X)$. One calls $\{U_i, h_i\}$ a D-representing *Cousin* II *distribution*. A meromorphic function $h \in \mathscr{M}^*(X)$ has divisor D (i.e. $\psi(h) = D$) if and only if $h_i h^{-1} \in \mathcal{O}^*(U_i)$ for all i.

A divisor D is called *positive*, denoted by $D \geq 0$, whenever there is a D-representing *Cousin* II *distribution* $\{U_i, h_i\}$ where $h_i \in \mathcal{O}(U_i)$ for all i. It follows that *a meromorphic function* $h \in \mathscr{M}^*(X)$ *is holomorphic if and only if* $(h) \geq 0$.

The classical *Cousin* II *problem*, which is also called *the multiplicative Cousin problem*, asks for a characterization of the principal divisors in $\mathscr{D}(X)$. The exact cohomology sequence immediately gives an answer:

A divisor $D \in \mathscr{D}(X)$ *is the divisor of a meromorphic function* $h \in \mathscr{M}^*(X)$ *if and only if* $\eta(D) = 0 \in H^1(X, \mathcal{O}^*)$.

One says that the Cousin II problem is *universally solvable* on X whenever ψ is surjective. The following is analogous to the situation with the Cousin I problem:

Theorem 2. *The Cousin* II *problem is universally solvable for a complex space* X *if and only if the natural homomorphism* $H^1(X, \mathcal{O}^*) \to H^1(X, \mathscr{M}^*)$ *is injective.*

In particular the Cousin II *problem is universally solvable for all spaces* X *with* $H^1(X, \mathcal{O}^*) = 0$.

In Section 4 we will look more closely at the group $H^1(X, \mathcal{O}^*)$ and will show in particular that it vanishes if $H^1(X, \mathcal{O}) = H^2(X, \mathbb{Z}) = 0$. □

We want now to give the divisor group $\mathscr{D}(X)$ a more geometric interpretation. For this let X be a *reduced* space which is *irreducible at every point* $x \in X$ (i.e. *locally irreducible*). Then every stalk $\mathscr{M}_x$ is the quotient field of the integral domain $\mathcal{O}_x$ and $\mathscr{M}_x^* = \mathscr{M}_x \setminus \{0\}$. A germ in $\mathcal{O}_x^*$ has a nowhere vanishing representation in some neighborhood of x. Every divisor D on X is therefore locally represented by a function f/g where $f, g \neq 0$ are uniquely determined up to *nowhere vanishing* holomorphic functions. The functions f and g determine well-defined hypersurfaces (i.e. $\{f = 0\}$, $\{g = 0\}$ which may be empty) of positive order. Counting the order of $\{g = 0\}$ negatively, one can therefore view every divisor $D \in \mathscr{D}(X)$ as an analytic hypersurface H in X where (at most countably many) irreducible components are counted with integral multiplicities. The family $\{H_i\}$ must be *locally-finite* (i.e. every relatively compact open set $U \subset X$ intersects only finitely many of the H_i's). If one additionally assumes that *every hypersurface in* X *is locally the first order zero set of some analytic function* (this is always true for manifolds), then the divisor group is canonically isomorphic to the *additive* group of all (even infinite) linear combinations $\sum_1^\infty n_i H_i$, $n_i \in \mathbb{Z}$, $n_i \neq 0$, where $\{H_i\}$ is any *locally-finite* family of irreducible analytic hypersurfaces in X with $H_i \neq H_j$ for $i \neq j$. We call the H_i's the *prime components* of D.

3. Poincaré Problem. In his work "Sur les fonctions de deux variables" Act. Math. **2**, p. 97–113, published in 1883, Poincaré had already shown that every meromorphic function on $\mathbb{C}^2$ is the ratio of two functions which are holomorphic on $\mathbb{C}^2$. Thus the field $\mathcal{M}(\mathbb{C}^2)$ is the quotient field of the ring $\mathcal{O}(\mathbb{C}^2)$. If X is a complex space where $\mathcal{M}(X)$ is the quotient field of $\mathcal{O}(X)$ with respect to the elements which are not zero divisors, then one says that Poincaré's Theorem holds.

To keep matters simple, we consider here only complex manifolds X. From Paragraph 2 we see that every divisor $D \in \mathcal{D}(X)$ is uniquely representable as a linear combination $\sum_1^\infty n_i H_i$ of its prime components. For every non-empty open set U in X one obtains the restriction $D|U \in \mathcal{D}(U)$ as follows: If $H_i \cap U = \sum_{j=1}^\infty H_{ij}$ is *the* decomposition of $H_i \cap U$ into irreducible components in U, then $D|U = \sum_{i,j=1}^\infty n_{ij} H_{ij}$ where $n_{ij} := n_i$. If $D, D' \in \mathcal{D}(X)$ and $D|U$, $D'|U$ have a common prime component, then, by the identity theorem for analytic sets (see Chapter A.3.5), D and D' have a common prime component.

The divisor $D = \sum_1^\infty n_i H_i$ is positive (i.e. $D \geq 0$) if and only if $n_i \geq 0$ for all i. Every divisor is uniquely representable as the difference of two positive divisors which have no common prime components:

$$D = D^+ - D^- \quad \text{with} \quad D^+ := \sum_{n_i \geq 0} n_i H_i \quad \text{and} \quad D^- := -\sum_{n_i < 0} n_i H_i.$$

After the above preparations the following is easy to prove.

Theorem 3. *Let X be a complex manifold on which the Cousin* II *problem is universally solvable. Then the following sharp form of the Poincaré Theorem holds on X:*

Every meromorphic function $h \in \mathcal{M}(X)$, $h \not\equiv 0$, is the quotient f/g of two holomorphic functions $f, g \in \mathcal{O}(X)$ whose germs $f_x, g_x \in \mathcal{O}_x$ in the (unique factorization) ring $\mathcal{O}_x$ are relatively prime for all $x \in X$.

Proof: We may assume that X is connected. Then $h \in \mathcal{M}^*(X)$ and $(h) \in \mathcal{D}(X)$ is well-defined. Let $(h) = D^+ - D^-$ with $D^+, D^- \geq 0$. By assumption there exists $g \in \mathcal{M}^*(X)$ with $(g) = D^- \geq 0$. Thus $g \in \mathcal{O}(X)$. Furthermore, for $f := gh \in \mathcal{M}^*(X)$, it follows that $(f) = (g) + (h) = D^+ \geq 0$ and f is likewise in $\mathcal{O}(X)$.

Now suppose there is a point $x_0 \in X$ where f_{x_0} and g_{x_0} have a non-unit common divisor. Then there exists a neighborhood U of x_0 with functions $\tilde{f}$, $\tilde{g}$, $p \in \mathcal{O}(U)$ so that

$$f_U = p\tilde{f}, \qquad g_U = p\tilde{g}$$

and

$$D^+ \mid U = (f_U) = (p) + (\tilde{f}), \qquad D^- \mid U = (g_U) = (p) + (\tilde{g}),$$

where $(p) \in \mathscr{D}(U)$ is positive and not the zero divisor. But this implies that $D^+ \mid U$ and $D^- \mid U$ (consequently D^+ and D^-) have a common prime component. Since this is not the case, we have the desired contradiction. □

In the next section we will see that the second Cousin problem is universally solvable on every Stein manifold X with $H^2(X, \mathbb{Z}) = 0$. From this and Theorem 3 we have the following: *The sharp form of Poincaré's Theorem holds for every Stein manifold X with $H^2(X, \mathbb{Z}) = 0$.*

The following is a consequence of Theorem A:

Theorem 4. *The Theorem of Poincaré holds for every Stein manifold X.*

Proof: Again let X be connected and $h \in \mathscr{M}^*(X)$. The $\mathcal{O}$-sheaves $\mathcal{O}$, $\mathcal{O}h$, and $\mathcal{O} + \mathcal{O}h$ are coherent subsheaves of $\mathscr{M}$.[1] Thus $\mathcal{O} \cap \mathcal{O}h$ is a coherent $\mathcal{O}$-sheaf (see Chapter A.2.3c). Since $\mathscr{M}_x$ is the quotient field of $\mathcal{O}_x$, it follows that $(\mathcal{O} \cap \mathcal{O}h)_x \neq 0$ for all $x \in X$. The $\mathcal{O}$-epimorphism $\varphi\colon \mathcal{O} \to \mathcal{O}h$, $f_x \to f_x h_x$, $x \in X$, determines a coherent ideal $\mathscr{I} := \varphi^{-1}(\mathcal{O} \cap \mathcal{O}h)$ with $\mathscr{I}_x \neq 0$ $x \in X$.

By Theorem A there is a global section $g \neq 0$ in $\mathscr{I}$ over X. For every such section we know that $g_x \neq 0$ for all $x \in X$ and $f := gh \in \mathcal{O}(X)$. □

The Poincaré problem has had tremendous influence on development of several complex variables. In order to solve the Poincaré question, Cousin, in his 1895 work "Sur les fonction de n variables complexes" Act. Math. **19,** 1–62, formulated the two problems which are named after him, and solved them in important special cases (e.g. product domains $B_1 \times \cdots \times B_m$ in $\mathbb{C}^m$). Even in the case of product domains, as Gronwall remarked in his 1917 work "On the expressibility of a uniform function of several complex variables as a quotient of two functions of entire character," Trans. Amer. Math. Soc. **18,** 50–64, it is necessary to make the additional assumption that with at most one exception all of the B_μ's must be simply connected (i.e. $H^2(X, \mathbb{Z}) = 0$, see the next section). In fact Gronwall gave the product domain $\mathbb{C}^* \times \mathbb{C}^* \subset \mathbb{C}^2$ as an example of a Stein manifold for which the Cousin II problem is *not universally solvable* and for which *Poincaré's Theorem in its sharp form does not hold.*

---

[1] In general *if* $h_1, \ldots, h_t \in \mathscr{M}(X)$, then $\mathscr{S} := \mathcal{O}h_1 + \cdots + \mathcal{O}h_t \subset M$ *is coherent.* To see this note that every point $x \in X$ has a neighborhood U so that $h_i \mid U = p_i/q$ with $p_i, q \in \mathcal{O}(U)$ and $q_u \neq 0$ for all $u \in U$. Multiplication by the common denominator q yields an $\mathcal{O}_U$-monomorphism $\sigma\colon \mathscr{S}_U \to \mathscr{S}_U$ and obviously

$$\operatorname{Im} \sigma = \mathcal{O}_U p_i + \cdots + \mathcal{O}_U p_t \subset \mathcal{O}_U.$$

Thus $\operatorname{Im} \sigma$ (and consequently $\mathscr{S}_U$) is coherent.

In the following table we summarize the 8 combinations of solvability/unsolvability of our 3 problems and whether or not each can happen (see [3], p. 192):

	Cousin I	Cousin II	Poincaré	Possible
1	+	+	+	+
2	+	+	−	−
3	+	−	+	+
4	+	−	−	+
5	−	+	+	+
6	−	+	−	−
7	−	−	+	+
8	−	−	−	+

Except for 4), all examples are realizable using domains in $\mathbb{C}^2$. Case 4) is ruled out in $\mathbb{C}^2$ by Theorem 1.4. As examples, case 5) is demonstrated by the domain

$$G := \{(z_1, z_2) \in \mathbb{C}^2 \mid 0 < |z_1| < 1,\ |z_2| < 1\} \cup \{(z_1, z_2) \in \mathbb{C}^2 \mid z_1 = 0,\ |z_2| < \tfrac{1}{2}\}$$

and case 7) is shown by the notched bicylinder

$$D^* := \{(z_1, z_2) \in \mathbb{C}^2 \mid |z_1| < 1,\ |z_2| < 1\} \setminus \{(z_1, z_2) \in \mathbb{C}^2 \mid (|z_1| - 1)^2 + (|z_2| - \tfrac{1}{2})^2 < \tfrac{1}{16}\}.$$

4. The Exact Exponential Sequence $0 \to \mathbb{Z} \to \mathcal{O} \to \mathcal{O}^* \to 0$. In this section X will again denote an arbitrary complex space. The cohomology group $H^1(X, \mathcal{O}^*)$, which by Theorem 2 plays the deciding role for determining the solvability of the Cousin II problem, is far more difficult to actually compute than is $H^1(X, \mathcal{O})$. We will see that $H^1(X, \mathcal{O}^*)$ contains *topological* information about X. The main tool for studying this group is the classical exponential map, the relevant properties of which are contained in the following:

Lemma. *For every complex space X there is an exponential map $\vartheta: \mathcal{O} \to \mathcal{O}^*$. The sequence*

$$0 \to \mathbb{Z} \to \mathcal{O} \xrightarrow{\vartheta} \mathcal{O}^* \to 1,$$

where $\mathbb{Z}$ denotes the constant sheaf of integers.

Proof: For every open set $U \supset X$ the algebra $\mathcal{O}(U)$ is Fréchet (see Section 4). For every $f \in \mathcal{O}(U)$ the series $\sum_0^\infty (f^\nu/\nu!)$ converges to an element $\exp f \in \mathcal{O}(U)$. It is easy to see that $\exp(f+g) = \exp f \cdot \exp g$ and in particular $\exp f \in \mathcal{O}^*(U)$. Thus the map

$$\vartheta_U\colon \mathcal{O}(U) \to \mathcal{O}^*(U), \qquad f \to \exp 2\pi i f,$$

is a homomorphism. The family $\{\vartheta_U\}$ determines a sheaf homomorphism $\vartheta\colon \mathcal{O} \to \mathcal{O}^*$.

Obviously $\ker \vartheta \supset \mathbb{Z}$. In order to show equality it is enough to show that every $f \in \mathfrak{m}_x$ with $\exp f = 1$ is the zero germ. As a consequence of $\exp f = 1$ we have $f = -\sum_{\nu=2}^\infty \frac{f^\nu}{\nu!} = f^2 \cdot g$ where $g := (1/2!) + (f/3!) + \cdots \in \mathcal{O}_x$. Thus $f = f^2 g = f^3 g^2 = \cdots = f^{j+1} \cdot g^j = \cdots$ and consequently $f \in \bigcap_{j=1}^\infty \mathfrak{m}_x^j = \{0\}$. This completes the verification that $\ker \vartheta = \mathbb{Z}$.

Every unit $e := 1 + f$, $f \in \mathfrak{m}_x$, has a logarithm

$$h := \log(1+f) := \sum_{\nu=1}^\infty \frac{(-1)^{\nu+1}}{\nu} f^\nu,$$

with $\exp h = e$. Thus ϑ is surjective. □

We will now exploit the exact cohomology sequence associated to the exponential sequence. One first observes that if $H^1(X, \mathbb{Z}) = 0$, then the mapping $\mathcal{O}(X) \to \mathcal{O}^*(X)$ induced from ϑ is surjective. This is just the classical theorem which states that on a (cohomologically) simply-connected manifold every non-vanishing holomorphic function has a logarithm. More important for us is the following piece of the exact cohomology sequence:

$$\cdots \longrightarrow H^1(X, \mathcal{O}) \longrightarrow H^1(X, \mathcal{O}^*) \xrightarrow{\delta} H^2(X, \mathbb{Z}) \longrightarrow H^2(X, \mathcal{O}) \quad \cdots$$

We see immediately that if $H^1(X, \mathcal{O}) = 0$, then $H^1(X, \mathcal{O}^*)$ is isomorphic to a subgroup of $H^2(X, \mathbb{Z})$. Hence the following is an immediate consequence of Theorem 2.

Theorem 5. *If X is a complex space with $H^1(X, \mathcal{O}) = 0$ and $H^1(X, \mathbb{Z}) = 0$, then the Cousin* II *problem is universally solvable on X.*

If X is a non-compact Riemann surface then $H^2(X, \mathbb{Z}) = 0$. Consequently the Cousin II problem is universally solvable on every non-compact Riemann surface (this is the generalized Weierstrass product theorem, see [6]).

It is already clear here that $H^2(X, \mathbb{Z})$ contains important information about the solvability of Cousin II problems. Even in the 1930's it was still believed that the fundamental group $\pi_1(X)$ would play a great role in these considerations (based for example on Gronwall's work). For example it is stated in [BT] p. 102 that it is completely open whether or not the Cousin I and Cousin II problems are always solvable for simply connected domains of holomorphy. In 1953 [35] Serre first gave an explicit example of such a domain where the Cousin II problem is not universally solvable. His example is

$$G := \{z \in \mathbb{C}^3 \mid |z_1^2 + z_2^2 + z_3^2 - 1| < 1\} \subset \mathbb{C}^3.$$

Clearly G is analytically isomorphic to the product of the unit disk and the affine quadric $Q := \{z \in \mathbb{C}^3 \mid z_1^2 + z_2^2 + z_3^2 = 1\}$. Since Q is retractible to the 2-sphere S^2, it follows that $\pi_1(G) = 0$ and $H^2(G, \mathbb{Z}) = \mathbb{Z}$. The Cousin II problem is not universally solvable in G. In fact $G \cap \{z_1 = iz_2\}$ has two disjoint components neither of which can be the divisor of a meromorphic function.

It is easy to generalize Theorem 5. For this purpose we compose the homomorphism δ in the above cohomology sequence with the homomorphism $\mathscr{D}(X) \xrightarrow{\eta} H^1(X, \mathcal{O}^*)$ from the cohomology sequence associated to $1 \to \mathcal{O}^* \to \mathscr{M}^* \to \mathscr{D} \to 0$. We thus obtain a group homomorphism

$$c: \mathscr{D}(X) \xrightarrow{\eta} H^1(X, \mathcal{O}^*) \xrightarrow{\delta} H^2(X, \mathbb{Z}), \qquad D \longmapsto \delta(\eta(D)),$$

which associates to every divisor D a 2-dimensional integral cohomology class $c(D) \in H^2(X, \mathbb{Z})$, the so-called *characteristic class* (*Chern class*) of D.

If X is an m-dimensional complex manifold, then $c(D)$ is dual to the $(2m-2)$-dimensional integral homology class which is determined by D (i.e. the hypersurface along with multiplicities). The following is just a consequence of the basic definitions:

Theorem 6. *If D is a principal divisor on X, then its Chern class $c(D)$ vanishes. If $H^1(X, \mathcal{O}) = 0$, then every divisor D with $c(D) = 0$ is principal.*

Proof: By Theorem 2 a divisor D is principal whenever $\eta(D) = 0$. Thus principal divisors satisfy $c(D) = \delta(\eta(D)) = 0$. If $H^1(X, \mathcal{O}) = 0$, then δ is injective. Consequently $c(D) = 0$ if and only if $\eta(D) = 0$. □

In Section 3 (Theorem 3) we will see that if X is an irreducible, reduced Stein space, then every cohomology class in $H^2(X, \mathbb{Z})$ is in fact the Chern class of a divisor. This implies the following:

If X is an irreducible, reduced Stein space, then the Cousin II *problem is universally solvable if and only if $H^2(X, \mathbb{Z}) = 0$.*

5. Oka's Principle. In the case of reduced complex spaces the structure sheaf $\mathcal{O}$ is a subsheaf of the sheaf $\mathscr{C}$ of germs of continuous functions. One defines $\mathscr{C}^*$ to be the sheaf of germs of nowhere vanishing continuous functions. Obviously it fol lows that $\mathcal{O}^* \subset \mathscr{C}^*$. Again we have an exact exponential sequence as well as the commutative diagram

$$\begin{array}{ccccccccc} 0 & \longrightarrow & \mathbb{Z} & \longrightarrow & \mathcal{O} & \longrightarrow & \mathcal{O}^* & \longrightarrow & 0 \\ & & \| & & \downarrow & & \downarrow & & \\ 0 & \longrightarrow & \mathbb{Z} & \longrightarrow & \mathscr{C} & \longrightarrow & \mathscr{C}^* & \longrightarrow & 0. \end{array}$$

Associated to this we have the following commutative diagram of exact cohomology sequences:

$$\begin{array}{ccccccccc} \cdots \longrightarrow & H^q(X, \mathcal{O}) & \longrightarrow & H^q(X, \mathcal{O}^*) & \longrightarrow & H^{q+1}(X, \mathbb{Z}) & \longrightarrow & H^{q+1}(X, \mathcal{O}) & \longrightarrow \cdots \\ & \downarrow & & \downarrow & & \| & & \downarrow & \\ \cdots \longrightarrow & H^q(X, \mathscr{C}) & \longrightarrow & H^q(X, \mathscr{C}^*) & \longrightarrow & H^{q+1}(X, \mathbb{Z}) & \longrightarrow & H^{q+1}(X, \mathscr{C}) & \longrightarrow \cdots \end{array}$$

Since the sheaf $\mathscr{C}$ is soft (Chapter A.4.2), it follows that $H^i(X, \mathscr{C}) = 0$ for all $i \geq 1$ (Chapter B.1.2). Thus we have the following:

Let X be a reduced complex space such that for some $q \geq 1$ $H^q(X, \mathcal{O}) = H^{q+1}(X, \mathcal{O}) = 0$ (e.g. X a Stein space). Then the injection $\mathcal{O}^* \to \mathscr{C}^*$ induces an isomorphism $H^q(X, \mathcal{O}^*) \simeq H^q(X, \mathscr{C}^*)$.

This is a rudimentary form of the important "Oka Principle" which can be vaguely stated as follows:

On a reduced Stein space X, problems which can be cohomologically formulated have only topological obstructions. In other words, such problems are holomorphically solvable if and only if they are continuously solvable.

Oka's famous theorem illustrates this principle:

Theorem of Oka ([31], 1939). *A Cousin II Distribution $\{U_i, h_i\}$ on a reduced Stein space X has a holomorphic solution if and only if it has a continuous solution.*

Proof: The obstruction to solving the problem is the cocycle $(g_{ij}) \in Z^1_a(\mathfrak{U}, \mathcal{O}^*)$, where $\mathfrak{U} = \{U_i\}$ and such that $g_{ij} := h_j h_i^{-1} \in \mathcal{O}^*(U_{ij})$ represents the cohomology class of the associated divisor D. If there is a *continuous solution* $s \in \mathscr{C}(X)$, then $s_i := h_i(s \mid U_i)^{-1} \in \mathscr{C}^*(U_i)$ and since $g_{ij} = s_j s_i^{-1}$, the cocycle (g_{ij}) is cohomologous to zero in $Z^1_a(\mathfrak{U}, \mathscr{C}^*)$. But $H^1(X, \mathcal{O}^*)$ and $H^1(X, \mathscr{C}^*)$ are canonically isomorphic. Thus $\eta(D) = 0$ and a holomorphic solution also exists. □

The literature on the Oka principle is voluminous. For a more detailed account we refer the reader to the report of O. Forster [15] and the list of references contained in it.

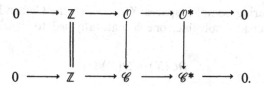

§ 3. Divisor Classes and Locally Free Analytic Sheaves of Rank 1

When considering complex spaces X on which the Cousin II problem is not necessarily universally solvable, one is naturally led to consider the quotient group

$$DC(X) := \mathscr{D}(X)/\mathrm{Im}\ \psi$$

of the group of divisors by the subgroup of principal divisors. One calls $DC(X)$ the *group of divisor classes* of X and the elements are called divisor classes.

The groups $DC(X)$ and $H^1(X, \mathcal{O}^*)$ turn out to be isomorphic to certain groups of isomorphy classes of analytic, locally-free sheaves of rank 1 over X. These identifications permit us to derive non-trivial statements about such sheaves on one hand and about the group of divisor classes on the other.

1. Divisors and Locally Free Sheaves of Rank 1. Given a divisor $D \in \mathscr{D}(X)$ we can associate to it an $\mathcal{O}$-subsheaf $\mathcal{O}(D)$ of the sheaf of germs of meromorphic functions: If $\{U_i, h_i\}$ is a Cousin II distribution for D on X, then $h_j h_i^{-1} \in \mathcal{O}^*(U_{ij})$ and consequently $(\mathcal{O}|U_{ij})h_i^{-1}|U_{ij} = (\mathcal{O}|U_{ij}) \cdot h_j^{-1}|U_{ij}$. Thus we have an $\mathcal{O}$-subsheaf $\mathcal{O}(D)$ of M with $\mathcal{O}(D)|U_i = (\mathcal{O}|U_i)h_i^{-1}$. Since $h_i \in \mathscr{M}^*(U_i)$, it follows that $(\mathcal{O}|U_i)h_i^{-1} \cong \mathcal{O}|U_i$. Thus $\mathcal{O}(D)$ is *locally-free* of rank 1. Obviously $\mathcal{O}(D)$ does not depend on the choice of the Cousin II distribution for D.

The collection $G(\mathscr{M})$ of all $\mathcal{O}$-subsheaves of $\mathscr{M}$ which are locally-free of rank 1 on X is a *semi-group*: For $\mathscr{L}, \mathscr{L}' \in G(\mathscr{M})$ it follows that $\mathscr{L} \cdot \mathscr{L}' \in G(\mathscr{M})$, where $\mathscr{L} \cdot \mathscr{L}'$ is the *product sheaf* with stalks $(\mathscr{L} \cdot \mathscr{L}')_x := \mathscr{L}_x \cdot \mathscr{L}'_x$. The sheaf $\mathscr{L} \cdot \mathscr{L}'$ is isomorphic to the tensor product $\mathscr{L} \otimes_{\mathcal{O}} \mathscr{L}'$.

For every $\mathscr{L} \in G(\mathscr{M})$ we have a covering $\{U_i\}$ of X so that $\mathscr{L}|U_i = (\mathcal{O}|U_i)f_i$ with $f_i \in \mathscr{M}^*(U_i)$. Since $(\mathcal{O}|U_{ij})(f_i|U_{ij}) = \mathscr{L}|U_{ij} = (\mathcal{O}|U_{ij})(f_j|U_{ij})$, it follows that $f_i f_j^{-1} \in \mathcal{O}^*(U_{ij})$. Thus $\{U_i, f_i^{-1}\}$ is a Cousin II distribution on X which represents a divisor $D \in \mathscr{D}(X)$ such that $\mathcal{O}(D) = L$. Consequently

$G(\mathscr{M})$ is a group; the mapping $\mathscr{D}(X) \to G(\mathscr{M})$, $D \to \mathcal{O}(D)$, is a group isomorphism:

$$\mathcal{O}(D + D') = \mathcal{O}(D) \cdot \mathcal{O}(D') \cong \mathcal{O}(D) \otimes_{\mathcal{O}} \mathcal{O}(D').$$

We see that $D = (h)$ for some $h \in \mathscr{M}^*(X)$ if and only if $\mathcal{O}(D) = \mathcal{O}h^{-1}$ (i.e. whenever h^{-1} is a section of $\mathcal{O}(D)$). Furthermore $\mathscr{L} \in G(\mathscr{M})$ has a section $s \in \mathscr{L}(X)$ with $\mathscr{L}_x = \mathcal{O}s_x$ for all $x \in X$ if and only if $\mathscr{L}$ is isomorphic to $\mathcal{O}$: An isomorphism $\varphi: \mathcal{O} \to \mathscr{L}$ determines such a section $s := \varphi(1) \in \mathscr{L}(X) \cap \mathscr{M}^*(X)$ and every such section determines an isomorphism $\mathcal{O} \to \mathscr{L}$ defined by $f_x \to f_x s_x$ Thus

a divisor D is principal if and only if $\mathcal{O}(D)$ is isomorphic to $\mathcal{O}$.

Two sheaves $\mathscr{L}, \mathscr{L}' \in G(\mathscr{M})$ are analytically isomorphic if and only i $\mathscr{L}' \cdot \mathscr{L}^{-1}$ is isomorphic to $\mathcal{O}$. Following the classical language, two divisors D, L are called *linearly equivalent* if $D' - D$ is a principal divisor. Thus the above considerations imply that

two divisors $D, D' \in \mathscr{D}(X)$ *are linearly equivalent if and only if their sheaves* $\mathcal{O}(D), \mathcal{O}(D')$ *are isomorphic.*

Introducing the group

$$LF(\mathscr{M}) := G(\mathscr{M})/\text{subsheaves which are isomorphic to } \mathcal{O},$$

the (*analytic*) *isomorphism classes of locally free subsheaves of rank 1 on X*, we can summarize as follows:

The isomorphism $\mathscr{D}(X) \to G(\mathscr{M})$, $D \to \mathcal{O}(D)$, *induces an isomorphism*

$$DC(X) \to LF(\mathscr{M})$$

of the group of divisor classes onto the group of isomorphy classes of sheaves in $G(\mathscr{M})$.

2. The Isomorphism $H^1(X, \mathcal{O}^*) \to LF(X)$. Using the exact cohomology sequence associated to $1 \to \mathcal{O}^* \to \mathscr{M}^* \to \mathscr{D} \to 0$, one has the natural isomorphism $DC(X) = \mathscr{D}(X)/\text{Im}\ \psi \simeq \text{Im}\ \eta \in H^1(X, \mathcal{O}^*)$. We identify the groups $DC(X)$ and $\text{Im}\ \eta$. From the results in the previous section we see that every divisor class in $DC(X) = \text{Im}\ \eta \in H^1(X, \mathcal{O}^*)$ determines an isomorphy class of locally-free sheaves of rank 1 over X. This construction can be carried out for *all* cohomology classes in $H^1(X, \mathcal{O}^*)$. For this we begin with a cover $\mathfrak{U} = \{U_i\}$ of X. For every cocycle $(g_{ij}) \in Z_a^1(\mathfrak{U}, \mathcal{O}^*)$ we have

$$g_{ij} \in \mathcal{O}^*(U_{ij}) \quad \text{and} \quad g_{ij}g_{jk} = g_{ik} \quad \text{on} \quad U_{ijk}.$$

We define an analytic sheaf automorphism Θ_{ij}: $\mathcal{O} | U_{ij} \to \mathcal{O} | U_{ij}$ by $f_x \to g_{ijx} \cdot f_x$. Thus $\Theta_{ij} \circ \Theta_{jk} = \Theta_{ik}$ on U_{ijk}. By Chapter A.0.9 there exists on X a *gluing* $L = (L, \vartheta_i)$ of the sheaves $\mathcal{O} | U_i$ with respect to the automorphisms Θ_{ij}, where ϑ_i: $\mathscr{L} | U_i \to \mathcal{O} | U_i$ is an isomorphism with $\Theta_{ij} = \vartheta_j \vartheta_i^{-1}$ (i.e. $g_{ij} = \vartheta_j \vartheta_i^{-1}(1)$). Since the maps ϑ_i can be interpreted as $\mathcal{O} | U_i$-isomorphisms, $\mathscr{L}$ is a locally-free analytic sheaf of rank 1. We call $\mathscr{L}$ the *sheaf glued by the cocycle* (g_{ij}).

If $D \in \mathscr{D}(X)$ is a divisor which is represented by a Cousin II distribution $\{U_i, h_i\}$, $h_i \in \mathscr{M}^*(U_i)$, then $\eta(D)$ is represented by the cocycle (g_{ij}), $g_{ij} := h_j \cdot h_i^{-1}$. The associated glued sheaf is isomorphic to $\mathcal{O}(D)$, because we chose $(\mathcal{O} | U_i)h_i^{-1}$ for the sheaf $\mathscr{L} | U_i$ and the isomorphism ϑ_i: $\mathscr{L} | U_i \to \mathcal{O} | U_i$ is given by $f_x \to h_{ix} \cdot f_x$.

If $\mathfrak{U}'$ is another cover of X and $(g'_{rs}) \in Z_a^1(U, \mathcal{O}^*)$ is a cocycle which glues together the sheaf $\mathscr{L}'$, then one verifies directly that $\mathscr{L}$ and $\mathscr{L}'$ are $\mathcal{O}$-isomorphic if and only if (g_{ij}) and (g'_{rs}) represent the same cohomology classes in $H^1(X, \mathcal{O}^*)$. Letting $LF(X)$ denote the *set of all isomorphy classes of analytic, locally-free sheaves of rank 1* over X, we have therefore defined an *injection*

$$\gamma: H^1(X, \mathcal{O}^*) \to LF(X)$$

which associates to a divisor class $\eta(D) \in H^1(X, \mathcal{O}^*)$ the isomorphy class of $\mathcal{O}(D)$.

The map γ is *surjective*: For every locally free sheaf $\mathscr{L}$ of rank 1 over X we have a cover $\mathfrak{U} = \{U_i\}$ of X and the isomorphisms $\vartheta: \mathscr{L} | U_i \to \mathcal{O} | U_i$. The family (g_{ij}) defined by $g_{ij} := \vartheta_j \vartheta_i^{-1}(1) \in \mathcal{O}(U_{ij})$ is a cocycle which glues together $\mathscr{L}$.

The tensor product $\mathscr{L} \otimes_{\mathcal{O}} \mathscr{L}$ of locally-free sheaves of rank 1 is again a locally-free sheaf of rank 1: If ϑ_i: $\mathscr{L} | U_i \to \mathcal{O} | U_i$ and ϑ_i': $\mathscr{L}' | U_i \to \mathcal{O} | U_i$ are the isomorphisms, then

$$\hat{\vartheta}_i := \vartheta_i \otimes \vartheta_i': \mathscr{L} | U_i \otimes_{\mathcal{O} | U_i} \mathscr{L}' | U_i \to \mathcal{O} | U_i \otimes_{\mathcal{O} | U_i} \mathcal{O} | U_i$$

are isomorphisms for which $\hat{\vartheta}_j \hat{\vartheta}_i^{-1} = \vartheta_j \vartheta_i^{-1} \otimes \vartheta_j' \vartheta_i'^{-1}$. Since $\mathcal{O} \otimes_{\mathcal{O}} \mathcal{O} = \mathcal{O} \cdot \mathcal{O} = \mathcal{O}$, it follows that $\hat{g}_{ij} = g_{ij} \cdot g_{ij}'$ is the associated cocycle. Thus $\mathscr{L} \otimes_{\mathcal{O}} \mathscr{L}'$ is isomorphic to the sheaf which is glued by the product cocycle. Since isomorphic sheaves yield isomorphic tensor products, a tensor product is likewise defined in $LF(X)$. In fact $\gamma(vv') = \gamma(v) \otimes \gamma(v')$ for all $v, v' \in H^1(X, \mathcal{O}^*)$.

The following is a summary of the above discussion.

Theorem 1. *The set $LF(X)$ of (analytic) isomorphism classes of locally-free sheaves of rank 1 over X is a group with respect to tensor product. The map γ: $H^1(X, \mathcal{O}^*) \to LF(X)$ is a group isomorphism and takes the group of divisor classes $DC(X) \subset H^1(X, \mathcal{O}^*)$ onto the subgroup $LF(\mathscr{M}) \subset LF(X)$ of isomorphism classes which are represented by $\mathcal{O}$-subsheaves of $\mathscr{M}$. For every divisor $D \in \mathscr{D}(X)$ the sheaf $\mathcal{O}(D)$ is in the isomorphism class $\gamma(\eta(D))$.*

As a corollary we note that

if X is a complex space with $H^1(X, \mathcal{O}) = H^2(X, \mathbb{Z}) = 0$, then every locally-free sheaf of rank 1 on X is free (i.e. isomorphic to the structure sheaf $\mathcal{O}$).

Proof: The exponential sequence implies that $H^1(X, \mathcal{O}^*) = 0$ and thus $LF(X)$ consists of one isomorphism class which contains $\mathcal{O}$. □

3. The Group of Divisor Classes on a Stein Space. The group $LF(\mathscr{M})$ is in general a *proper* subgroup of $LF(X)$. There is however a simple sufficient condition for an isomorphism class in $LF(X)$ to be in $LF(\mathscr{M})$:

Lemma. *Let X be reduced and $\mathscr{L}$ a locally-free analytic sheaf of rank 1 on X which has a section $s \in \mathscr{L}(X)$ whose zero set is nowhere dense in X. Then there is a positive divisor $D \in D(X)$ so that $\mathscr{L}$ is isomorphic to $\mathcal{O}(D)$.*

Proof: We may assume that L is glued together by isomorphisms ϑ_i: $\mathscr{L} | U_i \to \mathcal{O} | U_i$ with associated cocycle $(g_{ij}) \in Z_a^1(\mathfrak{U}, \mathcal{O}^*)$ defined by $g_{ij} = \vartheta_j \vartheta_i^{-1}(1)$. Thus the zero sets of $f_i := \vartheta_i(s | U_i) \in \mathcal{O}(U_i)$ are nowhere dense in U_i, and consequently $f_i \in \mathscr{M}^*(U_i)$. On U_{ij}

$$g_{ij} f_i = \vartheta_j \vartheta_i^{-1}(f_i) = \vartheta_j(s) = f_j.$$

Thus $g_{ij} = f_j f_i^{-1} \in \mathcal{O}^*(U_{ij})$ and $\{U_i, f_i\}$ is a Cousin II distribution on X whose associated divisor D is positive (recall $f_i \in \mathcal{O}(U_i)$). Since $\mathcal{L}$ and $\mathcal{O}(D)$ both have associated cocycle (g_{ij}), they are isomorphic. □

The following is now an immediate consequence of Theorem A.

Theorem 2. *If X is an irreducible, reduced Stein space, then*

$$LF(\mathcal{M}) = LF(X).$$

In other words, every locally-free, analytic sheaf of rank 1 on X is isomorphic to $\mathcal{O}(D)$ for some positive divisor $D \in \mathcal{D}(X)$.

Proof: By Theorem A there exists a section $s \not\equiv 0$ in $\mathcal{L}(X)$. Since X is irreducible, the zero set of s is nowhere dense in X and the claim follows from the lemma above. □

The isomorphism $\gamma\colon H^1(X, \mathcal{O}^*) \to LF(X)$ with $\gamma(DC(X)) = LM(X)$ translates the above theorem into the following:

Theorem 2′. *If X is an irreducible, reduced Stein space, then*

$$DC(X) = H^1(X, \mathcal{O}^*).$$

Every divisor $D \in \mathcal{D}(X)$ is linearly equivalent to a positive divisor.

The next result is now an immediate consequence of Theorem 2′:

Theorem 3. *If X is an irreducible, reduced Stein space, then every 2-dimensional, integral cohomology class in $H^2(X, \mathbb{Z})$ is the characteristic class $c(D)$ of a positive divisor $D \in \mathcal{D}(X)$.*

Proof: Let $v \in H^2(X, \mathbb{Z})$ be given. Since $H^2(X, \mathcal{O}) = 0$, the map $\delta\colon H^1(X, \mathcal{O}^*) \to H^2(X, \mathbb{Z})$ is surjective (i.e. there exists $u \in H^1(X, \mathcal{O}^*)$ with $\delta(u) = v$). By Theorem 2′ *there exists a positive divisor* $D \in \mathcal{D}(X)$ with $\eta(D) = u$. Consequently $c(D) = \delta\eta(D) = v$. □

In his works [36, 37] K. Stein had already dealt (in the language of homology) with the question of which cohomology classes in $H^2(X, \mathbb{Z}) = H_{2n-2}(X, \mathbb{Z})$ on a Stein manifold X can appear as characteristic classes of positive divisors. In his Habilitationsscrift (published in 1941), which signaled the fruitful entrance of methods from algebraic topology into complex analysis, he solved the problem for polycylinders. In [37], which was published 10 years later, the question for "infinitely divisible" elements of $H^2(X, \mathbb{Z})$ was answered in the positive for arbitrary Stein manifolds. Theorem 3 was proved by Serre [35] and gives the Stein result a final, optimal form.

§ 4. Sheaf Theoretical Characterization of Stein Spaces

By definition, every group $H^q(X, \mathscr{S})$, $q \geq 1$, vanishes when X is a Stein space, and $\mathscr{S}$ is a coherent sheaf on X. We will now give theorems which show that a space X is Stein if only *certain* cohomology groups $H^q(X, \mathscr{S})$ are zero.

1. Cycles and Global Holomorphic Functions. The following language is convenient:

Definition 1 (Cycle). *A map $\mathscr{o}: X \to \mathbb{Z}_0 := \{0, 1, 2, \ldots\}$ is called a cycle on X if its "support" $\operatorname{supp} \mathscr{o} := \{x \in X \mid \mathscr{o}(x) \neq 0\}$ is discrete in X (it would be more precise to call $\mathscr{o}$ a 0-dimensional, nonnegative cycle).*

Every cycle $\mathscr{o}$ determines the analytic ideal sheaf

$$\mathscr{Z}(\mathscr{o}) := \bigcup_{x \in X} \mathscr{Z}(\mathscr{o})_x, \qquad \mathscr{Z}(\mathscr{o})_x := \mathfrak{m}_x^{\mathscr{o}(x)} \qquad (\mathfrak{m}_x = \text{the maximal ideal in } \mathcal{O}_x).$$

It follows that $\mathscr{Z}(\mathscr{o})_x = \mathcal{O}_x$ for every $x \notin \operatorname{supp} \mathscr{o}$.

Theorem 1. *The ideal sheaf $\mathscr{Z}(\mathscr{o})$ is coherent.*

Proof: The "1 section" $1 \in \mathscr{Z}(\mathscr{o})(X \backslash \operatorname{supp} \mathscr{o}) = \mathcal{O}(X \backslash \operatorname{supp} \mathscr{o})$ generates $\mathscr{Z}(\mathscr{o})$ over $X \backslash \operatorname{supp} \mathscr{o}$. Let $p \in \operatorname{supp} \mathscr{o}$ and $r := \mathscr{o}(p) \geq 1$. There exists a neighborhood U of p in X with $U \cap \operatorname{supp} \mathscr{o} = \{p\}$ so that $(U, \mathcal{O}_U)$ is isomorphic to a complex subspace of a domain $B \subset \mathbb{C}^m$. We identify $(U, \mathcal{O}_U)$ with this subspace and let $\mathscr{J}$ be the coherent ideal sheaf in $\mathcal{O}_B$ so that $\mathcal{O}_U = (\mathcal{O}_B/\mathscr{J}) | U$. Let $z_1, \ldots, z_m$ be coordinates for $\mathbb{C}^m$ which are centered at p. Then the monomials

$$q_{i_1 \cdots i_m} := z_1^{i_1} \cdot \cdots \cdot z_m^{i_m}, \qquad i_1 + \cdots + i_m = r,$$

generate the ideal $\mathfrak{m}(\mathcal{O}_{B,p})^r$ and hence their equivalence classes $\bar{q}_{i_1 \cdots i_m} \in \mathcal{O}_B/\mathscr{J}$ generate the ideal $\mathfrak{m}(\mathcal{O}_B/\mathscr{J})_p^r = \mathfrak{m}(\mathcal{O}_{U,p})^r = \mathfrak{m}_p^{\mathscr{o}(p)}$. Since the monomials $q_{i_1 \cdots i_m}$ generate all stalks of $\mathcal{O}_B$ over $B \backslash p$, the functions $\bar{q}_{i_1 \cdots i_m} | U \in \mathcal{O}_U(U)$ generate every stalk $\mathcal{O}_{U,x}$, $x \in U \backslash p$. Since $\mathscr{Z}(\mathscr{o})_x = \mathcal{O}_{U,x}$ for all $x \in U \backslash p$ (recall $U \cap \operatorname{supp} \mathscr{o} = p$), the functions $\bar{q}_{i_1 \cdots i_m} | U$ therefore generate the sheaf $\mathscr{Z}(\mathscr{o}) | U$. □

The following is quite useful.

Theorem 2 (Existence Criterion). *Let $\mathscr{o}$ be a cycle on X so that $\mathcal{O}(X) \to (\mathcal{O}/\mathscr{Z}(\mathscr{o}))(X)$ is surjective and suppose that to every point $p \in \operatorname{supp} \mathscr{o}$ there is arbitrarily assigned a germ $g_p \in \mathcal{O}_p$. Then there exists a function $f \in \mathcal{O}(X)$ so that*

$$f_p - g_p \in \mathfrak{m}_p^{\mathscr{o}(p)} \quad \text{for all} \quad p \in \operatorname{supp} \mathscr{o}.$$

Proof: Let $\rho: \mathcal{O} \to \mathcal{O}/\mathscr{Z}(\mathscr{o})$ be the quotient epimorphism. Since $\operatorname{supp}(\mathcal{O}/\mathscr{Z}(\mathscr{o})) = \operatorname{supp} \mathscr{o}$, a section $s \in (\mathcal{O}/\mathscr{Z}(\mathscr{o}))(X)$ is defined by $s(p) := g_p$ for

$p \in \operatorname{supp} \mathfrak{o}$ and $s(p) := 0$ otherwise. By assumption the induced map $\rho_*: \mathcal{O}(X) \to (\mathcal{O}/\mathcal{Z}(\mathfrak{o}))(X)$ is an epimorphism. Consequently there is an $f \in \mathcal{O}(X)$ so that $\rho_*(f) = s$. Since $\rho(f_p) = \rho(g_p)$, it follows that $f_p - g_p \in (\operatorname{Ker} \rho)_p = \mathcal{Z}(\mathfrak{o})_p = \mathfrak{m}_p^{\mathfrak{o}(p)}$ for all $p \in \operatorname{supp} \mathfrak{o}$. □

If every point of supp $\mathfrak{o}$ is *non-singular*, then the epimorphism condition in Theorem 2 guarantees that one can always find a holomorphic function on X whose *Taylor series* at every point $p \in \operatorname{supp} \mathfrak{o}$ with respect to some local coordinate at p is arbitrarily prescribed up to order $\mathfrak{o}(p)$.

We now consider complex spaces X which have at least one of the following properties:

(S) *For every coherent ideal $\mathcal{I} \subset \mathcal{O}$ it follows that $H^1(X, \mathcal{I}) = 0$.*

(S′) *The section functor is exact on the category of coherent analytic sheaves on X (i.e. every exact sequence $0 \to \mathcal{S}' \to \mathcal{S} \to \mathcal{S}'' \to 0$ induces an exact $\mathcal{O}(X)$-sequence $0 \to \mathcal{S}'(X) \to \mathcal{S}(X) \to \mathcal{S}''(X) \to 0$).*

Since $\mathcal{Z}(\mathfrak{o})$ is coherent (Theorem 1), every space having property (S) or (S′) satisfies the surjectivity condition of Theorem 2 (i.e. $\mathcal{O}(X) \to (\mathcal{O}/\mathcal{Z}(\mathfrak{o}))(X)$ is surjective). We have two immediate consequences of this:

Consequence 1. *Let X be a complex which has property (S) or (S′). Let $(x_n)_{n\geq 0}$ be a discrete sequence in X and $(c_n)_{n\geq 0}$ an arbitrary sequence of complex numbers. Then there exists a holomorphic function $f \in \mathcal{O}(X)$ with $f(x_n) = c_n$, $n \geq 0$.*

Proof: Let $\mathfrak{o}(x) := 1$ for $x = x_n$, $n \geq 0$, and $\mathfrak{o}(x) := 0$ otherwise. Then, applying Theorem 2 for $g_{x_n} := c_n \in \mathcal{O}_{x_n}$, there exists $f \in \mathcal{O}(X)$ with $f_{x_n} - c_n \in \mathfrak{m}_{x_n}$. In other words $f(x_n) = c_n$ for all $n \geq 0$. □

Consequence 2. *Let X be a complex space which satisfies (S) or (S′). Let $e := \dim_{\mathbb{C}} \mathfrak{m}_p/\mathfrak{m}_p^2$ be the embedding dimension*[2] *of X at p. Then there are e holomorphic functions $f_1, \ldots, f_e \in \mathcal{O}(X)$ whose germs $f_{1p}, \ldots, f_{ep} \in \mathcal{O}_p$ form a generating system for $\mathfrak{m}_p$ as an $\mathcal{O}_p$-module.*

Proof: Let $g_{1p}, \ldots, g_{ep} \in \mathfrak{m}_p$ be germs which generate $\mathfrak{m}_p$. Applying Theorem 2 with $\mathfrak{o}(p) := 2$ and $\mathfrak{o}(x) := 0$ otherwise, there exist functions $f_i \in \mathcal{O}(X)$ with $f_{ip} - g_{ip} \in \mathfrak{m}_p^2$ for $1 \leq i \leq e$. Thus the equivalence classes $\bar{f}_{1p}, \ldots, \bar{f}_{ep} \in \mathfrak{m}_p/\mathfrak{m}_p^2$ generate the $\mathbb{C}$-vector space $\mathfrak{m}_p/\mathfrak{m}_p^2$ and consequently $f_{1p}, \ldots, f_{ep}$ generate $\mathfrak{m}_p$ as an $\mathcal{O}_p$-module (see the footnote on p. 101). □

If p is a *non-singular* point of X, then the embedding dimension at p is the same as the dimension of X at p. Thus Consequence 2 says that for spaces fulfilling (S) or (S′) Stein's global coordinates axiom (see the introduction) holds at all non-singular points of X.

---

[2] The embedding dimension of X at p is the smallest integer $e \geq 0$ so that $\mathfrak{m}_p$ is generated as an $\mathcal{O}_p$-module by e germs in $\mathfrak{m}_p$ (see [AS], Chapter II.3).

2. Equivalent Criteria for a Stein Space. It is now easy to prove that our weakened axioms are equivalent to Stein's original ones. As a matter of fact, the following could be considered the main theorem of this book.

Theorem 3 (Equivalent Criteria for a Stein Space). *The following statements about a complex space X (with countable topology) are equivalent:*

i) *X is holomorphically complete (i.e. weakly holomorphically convex, and every compact analytic set in X is finite).*
ii) *X is Stein.*
iii) *If $\mathscr{I}$ is a coherent ideal contained in $\mathcal{O}$, then $H^1(X, \mathscr{I}) = 0$ (property (S)).*
iv) *The section functor is exact on the category of coherent analytic sheaves (property (S')).*
v) *X is holomorphically convex, holomorphically separable, and to every point $x_0 \in X$ there are e functions $f_1, \ldots, f_e$ where e is the embedding dimension of X at x_0 and $f_{1x_0}, \ldots, f_{ex_0}$ generate the maximal ideal $\mathfrak{m}_{x_0} \subset \mathcal{O}_{x_0}$.*

Proof: i)$\Rightarrow$ii): This is the fundamental theorem of Chapter IV. ii)$\Rightarrow$iii) and ii)$\Rightarrow$iv): Clear.

iii)$\Rightarrow$v) and iv)$\Rightarrow$v): If $D = \{x_n\}_{n \geq 0}$ is a discrete set in X then by Consequence 1 there exists a function $h \in \mathcal{O}(X)$ with $h(x_n) = n$ and in particular $|h|_D = \infty$. Therefore (by Theorem IV.2.4) X is holomorphically convex. Given $x_0, x_1 \in X$ with $x_0 \neq x_1$, Consequence 1 guarantees the existence of $f \in \mathcal{O}(X)$ with $f(x_0) = 0$ and $f(x_1) = 1$. Thus X is holomorphically separable. The last statement in v) follows from Consequence 2.

v)$\Rightarrow$i): Holomorphic convexity implies weak holomorphic convexity and holomorphic separability implies that every compact analytic set in X is finite. □

Property (S) was first discussed by Serre (see [35], p. 53). It should be remarked that it is enough to require $H^1(X, \mathscr{I}) = 0$ for ideals which coincide with $\mathcal{O}$ except on a discrete set.

We now mention still another consequence of $H^1(X, \mathscr{I}) = 0$:

Theorem 4. *Let $(Y, \mathcal{O}_Y)$ be a closed complex subspace of a Stein space $(X, \mathcal{O}_X)$. Then every function holomorphic on Y is the restriction of a function which is holomorphic on X.*

Proof: Let $\mathscr{I} \subset \mathcal{O}_X$ be the coherent ideal associated to Y. Then $Y = \operatorname{supp}(\mathcal{O}_X/\mathscr{I})$ and $\mathcal{O}_Y = (\mathcal{O}_X/\mathscr{I})|Y$. Since $H^1(X, \mathscr{I}) = 0$, the homomorphism $\mathcal{O}_X(X) \to \mathcal{O}_X/\mathscr{I}(X)$ is surjective and the claim follows from the canonical identification of $\mathcal{O}_Y(Y)$ with $(\mathcal{O}_X/\mathscr{I})(X)$. □

3. The Reduction Theorem. Here we consider complex spaces $X = (X, \mathcal{O}_X)$ and their reductions $\operatorname{red} X = (X, \mathcal{O}_{\operatorname{red} X})$. The kernel of the canonical $\mathcal{O}_X$-homomorphism $\rho: \mathcal{O}_X \to \mathcal{O}_{\operatorname{red} X}$ is the nilradical $\mathcal{N} := \mathfrak{n}(\mathcal{O}_X)$ of $\mathcal{O}_X$. The follow ing is an immediate consequence of the fact that $h \in \mathcal{O}_X(X)$ has the same complex value at every point in X as the reduced function $\rho_*(h) \in \mathcal{O}_{\text{red } X}(X)$.

If X has any of the following properties, then red *X has them too: weak holomorphic convexity, holomorphic convexity, holomorphically spreadable, Stein.*

The converse of this statement is not true in general. In fact Schuster has given examples of complex spaces which are not holomorphically convex (resp. holomorphically separable), but whose reductions are holomorphically convex (resp. holomorphically) separable (see [34], p. 285). However, the following is easy to prove.

If $\rho_*: \mathcal{O}_X(X) \to \mathcal{O}_{\text{red } X}(X)$ is surjective, then red *X will have any of the following properties if and only if X does: weak holomorphic convexity, holomorphic convexity, holomorphically spreadable, Stein.*

We now give a beautiful application of Theorem B:

If red *X is Stein, then $\rho_*: \mathcal{O}_X(X) \to \mathcal{O}_{\text{red } X}(X)$ is surjective.*

Proof: We consider the $\mathcal{O}_X$-epimorphisms

$$\rho_i: \mathcal{O}_X \to \mathcal{H}_i \quad \text{with} \quad \mathcal{H}_i := \mathcal{O}_X/\mathcal{N}^i, \qquad i = 1, 2, \ldots,$$

where $\mathcal{N}^i$ is the i-fold product of $\mathcal{N}$ with itself. Note that $\mathcal{H}_1 = \text{red } \mathcal{O}_X$ and $\rho_1 = \rho$. Since $\mathcal{N}^i \supset \mathcal{N}^{i+1}$, there are $\mathcal{O}_X$-epimorphisms

$$\varepsilon_i: \mathcal{H}_{i+1} \to \mathcal{H}_i \quad \text{with} \quad \text{Ker } \varepsilon_i = \mathcal{N}^i/\mathcal{N}^{i+1}$$

and

$$\varepsilon_i \rho_{i+1} = \rho_i, \qquad i = 1, 2, \ldots.$$

At the section level the following diagram is commutative.

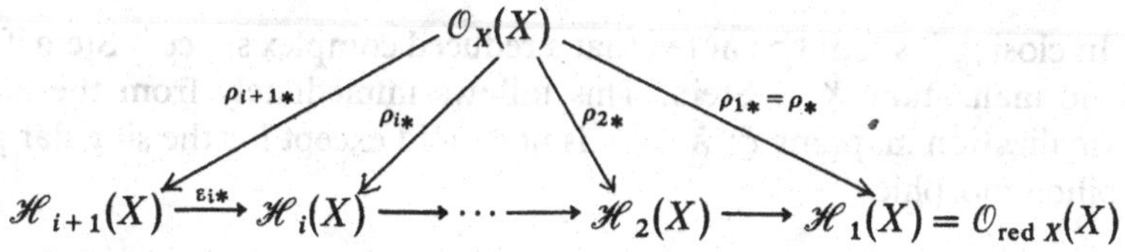

Since $\mathcal{N}$ is coherent, all of the products $\mathcal{N}^i$ are coherent $\mathcal{O}_X$-ideals (Chapter A.2.3). Since $\mathcal{N} \cdot \text{Ker } \varepsilon_i = \mathcal{N}(\mathcal{N}^i/\mathcal{N}^{i+1}) = 0$, it follows that $\mathcal{K}er\ \varepsilon_i$ is a coherent $\mathcal{O}_{\text{red } X}$-sheaf (Chapter A.2.4). Thus, since red X is Stein, $H^1(X, \mathcal{K}er\ \varepsilon_i) = 0$ and consequently $\varepsilon_{i*}: \mathcal{H}_{i+1}(X) \to \mathcal{H}_i(X)$ is surjective. Let $h_1 \in \mathcal{H}_1(X)$ be given. We successively choose functions $h_i \in \mathcal{H}_i(X)$ so that $\varepsilon_i(h_i) = h_{i-1}$, $i = 2, 3, \ldots$. These will allow us to determine a ρ_*-preimage $h \in \mathcal{O}_X(X)$ of h_1.

The sets $X_i := \{x \in X \mid \mathcal{N}^i_x = 0\}$ are open in X and $X_1 \subset X_2 \subset \cdots$. For every $x \in X$ there exists $i(x) \geq 1$ with $\mathcal{N}^{i(x)}_x = 0$. Thus $X = \bigcup_1^\infty X_i$. It follows from the definitions of X_i, $\mathcal{H}_i$ and ρ_i that

$$\mathcal{O}_{X_i} = \mathcal{H}_i | X_i, \qquad \rho_i | X_i = \text{identity and } h_i | X \in \mathcal{O}_X(X_i).$$

Since $\mathcal{K}er\ \varepsilon_i | X_i = 0$ and $\mathcal{H}_{i+1} | X_i = \mathcal{O}_{X_i}$, it follows that

$$\varepsilon_i \mid X_i \colon \mathcal{H}_{i+1} \mid X_i \to \mathcal{H}_i \mid X_i$$

is likewise the identity map $\mathcal{O}_{X_i} \to \mathcal{O}_{X_i}$. Hence

$$h_{i+1} | X_i = h_i | X_i, \qquad i = 1, 2, \ldots.$$

Consequently the family $\{h_i\}$ determines a section

$$h \in \mathcal{O}_X(X) \quad \text{with} \quad h | X_i = h_i | X_i, \qquad i = 1, 2, \ldots.$$

Since $\rho_* = \varepsilon_{1*}\varepsilon_{2*} \cdots \varepsilon_{i-1*}\rho_{i*}$ and $\rho_{i*}(h) | X_i = h | X_i = h_i | X_i$, it follows that

$$\rho_*(h) | X_i = \varepsilon_{1*}\varepsilon_{2*} \cdots \varepsilon_{i-1*}(h_i) | X_i = h_1 | X_i, \qquad i = 1, 2, \ldots,$$

and consequently $\rho_*(h) = h_1$. □

Remark: We have obviously just shown that if X is a complex space whose cohomology groups $H^1(X, \mathcal{N}^i/\mathcal{N}^{i+1})$ vanish $1 \leq i \leq \infty$, then the module of sections $\mathcal{O}_X(X)$ is the inverse projective limit $\lim\ (\mathcal{O}_X/\mathcal{N}^i)(X)$, and in this case $\rho_*\colon \mathcal{O}_X(X) \to \mathcal{O}_{\text{red}\,X}(X)$ is always surjective. □

As a result of the above, we now have the following remark:

Theorem 5 (Reduction Theorem). *A complex space X is Stein if and only if its reduction* red *X is Stein.*

In closing it should be noted that a reduced complex space is Stein if and only if its normalization $\tilde{X}$ is Stein. This follows immediately from the fact that the normalization mapping $\xi\colon \tilde{X} \to X$ is finite and except for the singular points in X is biholomorphic. □

4. Differential Forms on Stein Manifolds. For every complex manifold X, the sheaf Ω^p of germs of holomorphic p-forms is coherent on X (see Chapter II.2.2). In the Stein case $H^q(X, \Omega^p) = 0$ for all $p \geq 0$, $q \geq 1$. Thus the following is a simple consequence of Chapter II.4.2.

Theorem 6. *Let X be a Stein manifold and $p \geq 0$, $q \geq 1$. Then for every (p, q)-form $\varphi \in \mathcal{A}^{p,q}(X)$ with $\bar{\partial}\varphi = 0$ there exist $\psi \in \mathcal{A}^{p,q-1}(X)$ with $\varphi = \bar{\partial}\psi$.*

The following is a corollary of the results in Chapter II.4.3:

Theorem 7. *For every Stein manifold X there are natural $\mathbb{C}$-isomorphisms*

$$H^0(X, \mathbb{C}) \simeq \operatorname{Ker}(d \mid \mathcal{O}(X)),$$
$$H^q(X, \mathbb{C}) \simeq \operatorname{Ker}(d \mid \Omega^q(X))/d\Omega^{q-1}(X), \qquad q \geq 1.$$

Since by the formal de Rham Theorem (Chapter II.1.8) it follows that there are always natural $\mathbb{C}$-isomorphisms

$$H^0(X, \mathbb{C}) \simeq \operatorname{Ker}(d \mid \mathscr{E}(X)), \qquad H^q(X, \mathbb{C}) \simeq \operatorname{Ker}(d \mid \mathscr{A}^q(X))/d\mathscr{A}^{q-1}(X), \qquad q \geq 1,$$

in the Stein case we have the commutative diagram

$$\text{(D)} \qquad \begin{array}{ccccccccc} 0 & \longrightarrow & d \mid \Omega^{q-1}(X) & \longrightarrow & \operatorname{Ker}(d \mid \Omega^q(X)) & \longrightarrow & H^q(X, \mathbb{C}) & \longrightarrow & 0 \\ & & \downarrow & & \downarrow & & \| & & \\ 0 & \longrightarrow & d \mid \mathscr{A}^{q-1}(X) & \longrightarrow & \operatorname{Ker}(d \mid \mathscr{A}^q(X)) & \xrightarrow{\pi_q} & H^q(X, \mathbb{C}) & \longrightarrow & 0 \end{array}$$

with exact rows, where the vertical maps are the natural inclusions. This situation has the following consequence:

Theorem 8. *Let X be a Stein manifold and $\alpha \in \mathscr{A}(X)$ a differentiable differential form whose differential $d\alpha$ is a holomorphic differential form. Then there exists a differentiable form $\beta \in \mathscr{A}(X)$ so that $\alpha - d\beta$ is holomorphic.*[3]

Proof: It is enough to prove the theorem for forms $\alpha \in \mathscr{A}^r(X)$, $1 \leq r < \infty$. By assumption $d\alpha \in \operatorname{Ker}(d \mid \Omega^{r+1}(X))$. Since $\pi_{r+1}(d\alpha) = 0 \in H^{r+1}(X, \mathbb{C})$, the diagram (D) guarantees the existence of a form $\delta \in \Omega^r(X)$ with $d\delta = d\alpha$. The form $\alpha - \delta \in \operatorname{Ker}(d \mid \mathscr{A}^r(X))$ determines a cohomology class $\pi_r(\alpha - \delta) \in H^r(X, \mathbb{C})$. We choose a holomorphic r-form $\varepsilon \in \operatorname{Ker}(d \mid \Omega^r(X))$ which determines the same class. Then $\alpha - \delta - \varepsilon \in \operatorname{Ker}(d \mid \mathscr{A}^r(X))$ and $\pi_r(\alpha - \delta - \varepsilon) = 0$. Thus there is a differentiable $(r-1)$-form $\beta \in \mathscr{A}^{r-1}(X)$ with $d\beta = \alpha - \delta - \varepsilon$. Consequently $\alpha - d\beta = \delta + \varepsilon \in \Omega^r(X)$. □

Corollary to Theorem 8. *Let X be a Stein manifold and α any d-closed differentiable differential form on X. Then there exists a holomorphic differential form γ on X so that $\alpha - \gamma$ is a d-exact (differentiable) differential form.*

---

[3] For forms $\alpha \in \mathscr{A}^0(X)$ this theorem says that every differentiable function α with a holomorphic differential $d\alpha$ is in fact holomorphic: $\alpha \in \mathcal{O}(X)$. This statement is trivially true for any complex manifold (see Chapter II.2.6).

This corollary means for example that on a Stein manifold there exist holomorphic differential forms with arbitrarily prescribed periods.

The "Stein" assumption in Theorems 6, 7, 8 was made in order to guarantee that $H^q(X, \Omega^p) = 0$. It is not known if a complex manifold with $H^q(X, \Omega^p) = 0$ for all $p \geq 0$ and $q \geq 1$ is necessarily Stein.

5. Topological Properties of Stein Spaces. If X is any m-dimensional complex manifold, then $\Omega^q = 0$ for all $q > m$. Thus the following is an immediate consequence of Theorem 7:

Theorem 9. *If X is an m-dimensional Stein manifold, then*

$$H^q(X, \mathbb{C}) = 0 \quad \textit{for all} \quad q > m.$$

This theorem, given by Serre in 1953, is a purely topological, necessary condition for a complex manifold, in particular for a domain in $\mathbb{C}^m$, to be Stein. The following is a homological reformulation of this.

Theorem 9'. *If X is an m-dimensional, Stein manifold, then the integral homology group $H_q(X, \mathbb{Z})$ is a torsion group for all $q \geq m$.*

Proof: By general theorems from algebraic topology $H^q(X, \mathbb{C})$ is isomorphic to the group $\mathrm{Hom}(H_q(X, \mathbb{Z}), \mathbb{C})$ for every q. In the case when $H^q(X, \mathbb{C}) = 0$ the group cannot contain a free element, because it would give a non-trivial homomorphism $H_q(X, \mathbb{Z}) \to \mathbb{C}$. □

Until 1958 the problem of whether or not $H_q(X, \mathbb{Z})$ for $q > m$ could contain non-trivial torsion elements remained open. At that time A. Andreotti and T. Frankel [1], using the embedding theorem for Stein manifolds and Morse theory, showed that this is indeed not possible:

If X is an m-dimensional Stein manifold, then

$$H_q(X, \mathbb{Z}) = 0 \quad \textit{for} \quad q > m \quad \textit{and} \quad H_m(X, \mathbb{Z}) \quad \textit{is} \quad \textit{free}.$$

Again using the methods of Morse theory, J. Milnor (see [28], p. 39) sharpened this result:

Every complex m-dimensional Stein manifold is homotopy equivalent to a real m-dimensional CW-complex.

This statement can still be greatly improved:

In every m-dimensional Stein manifold X there is a real m-dimensional, closed CW-complex K which is a "strong deformation retract" of X. In other words, there is a continuous map $f\colon X \times [0, 1] \to X$ with the following property:

$$f(x, 0) = x \textit{ for all } x \in X, \quad f(p, t) = p \textit{ for all } (p, t) \in K \times [0, 1], \; f(X \times \{1\}) = K$$

It is natural to ask which CW-complexes can arise as retracts of Stein manifolds. In this regard one can show the following: (see [17], p. 468ff.):

Tube Theorem. *Every paracompact, real m-dimensional, real-analytic manifold R has a Stein tubular neighborhood X. In other words,*

1) *X is a complex m-dimensional Stein manifold and R is a real-analytic submanifold of X;*
2) *R is a strong deformation retract of X.*

It is also quite reasonable to ask if the lower homology groups $H_q(X, \mathbb{Z})$, $1 \leq q < m$, of an m-dimensional Stein manifold can be arbitrarily prescribed. In 1959 K. J. Ramspott (Existenz von Holomorphiegebieten zu vorgegebener erster Bettischer Gruppe, Math. Ann. **138**, 342–355) showed that quite a bit can in fact be prescribed:

For every countable, torsion free, abelian group B there exists a Stein domain X in $\mathbb{C}^2$ whose first Betti group (i.e. the quotient group of $H_1(X, \mathbb{Z})$ by its torsion group) is isomorphic to B.

Furthermore Narasimhan [29] showed the following:

For every countable abelian group G and every natural number $q \geq 1$ there is a Stein domain (even a Runge domain) X in C^{2q+3} so that $H_q(X, \mathbb{Z})$ is isomorphic to G.

Using complex analytic methods, the theorem of Andreotti-Frankel was generalized in [29]:

If X is an m-dimensional Stein space, then

$$H_q(X, \mathbb{Z}) = 0 \text{ for } q > m \text{ and } H_m(X, \mathbb{Z}) \text{ is torsion free.}$$

§ 5. A Sheaf Theoretical Characterization of Stein Domains in $\mathbb{C}^m$

A domain B in $\mathbb{C}^m$, $1 \leq m < \infty$, is Stein if and only if it is holomorphically convex. In this section we will show that such domains have a particularly simple sheaf theoretical characterization.

1. An Induction Principle. Our beginning point is the following classical result:

Lemma (Simultaneous Continuability, see [BT], p. 121). *Let $B \subset \mathbb{C}^m$ be a domain which is not holomorphically convex. Then there is a point $p \in B$ and a polycylinder Δ about p with $\Delta \not\subset B$ so that every $f \in \mathcal{O}(B)$ has a Taylor series on Δ which converges compactly on Δ to a function $F \in \mathcal{O}(\Delta)$. Let W be the connected component of $B \cap \Delta$ which contains p. Then $f \mid W = F \mid W$.*

Proof: Since B is not holomorphically convex, there exists a compact set $K \subset B$ whose hull $\hat{K}$ relative to B is not compact in B. Let $\Delta_t(c)$ denote the polydisk

$$\{(z_1, \ldots, z_m) \in \mathbb{C}^m \mid |z_\mu - c_\mu| < t, 1 \leq \mu \leq m\}$$

of radius t about c in $\mathbb{C}^m$. There exists a real number $r > 0$ so that $B' := \bigcup_{a \in K} \Delta_r(a)$ is a *relatively compact* subset in B. Consequently $|f|_{B'} < \infty$ for every $f \in \mathcal{O}(B)$. For such a function, the Taylor series of f converges on $\Delta_r(a)$ for all a and the Cauchy inequalities yield.

$$(*) \quad |f_{\mu_1 \cdots \mu_m}|_K \leq \frac{|f|_{B'}}{r^{\mu_1 + \cdots + \mu_m}}, \quad \text{where } f_{\mu_1 \cdots \mu_m} := \frac{1}{\mu_1! \cdots \mu_m!} \frac{\partial^{\mu_1 + \cdots + \mu_m} f}{\partial z_1^{\mu_1} \cdots \partial z_m^{\mu_m}} \in \mathcal{O}(B).$$

Since $\hat{K}$ is bounded in $\mathbb{C}^m$, there exists $p \in \hat{K}$ such that $\Delta := \Delta_r(p)$ is not contained in B. We consider the Taylor series of f about p (= the origin):

$$\sum_0^\infty f_{\mu_1 \cdots \mu_m}(p) z_1^{\mu_1} \cdots z_m^{\mu_1}.$$

Since $p \in \hat{K}$, it follows that $|f_{\mu_1 \cdots \mu_m}(p)| \leq |f_{\mu_1 \cdots \mu_m}|_K$. Applying $(*)$, we see that the above power series converges on Δ to $F \in \mathcal{O}(\Delta)$. The identity principle implies that f and F coincide on the connected component of $B \cap \Delta$ which contains p. □

We now use the above lemma to prove a theorem which will in turn allow us to give an induction argument for a sheaf theoretical characterization of Stein domains.

Theorem 1. *A domain B in $\mathbb{C}^m$ is Stein if and only if at least one of the following conditions is fulfilled:*

*) *For every $(m-1)$-dimensional, analytic plane $H \subset \mathbb{C}^m$ the intersection $B \cap H \subset H \simeq \mathbb{C}^{m-1}$ is Stein and the restriction $\mathcal{O}_B(B) \to \mathcal{O}_{B \cap H}(B \cap H)$ is surjective.*

**) *For every complex line $E \subset \mathbb{C}^m$, the restriction $\mathcal{O}_B(B) \to \mathcal{O}_{B \cap E}(B \cap E)$ is surjective.*

Proof: If B is Stein, then every subspace $B \cap H$ is Stein and $\mathcal{O}_B(B) \to \mathcal{O}_{B \cap H}(B \cap H)$ is surjective by Theorem 4.4. Thus $(*)$ holds.

$*) \Rightarrow **)$: Let E be given and choose a hyperplane $H \subset \mathbb{C}^m$ which contains E. Then $\mathcal{O}_B(B) \to \mathcal{O}_{B \cap E}(B \cap E)$ is the composition of $\mathcal{O}_B(B) \to \mathcal{O}_{B \cap H}(B \cap H)$ and $\mathcal{O}_{B \cap H}(B \cap H) \to \mathcal{O}_{B \cap E}(B \cap E)$. The first map is surjective by assumption and the second is surjective since $B \cap H$ is Stein.

It remains to show that $(**)$ implies the holomorphic convexity of B. Suppose this is not the case and choose p, Δ, and W as in the lemma. Let $q \in \Delta \backslash B$ and take

E to be the complex line through p and q. On the *real* segment $pq \subset E \cap \Delta$ from p to q there exists a *first* point $y \notin W$. Thus $y \in \partial(W \cap E) \cap \Delta$. We now choose a function $\tilde{f}$ on $E \simeq \mathbb{C}^1$ which is holomorphic on $E \setminus y$ and has a pole at y. By assumption there exists a function $f \in \mathcal{O}_B(B)$ with $f \mid B \cap E = \tilde{f} \mid B \cap E$. By the lemma there exists $F \in \mathcal{O}_\Delta(\Delta)$ with $F \mid W = f \mid W$. But F is holomorphic at $y \in \Delta$. Consequently $\tilde{f}$ is bounded along all sequences approaching $y \in \partial(W \cap E)$. This is contrary to $\tilde{f}$ having a pole at y. Thus B is holomorphically convex and consequently is Stein. □

2. The Equations $H^1(B, \mathcal{O}_B) = \cdots = H^{m-1}(B, \mathcal{O}_B) = 0$. We now apply Theorem 1 in order to prove the following:

Theorem 2. *Let B be a domain in $\mathbb{C}^m$. Then the following are equivalent:*

i) *B is Stein.*
ii) $H^1(B, \mathcal{O}_B) = \cdots = H^{m-1}(B, \mathcal{O}_B) = 0$.

Proof: It is enough to prove ii) ⇒ i). We proceed by induction on the dimension m. For $m = 1$ it is clear. We will use (∗) for the induction step when $m > 1$. Let $H \subset \mathbb{C}^m$ be an $(m-1)$-dimensional analytic plane which intersects B. Thus $B' := B \cap H$ is a non-empty domain in $H \simeq \mathbb{C}^{m-1}$. We choose a *linear* function l on $\mathbb{C}^m$ which vanishes on H. Thus $\mathcal{O}_{B'} \simeq (\mathcal{O}_B / l\mathcal{O}_B) \mid B'$ and $H^q(B', \mathcal{O}_{B'}) \simeq H^q(B, \mathcal{O}_B / l\mathcal{O}_B)$ for all $q \geq 0$.

Thus we have the exact sequence

$$0 \longrightarrow \mathcal{O}_B \xrightarrow{\lambda} \mathcal{O}_B \longrightarrow \mathcal{O}_B / l\mathcal{O}_B \longrightarrow 0,$$

where λ is defined by $h_z \to l_z h_z$ for $h_z \in \mathcal{O}_z$, $z \in B$. The associated exact cohomology sequence is

$$\mathcal{O}_B(B) \to \mathcal{O}_{B'}(B') \to H^1(B, \mathcal{O}_B) \to \cdots$$
$$\to H^q(B, \mathcal{O}_B) \to H^q(B', \mathcal{O}_{B'}) \to H^{q+1}(B, \mathcal{O}_B) \to \cdots.$$

From this we read off the fact that $\mathcal{O}_B(B) \to \mathcal{O}_{B'}(B')$ is surjective and that $H^1(B', \mathcal{O}_{B'}) = \cdots = H^{m-2}(B', \mathcal{O}_{B'}) = 0$. This along with the induction assumption implies that B' is Stein. □

Remark: The assumption in Theorem 2 that B is a domain in $\mathbb{C}^m$ is quite important. There certainly are manifolds X which are not Stein and all of whose cohomology groups $H^q(X, \mathcal{O})$, $q \geq 1$, vanish. For example every projective space $\mathbb{P}_m$ is such a manifold. Nevertheless in 1966 in "On sheaf cohomology and envelopes of holomorphy," Ann. Math. **84**, 102–118, H. B. Laufer proved the following generalization of Theorem 2:

Every subdomain B of an m-dimensional, Stein manifold whose cohomology groups $H^\mu(B, \mathcal{O})$, $1 \le \mu < m$, all vanish is Stein.

The proof uses among other things the fact that every point $p \in X$ is the simultaneous zero set of m-functions in $\mathcal{O}(X)$.

The condition $(*)$ in Theorem 1 can be further exploited:

Let B be a domain in $\mathbb{C}^m$ for which the Cousin I *problem is universally solvable. Then for every analytic, $(m-1)$-dimensional plane $H \subset \mathbb{C}^m$ which intersects B, the restriction $\mathcal{O}_B(B) \to \mathcal{O}_{B \cap H}(B \cap H)$ is surjective.*

Proof (also see [3], p. 183–4): Let $H = \{(z_1, \ldots, z_m) \in \mathbb{C}^m \mid z_1 = 0\}$. For every point $z \in B$ we choose an open polydisk neighborhood $U_z \subset B$ so that $U_z \cap H = \emptyset$ whenever $z \notin H$. Let $g(z_2, \ldots, z_n) \in \mathcal{O}_{B \cap H}(B \cap H)$ be given and let

$$f^z(z_1, \ldots, z_m) := \frac{g(z_2, \ldots, z_m)}{z_1} \,\Big|\, U_z \in \mathcal{M}_B(U_z) \quad \text{if} \quad z \in H$$

and

$$f^z(z_1, \ldots, z_m) := 0 \in \mathcal{O}_B(U_z) \quad \text{for} \quad z \in H.$$

Since $f^z - f^{z'}$ is holomorphic on $U_z \cap U_{z'}$ for all $z, z' \in B$, the family $\{U_z, f^z\}$ is a Cousin I distribution on B. Thus by assumption there is a function F which is meromorphic on B satisfying $r^z := F \mid U_z - f^z \in \mathcal{O}_B(U_z)$. We set $G := z_1 F$ and note that G is holomorphic on $B \setminus H$ since F is holomorphic there. Moreover for all $z \notin H$

$$G \mid U_z = z_1 r^z + g \mid U_z \in \mathcal{O}_B(U_z).$$

Hence $G \in \mathcal{O}_B(B)$ and $G \mid B \cap H = g$. Consequently $\mathcal{O}_B(B) \to \mathcal{O}_{B \cap H}(B \cap H)$ is surjective. □

The following is now an immediate corollary:

Theorem 3. *The following statements about a domain $B \subset \mathbb{C}^m$ are equivalent:*

i) *B is Stein.*
ii) *The Cousin* I *problem is universally solvable on B, and for every $(m-1)$-dimensional plane H which meets B the intersection $B \cap H$ is Stein.*

Proof: i) ⇒ ii): This is a direct consequence of Theorem 1.1b and Theorem 1.16. ii) ⇒ i): By the remarks directly above, it is clear that B has property $(*)$ of Theorem 1. Thus B is Stein. □

When $m = 2$, Theorem 3 says that a domain $B \subset \mathbb{C}^2$ is Stein if and only if the Cousin problem is universally solvable (Cartan [CAR], 1934).

3. Representation of 1. If $f_1, \ldots, f_l$ are holomorphic functions on a complex space X, then in order for there to exist functions $g_i \in \mathcal{O}(X)$, $1 \le i \le l$, with $1 = \sum_{i=1}^{l} g_i f_i$, the set $\{f_1 = \cdots = f_l = 0\}$ must be empty. The converse is true for Stein spaces. For a proof of this, we begin with a preparatory theorem:

Theorem 4. *Let $\mathcal{S}$ be a coherent sheaf on a Stein space X and take $\mathcal{S}'$ to be an $\mathcal{O}$-subsheaf which is generated by finitely many sections $s_1, \ldots, s_l$ in the $\mathcal{O}(X)$-module $\mathcal{S}(X)$. Then $s_1, \ldots, s_l$ generate the $\mathcal{O}(X)$-module $\mathcal{S}'(X)$. In particular if every stalk $\mathcal{S}_x$ is generated as an $\mathcal{O}_x$-module by $s_{1x}, \ldots, s_{lx} \in \mathcal{S}_x$, then $s_1, \ldots, s_l$ generate the $\mathcal{O}(X)$-module $\mathcal{S}(X)$.*

Proof: Define on X the $\mathcal{O}$-homomorphism $\sigma: \mathcal{O}^l \to \mathcal{S}$ by $(f_{1x}, \ldots, f_{lx}) \to \sum_{i=1}^{l} f_{ix} s_{ix}$. Since X is Stein, the induced $\mathcal{O}(X)$-homomorphism $\mathcal{O}^l(X) \to \mathcal{S}'(X)$ is surjective. If $s_{1x}, \ldots, s_{lx}$ generate $\mathcal{S}_x$ for all $x \in X$, then $\mathcal{S}'(X) = \mathcal{S}(X)$. □

The main result of this paragraph follows as an application of Theorem 4:

Theorem 5 (Representation of 1 by everywhere locally relatively prime functions). *Let X be a Stein space and $f_1, \ldots, f_l \in \mathcal{O}(X)$ be holomorphic on X with $\{x \in X \mid f_1(x) = \cdots = f_l(x) = 0\} = \emptyset$. Then there exist functions $g_1, \ldots, g_l \in \mathcal{O}(X)$ so that*

$$1 = \sum_{i=1}^{l} g_i f_i.$$

Proof: The ideal $\mathcal{I} := \mathcal{O}f_1 + \cdots + \mathcal{O}f_l$ is coherent. Since $f_1, \ldots, f_l$ have no common zeros, $\mathcal{I}_x = \mathcal{O}_x$ for all $x \in X$. It follows from Theorem 4 that $\mathcal{I}(X) = \mathcal{O}(X)$, and as a result $1 \in \mathcal{I}(X) = \mathcal{O}(X)f_1 + \cdots + \mathcal{O}(X)f_l$. □

The following partial converse shows that the conclusion of Theorem 4 is quite strong:

Theorem 6. *The following statements about a domain $B \subset \mathbb{C}^m$ are equivalent:*

i) *B is Stein.*
ii) *If $f_1, \ldots, f_l \in \mathcal{O}(B)$ are such that $\{z \in B \mid f_1(z) = \cdots = f_l(z) = 0\} = \emptyset$, then there exist $g_1, \ldots, g_l \in \mathcal{O}(B)$ with $1 = \sum_{i=1}^{l} g_i f_i$.*

Proof: It is enough to show that ii) implies the holomorphic convexity of B. Let D be a discrete set in B. We may assume that D has an accumulation point $c = (c_1, \ldots, c_m) \in \mathbb{C}^m$, because if not, then one of the coordinate functions $z_1, \ldots, z_m$ would be unbounded on D. Since $c \notin B$, it follows that the m functions $z_\mu - c_\mu$, $1 \le \mu \le m$, have no common zeros on B. Thus there exist functions $g_1, \ldots,$

$g_m \in \mathcal{O}(B)$ with $1 = \sum_{\mu=1}^{m} g_\mu(z_\mu - c_\mu)$. At least one of the g_μ's must be unbounded on D, as otherwise letting z tend to c would show that $1 = 0$. □

4. The Character Theorem. We now prove a theorem which is closely related to Theorem 6. A $\mathbb{C}$-algebra homomorphism $\chi\colon \mathcal{O}(X) \to \mathbb{C}$ is called a (complex) *character*. As usual we let $z_1, \ldots, z_m$ be holomorphic coordinates in $\mathbb{C}^m$. The following is due to J. Igusa [23]:

Theorem 7 (Character Theorem). *The following statements about a domain B in $\mathbb{C}^m$ are equivalent:*

i) *B is Stein.*
ii) *For every character $\chi\colon \mathcal{O}(B) \to \mathbb{C}$, it follows that $(\chi(z_1), \ldots, \chi(z_n)) \in B$.*
iii) For every character $\chi\colon \mathcal{O}(B) \to \mathbb{C}$ there is a point $b \in B$ so that for all $f \in \mathcal{O}(B)$ it follows that $\chi(f) = f(b)$.

Proof: i)⇒ii): Suppose $(\chi(z_1), \ldots, \chi(z_m)) \notin B$. Then the m functions $z_\mu - \chi(z_\mu) \in \mathcal{O}(B)$, $1 \le \mu \le m$, have no common zeros in B. Thus by Theorem 6 there exist functions $g_1, \ldots, g_m \in \mathcal{O}(B)$ with $1 = \sum_{\mu=1}^{m} g_\mu(z_\mu - \chi(z_\mu))$. This yields the following contradiction:

$$1 = \chi(1) = \sum_{\mu=1}^{m} \chi(g_\mu) \cdot \chi(z_\mu - \chi(z_\mu)) = \sum_{\mu=1}^{m} \chi(g_\mu) \cdot 0 = 0.$$

ii)⇒i): Suppose B is not holomorphically convex. Then the lemma in Section 1 on simultaneous continuability guarantees the existence of a point $p = (p_1, \ldots, p_n) \in B$ and a polycylinder Δ about p with $\Delta \not\subset B$ so that every $f \in \mathcal{O}(B)$ has a power series expansion at p which converges compactly to $F \in \mathcal{O}(\Delta)$. The map $\mathcal{O}(B) \to \mathcal{O}(\Delta)$, $f \to F$, is a $\mathbb{C}$-algebra homomorphism. Thus for every $c = (c_1, \ldots, c_m) \in \Delta$ the map $\chi_c\colon \mathcal{O}(B) \to \mathbb{C}, f \to F(c)$, is a character. Since $p_\mu + (z_\mu - p_\mu)$ is the Taylor series of z_μ about p, it follows that $\chi_c(z_\mu) = c_\mu$ and consequently $c \in B$ by ii). This contracts $\Delta \not\subset B$. Hence B is holomorphically convex.

i) ∧ ii) ⇒ iii): Let χ be given and $b := (\chi(z_1), \ldots, \chi(z_m)) \in B$. If there exists $f \in \mathcal{O}(B)$ with $\chi(f) \neq f(b)$, then $f - \chi(f), z_1 - \chi(z_1), \ldots, z_m - \chi(z_m)$ have no common zeros in B. Applying Theorem 6, there exist functions $g, g_1, \ldots, g_m \in \mathcal{O}(B)$ so that

$$1 = g \cdot (f - \chi(f)) + \sum_{\mu=1}^{m} g_\mu \cdot (z_\mu - \chi(z_\mu)).$$

Since $\chi(h - \chi(h)) = 0$ for all $h \in \mathcal{O}(B)$, we have the following contradiction:

$$1 = \chi(1) = \chi(g) \cdot 0 + \sum_{\mu=1}^{m} \chi(g_\mu) \cdot 0 = 0$$

iii)⇒ii): This is trivial, because $(\chi(z_1), \ldots, \chi(z_m)) = b$. □

The following is a beautiful consequence of Theorem 4:

Theorem 8. *If $B \subset \mathbb{C}^m$ is Stein and $b = (b_1, \ldots, b_m) \in B$, then every $f \in \mathcal{O}(B)$ can be represented in the form*

$$f = f(b) + \sum_{\mu=1}^{m} f_\mu \cdot (z_\mu - b_\mu),$$

where $f_\mu \in \mathcal{O}(B)$, $1 \leq \mu \leq m$.

Proof: Let $\mathscr{I}$ be the ideal in $\mathcal{O}_B$ which is generated by the sections $z_1 - b_1, \ldots, z_m - b_m \in \mathcal{O}(B)$. Then $\mathscr{I}$ is coherent. Since B is Stein, Theorem 4 implies that

$$\mathscr{I}(B) = \mathcal{O}(B) \cdot (z_1 - b_1) + \cdots + \mathcal{O}(B) \cdot (z_m - b_m).$$

But $\mathscr{I}_b = \mathfrak{m}(\mathcal{O}_b)$ and $\mathscr{I}_z = \mathcal{O}_z$ for $z \neq b$. Thus $f - f(b) \in \mathscr{I}(B)$ for all $f \in \mathcal{O}(B)$. □

Theorem 8 says that every *character ideal* Ker χ_b, $b \in B$, where $\chi_b \colon \mathcal{O}(B) \to \mathbb{C}$ is defined by $f \to f(b)$, is generated by the m elements $z_1 - \chi_b(z_1), \ldots, z_m - \chi_b(z_m)$. Generalizations of this as well as of the character theorem can be found in Section 7.

§ 6. The Topology on the Module of Sections of a Coherent Sheaf

The goal here is to make the $\mathbb{C}$-vector space $\mathscr{S}(X)$ of global sections of a coherent sheaf $\mathscr{S}$ into a Fréchet space. It will be shown that this Fréchet topology on $\mathscr{S}(X)$ is uniquely determined by certain natural demands. The techniques of construction involve analytic blocks. If X is *reduced*, then this Fréchet topology on $\mathcal{O}(X)$ turns out to be the topology of compact convergence.

The results of this section have important applications in Chapter VI. In fact some of them were already used in Chapter IV.2.5.

0. Fréchet Spaces. Usually a $\mathbb{C}$-vector space V is called a Fréchet space whenever it is *locally convex, metrizable, and complete*. The following definition is more convenient for function theoretical purposes.

Definition 1 (Fréchet space). *A topological $\mathbb{C}$-vector space V is called a Fréchet space if there is a sequence $|\ \ |_\nu$, $\nu = 1, 2, \ldots$, of semi-norms on V so that*

$$(*) \qquad d(v, w) := \sum_{\nu=1}^{\infty} 2^{-\nu} \frac{|v - w|_\nu}{1 + |v - w|_\nu}, \qquad v, w \in V,$$

defines a complete metric d on V which induces the given topology.

Remark: It is always the case that (*) defines a translation invariant *pseudometric* on V. Thus d is a metric if and only if

$$|v|_\nu = 0 \quad \text{for all} \quad \nu \geq 1 \Rightarrow v = 0.$$

The following properties are easy to verify:

If U is a closed subspace of a Fréchet space V, then U with induced topology and V/U with the quotient topology are Fréchet spaces.

If $\{V_i\}_{i \in \mathbb{N}}$ is a sequence of Fréchet spaces, then $\prod_{i \in \mathbb{N}} V_i$ with the product topology is a Fréchet space.

On a given $\mathbb{C}$-vector space it is impossible to have two Fréchet topologies one of which is finer than the other. One can even say more:

Banach Open Mapping Theorem. *Every $\mathbb{C}$-linear, continuous map $\Psi: V \to W$ of a Fréchet space V onto a Fréchet space W is open.*

A proof of this can be found in any standard functional analysis textbook.

1. The Topology of Compact Convergence. The following is well-known:

Theorem 2. *If X is a complex manifold (with countable topology), then the vector space $\mathcal{O}(X)$ of holomorphic functions on X is a Fréchet space (even a Fréchet algebra) with respect to the topology of compact convergence. If $\{U_\nu\}_{\nu \geq 1}$ is a countable open cover of X such that every $\bar{U}$ is compact, then this Fréchet topology is given by the sequence*

$$|\ \ |_\nu: \mathcal{O}(X) \to \mathbb{R}, \qquad |f|_\nu := |f|_{\bar{U}_\nu} = \max_{x \in \bar{U}_\nu} |f(x)|, \qquad f \in \mathcal{O}(X),$$

of sup-norms.

The proof is exactly the same as that for domains in $\mathbb{C}$.

In general the topology of compact convergence is not describable by one (or finitely many) semi-norms:

Let X be a locally-compact topological space and A an $\mathbb{R}$-subalgebra of the algebra $\mathscr{C}(X)$ of all complex-valued continuous functions on X. Suppose that there is an unbounded function $u \in A$. Then there is no $\mathbb{R}$-vector space norm on A which induces the topology of compact convergence.

Proof: Suppose that there exists such a norm $|\ \ |$. The map $A \to A$, $a \to ua$, is a continuous $\mathbb{R}$-linear map and is therefore bounded. That is, there exists $M \in \mathbb{R}$ so that $|ua| \leq M|a|$ for all $a \in A$. So for every $r > 0$ it follows that $|ru \cdot a| \leq rM|a|$. We choose r so that $\varepsilon := rM < 1$. The function $v := ru \in A$ is likewise unbounded on X, and thus $|va| \leq \varepsilon |a|$ for all aA. By induction one sees that

$$|v^n| \leq \varepsilon^{n-1}|v| \quad \text{for all} \quad n = 1, 2, \ldots.$$

Since $\varepsilon < 1$, it follows that v^n converges to 0. This contradicts the fact that v is unbounded. □

As a corollary of this theorem we note that the space $\mathcal{O}(X)$ of all holomorphic functions on a *non-compact, holomorphically convex space X* does not carry a norm which induces the topology of compact convergence.

The topology of compact convergence is compatable with the sequence topology in the convergent power series ring (see [AS], p. 58):

Theorem 3 (Compatability Theorem). *If X is a complex manifold and $\mathcal{O}(X)$ carries the topology of compact convergence, then every restriction map*

$$\mathcal{O}(X) \to \mathcal{O}_x, \quad f \mapsto f_x, \quad x \in X,$$

is continuous whenever $\mathcal{O}_x$ is equipped with the (convergent power series) sequence topology.

Proof: Every point $x \in X$ lies in a compact polycylinder $\Delta \subset X$. The restriction $\mathcal{O}(X) \to \mathcal{O}(\Delta)$ is continuous when we equip $\mathcal{O}(\Delta)$ with the sup-norm $| \quad |_\Delta$. The restriction $\mathcal{O}(\Delta) \to \mathcal{O}_x$ is likewise continuous (see [AS], p. 58). □

For every natural number $l \geq 1$ we furnish the $\mathbb{C}$-vector space

$$\mathcal{O}^l(X) \simeq \prod_1^l \mathcal{O}(X)$$

with the product topology. If $\mathcal{J}$ is a coherent subsheaf of $\mathcal{O}^l$, then the module of sections $\mathcal{J}(X)$ is *closed* in $\mathcal{O}^l(X)$ (see Section 4.1) and is consequently a Fréchet subspace of $\mathcal{O}^l(X)$. Thus the quotient space $\mathcal{O}^l(X)/\mathcal{J}(X)$ is a Fréchet space.

2. The Uniqueness Theorem. Now let X be an arbitrary complex space and $\mathcal{S}$ a coherent $\mathcal{O}$-sheaf on X. There is no obvious notion of compact convergence for sequences of sections $s_\nu \in \mathcal{S}(X)$. In order to find a Fréchet topology for $\mathcal{S}(X)$ we think along the lines of Theorem 3. Every stalk $\mathcal{S}_x$, $x \in X$, carries the natural sequence topology (see [AS], p. 86ff). This topology is always Hausdorff. This, along with the theorem of Banach (i.e. the "open mapping theorem"), shows that the compatibility property of Theorem 3 is already quite significant:

Theorem 4 (Uniqueness Theorem). *If there is a Fréchet topology on $\mathcal{S}(X)$ so that the restriction mappings $\mathcal{S}(X) \to \mathcal{S}_x$, $s \to s_x$, $x \in X$, are continuous, where $\mathcal{S}_x$ carries the sequence topology, then it is unique.*

Proof: Let two such topologies on $\mathcal{S}(X)$ be given. We denote them by V and W. Furnishing $V \times W$ with the product topology, the projections $V \times W \to V$ and $V \times W \to W$ are continuous and the diagonal $\Delta \subset V \times W$ is mapped bijectively in both cases. It is enough to show that Δ is *closed* in $V \times W$, because it will then be a Fréchet space with the induced topology and, by the theorem of Banach, will therefore be homeomorphic to both V and W.

Since the restrictions $V \to S_x$ and $W \to S_x$ are continuous, it follows that if we equip $\mathcal{S}_x \times \mathcal{S}_x$ with the product topology, then

$$\lambda_x \colon V \times W \to \mathcal{S}_x \times \mathcal{S}_x, \qquad (v, w) \mapsto (v_x, w_x)$$

is continuous. Since $\mathscr{S}_x$ is *Hausdorff*, the diagonal Δ_x is closed in $\mathscr{S}_x \times \mathscr{S}_x$. Thus for all $x \in X$ the preimage $\lambda_x^{-1}(\Delta_x)$ is closed in $V \times W$. But $\Delta = \bigcap_{x \in X} \lambda_x^{-1}(\Delta_x)$. So Δ is also closed. □

3. The Existence Theorem. In this paragraph we construct a topology on $\mathscr{S}(X)$ for any given coherent sheaf $\mathscr{S}$ on a complex space X. This topology is compatible with the sequence topology on every $\mathscr{S}_x$, $x \in X$. The following simple lemma shows that this comes down to a local problem.

Lemma. *Let $(X_\nu)_{\nu \geq 1}$ be an open cover of X such that every $\mathscr{S}(X_\nu)$ carries a Fréchet topology with every restriction $\mathscr{S}(X_\nu) \to \mathscr{S}_x$, $x \in X_\nu$, continuous, $\nu > 1$. Define the $\mathbb{C}$-linear map ι, by*

$$\iota\colon \mathscr{S}(X) \to V := \prod_{\nu=1}^{\infty} \mathscr{S}(X_\nu), \qquad s \to (s \,|\, X_\nu)_{\nu \geq 1}.$$

Then ι maps $\mathscr{S}(X)$ bijectively onto a closed subspace of the Fréchet space V. The Fréchet topology on $\mathscr{S}(X)$ which is induced by ι is such that all of the restrictions $\mathscr{S}(X) \to \mathscr{S}_x$, $x \in X$, are continuous.

Proof: By results in Paragraph 0, the space V is Fréchet with the product topology. The mapping ι is $\mathbb{C}$-linear, and, since $\bigcup_1^\infty X_\nu = X$, it is also injective. The image space $\operatorname{Im} \iota \subset V$ consists of all sequences $(s_\nu)_{\nu \geq 1}$, $s_\nu \in \mathscr{S}(X)$, for which $s_{\alpha x} = s_{\beta x}$ whenever $x \in X_\alpha \cap X_\beta$. In order to give a better description of this set, we consider for every point $x \in X_\nu$ the $\mathbb{C}$-linear map $\eta_{\nu x}\colon V \to \mathscr{S}_x$ which is defined by composition of the projection $V \to \mathscr{S}(X_\nu)$ with the restriction $\mathscr{S}(X_\nu) \to \mathscr{S}_x$. It is clear that $\eta_{\nu x}$ is continuous. Setting $I := \{(\mu, \nu, x) \,|\, x \in X_\mu \cap X_\nu\}$, it follows that for every triple $(\mu, \nu, x) \in I$ the set

$$L(\mu, \nu, x) := \{v \in V \,|\, \eta_{\mu x}(v) = \eta_{\nu x}(v)\}$$

is a closed $\mathbb{C}$-subspace of V.[4]
Now

$$\operatorname{Im} \iota = \bigcap_{(\mu,\nu,x) \in I} L(\mu, \nu, x).$$

Thus $\operatorname{Im} \iota$ is closed in V and is consequently a Fréchet space. We transport the Fréchet topology on $\operatorname{Im} \iota$ back to $\mathscr{S}(X)$ by ι^{-1}. Since the restriction $\zeta_x\colon \mathscr{S}(X) \to \mathscr{S}_x$ is just $\zeta_x = \eta_{\nu x} \circ \iota$ for $x \in X_\nu$, it follows that all of the restrictions $\mathscr{S}(X) \to \mathscr{S}_x$ are continuous. □

[4] If $\eta_i\colon A \to B$, $i = 1, 2$, are continuous maps between Hausdorff spaces, then the set $\{a \in A \,|\, \eta_1(a) = \eta_2(a)\}$ is closed.

It remains to show that every point $p \in X$ has a neighborhood W so that $\mathcal{S}(W)$ carries a Fréchet topology which is compatible with the sequence topology on $\mathcal{S}_x$, $x \in W$. Every point $p \in X$ possesses an open neighborhood $U \subset X$ and a *holomorphic embedding* $\pi: U \to V$ of U into a domain $V \subset \mathbb{C}^m$. We choose a compact block $Q \subset V$ with $\pi(p) \in \mathring{Q}$ and set $P := \pi^{-1}(Q)$. Then (P, π) is an analytic block in U with $p \in P^0$, where $P^0 := \pi^{-1}(\mathring{Q})$ is the analytic interior of P. Whenever we have this situation, we call (P, π) a *block neighborhood* of p.

Lemma. *Let $\mathcal{S}$ be coherent on X and (P, π) be a block neighborhood. Then there is a Fréchet topology on $\mathcal{S}(P^0)$ so that all of the restrictions $\mathcal{S}(P^0) \to \mathcal{S}_x$, $x \in P^0$, are continuous.*

Proof: There exists $l \geq 0$ and a $\mathcal{O}$-epimorphism $\varepsilon: \mathcal{O}^l | Q \to \pi_*(\mathcal{S} | U) | Q$. We restrict ε to $\mathring{Q}$. Since $\mathring{Q}$ is Stein and the kernel sheaf $\mathcal{J} := \mathcal{K}er\, \varepsilon | \mathring{Q}$ is coherent, we obtain an exact sequence

$$0 \longrightarrow \mathcal{J}(\mathring{Q}) \longrightarrow \mathcal{O}^l(\mathring{Q}) \xrightarrow{\varepsilon_*} \pi_*(\mathcal{S} | U)(\mathring{Q}) \simeq \mathcal{S}(P^0) \longrightarrow 0.$$

Thus $\mathcal{S}(P^0)$ is algebraically isomorphic to $\mathcal{O}^l(\mathring{Q})/\mathcal{J}(\mathring{Q})$ which carries a natural Fréchet quotient topology. The isomorphism induces a Fréchet topology on $\mathcal{S}(P^0)$.

We have to show that every restriction $\zeta_x: \mathcal{S}(P^0) \to \mathcal{S}_x$, $x \in P^0$, is continuous. Let $z := \pi(x)$. We note that the injectivity of π implies that $\pi^{-1}(z) = x$. It is obvious that the following diagram of natural $\mathbb{C}$-linear maps is commutative:

$$\begin{array}{ccc} \mathcal{O}^l(\mathring{Q}) & \xrightarrow{\varepsilon_*} & \pi_*(\mathcal{S} | U)(\mathring{Q}) \simeq \mathcal{S}(P^0) \\ \downarrow \omega_z & & \downarrow \zeta_x \\ \mathcal{O}^l_z & \xrightarrow{\varepsilon_z} & \pi_*(\mathcal{S} | U)_z \simeq \mathcal{S}_x \end{array}$$

Now ε_* is open and ε_z is a continuous $\mathcal{O}_z$-epimorphism. By the results of Paragraph 1, the restriction ω_z is also continuous. Since $\mathcal{O}_x$ is isomorphic to a quotient algebra of $\mathcal{O}_z$, the sequence topology of the $\mathcal{O}_z$-module $\pi_*(\mathcal{S} | U)_z$ agrees with the sequence topology of the $\mathcal{O}_x$-module $\mathcal{S}_x$. Thus ζ_x is continuous for all $x \in P^0$. □

The main theorem is now an immediate consequence of the two lemmas:

Theorem 5 (Existence Theorem). *Let X be a complex space (with countable topology) and $\mathcal{S}$ a coherent sheaf on X. Then there exists a Fréchet topology on $\mathcal{S}(X)$ so that all of the restrictions $\mathcal{S}(X) \to \mathcal{S}_x$, $x \in X$, are continuous.*

Remark: In the proof of the second lemma, Theorem B for *open* blocks (namely $H^1(\mathring{Q}, \mathcal{J}) = 0$) was used. It should be noted that one can get away with using only Theorem B for compact blocks. As above, one uses the Fréchet space $F := \mathcal{O}^l(\mathring{Q})/\mathcal{J}(\mathring{Q}) \subset S(P^0)$—one doesn't yet know equality! Since $\varepsilon_*: \mathcal{O}^l(Q) \to \pi_*(\mathcal{S}\,|\,U)(Q) \simeq \mathcal{S}(P)$ is surjective, one can construct a $\mathbb{C}$-linear map $\sigma: \mathcal{S}(P) \to F$ with $\operatorname{Ker} \sigma \subset \{s \in \mathcal{S}(P)\,|\,s\,|\,P^0 = 0\}$. Further one can obtain continuous, linear maps $\varphi_x: F \to \mathcal{S}_x, x \in P$, so that $\varphi_x \cdot \sigma$ is always the restriction. The proof of the existence theorem now runs essentially like that above if one chooses in the first lemma a sequence of block neighborhoods $(P_\nu, \pi_\nu)_{\nu \leq 1}$ where $\bigcup_{\nu=1}^{\infty} P_\nu^0 = X$ and V is the corresponding product of Fréchet spaces $F_\nu \subset \mathcal{S}(P_\nu^0)$. □

We call the Fréchet topology which is characterized by Theorems 4 and 5 the *canonical topology* on $\mathcal{S}(X)$, and always think of $\mathcal{S}(X)$ as equipped with it. The following is a corollary of Theorems 4 and 5:

If X is a complex manifold, then the canonical topology on $\mathcal{O}(X)$ is the topology of compact convergence.

There is an important generalization of Theorem 8 which is quite useful for studying convergence:

Let $(s_\nu)_{\nu \geq 0}$ be a sequence of sections $s_\nu \in \mathcal{S}(X)$. Suppose that for every $x \in X$ there is a neighborhood U_x so that $(s_\nu\,|\,U_x)_{\nu \geq 1}$ converges in the canonical topology of $\mathcal{S}(U_x)$ to a section $s^{(x)} \in \mathcal{S}(U_x)$. Then $(s_\nu)_{\nu \geq 1}$ converges in the canonical topology of $\mathcal{S}(X)$ to $s \in \mathcal{S}(X)$ so that $s\,|\,U_x = s^{(x)}$ for all $x \in X$.

This follows immediately from the first lemma. Since X has countable topology, one can obtain the sequence X_ν from the U_x's.

4. Properties of the Canonical Topology. We begin with some remarks about arbitrary complex spaces:

Theorem 6. *The canonical topology on the module of sections in an arbitrary coherent sheaf has the following properties:*

a) *If $U \subset X$ is open, then the restriction map $\rho_U: \mathcal{S}(X) \to \mathcal{S}(U)$ is continuous.*
b) *If $\alpha: \mathcal{S} \to \mathcal{T}$ is an analytic homomorphism between coherent sheaves over X, then the induced map $\alpha_X: \mathcal{S}(X) \to \mathcal{T}(X)$ is continuous.*
c) *If $f: X \to Y$ is a finite holomorphic map, then for every coherent sheaf $\mathcal{S}$ on X the isomorphism $i: f_*(\mathcal{S})(Y) \xrightarrow{\sim} \mathcal{S}(X)$ is a homeomorphism.*

Proof: In all cases we procede as in the proof of the unicity theorem (Theorem 3). We set $V := \mathcal{S}(X)$ and define in the various cases appropriate sets:

a) $W := \mathcal{S}(U)$, $\quad C := \text{Graph } \rho_U = \{(s, s\,|\,U)\} \subset V \times W$
b) $W := \mathcal{T}(X)$, $\quad C := \text{Graph } \alpha_X = \{(s, \alpha_X(s))\} \subset V \times W$
c) $W := f_*(\mathcal{S})(Y)$, $\quad C := \text{Graph } i^{-1} = \{(s, i^{-1}(s))\} \subset V \times W$.

In each case C is proved to be a closed subspace of the Fréchet space $V \times W$ by realizing it as the intersection of closed sets:

a) $C = \bigcap_{x \in U} \{(v, w) \in V \times W \mid w_x = v_x\}$

b) $C = \bigcap_{x \in X} \{(v, w) \in V \times W \mid w_x = \alpha_x(v_x)\}$

c) $C = \bigcap_{y \in Y} \{(v, w) \in V \times W \mid \hat{f}_x(w_y) = v_x$, whenever $f(x) = y\}$.

The point is that the restrictions to stalks are continuous! The continuous projection of $V \times W$ onto V (on V and on W in case c)) induces a bijective map from the Fréchet space C. By the Banach open mapping theorem it is a homeomorphism. □

In the situation of Theorem 6b), the α_X-image of $\mathcal{S}(X)$ is *not in general closed* in $\mathcal{T}(X)$. This is due to the fact that not every global section in $\alpha(\mathcal{S}) \subset \mathcal{T}$ is the image of a section in $\mathcal{S}(X)$.

Closedness Theorem. *If $\mathcal{S}$ is a coherent subsheaf of the coherent sheaf $\mathcal{T}$, then $\mathcal{S}(X)$ is a closed subspace of $\mathcal{T}(X)$.*

Proof: All of the restrictions $\mathcal{T}(X) \to \mathcal{T}_x$ are continuous. Since $\mathcal{S}_x$ is closed in $\mathcal{T}_x$, it follows that $\mathcal{T}(X) \cap \mathcal{S}_x$ is closed in $\mathcal{T}(X)$. Thus $\bigcap_{x \in X} (\mathcal{T}(X) \cap \mathcal{S}_x)$ is likewise closed in $\mathcal{T}(X)$. □

It is now possible to prove several interesting corollaries.

Corollary 1. *If X is Stein and $\alpha\colon \mathcal{S} \to \mathcal{T}$ is an analytic homomorphism between two coherent sheaves over X, then $\alpha_X(\mathcal{S}(X))$ is closed in $\mathcal{T}(X)$.*

Proof: Since X is Stein, the induced homomorphism $\mathcal{S}(X) \to \alpha(\mathcal{S})(X)$ is surjective. Thus $\alpha_X(\mathcal{S}(X)) = \alpha(\mathcal{S})(X)$, and the claim follows from the Closedness Theorem. □

Corollary 2. *Let $\mathcal{S}$ be a coherent sheaf on a Stein space X, and let $s_1, \ldots, s_l \in \mathcal{S}(X)$ be finitely many sections. Then $\mathcal{O}(X)s_1 + \cdots + \mathcal{O}(X)s_l$ is a closed $\mathcal{O}(X)$-submodule of $\mathcal{S}(X)$.*

Proof: By Theorem 5.4 we know that $\mathcal{O}(X)s_1 + \cdots + \mathcal{O}(X)s_l = \mathcal{S}'(X)$, where $\mathcal{S}'$ is the coherent $\mathcal{O}$-subsheaf of $\mathcal{S}'$ which is generated by $s_1, \ldots, s_l$. The Closedness Theorem says that $\mathcal{S}'(X)$ is closed in $\mathcal{S}(X)$. □

We see in particular that

If X is Stein, then every finitely generated ideal $\mathcal{I}$ in $\mathcal{O}(X)$ is closed in $\mathcal{O}(X)$.

In closing we note without proof the following:

Approximation Theorem: *Let X be Stein and (P, π) be an analytic block in X. Then for every coherent $\mathcal{O}$-sheaf $\mathcal{S}$, the image* $\operatorname{Im} \rho$ *of the restriction map $\rho: \mathcal{S}(X) \to \mathcal{S}(P^0)$ is dense in $\mathcal{S}(P^0)$.*

5. The Topologies for $C^q(\mathfrak{U}, \mathcal{S})$ *and* $Z^q(\mathfrak{U}, \mathcal{S})$. Let $\mathcal{S}$, $\mathcal{S}'$ be coherent on X and $\mathfrak{U} = \{U_i\}$ be an open (countable) cover of X. Then every $\mathbb{C}$-vector space $\mathcal{S}(U_{\iota_0 \cdots \iota_q})$, $U_{\iota_0 \cdots \iota_q} = U_{\iota_0} \cap \cdots \cap U_{\iota_q}$, is a Fréchet space with the canonical topology. The $\mathbb{C}$-vector space

$$C^q(\mathfrak{U}, \mathcal{S}) = \prod_{\iota_0, \ldots \iota_q} \mathcal{S}(U_{\iota_0 \cdots \iota_q})$$

of q-cochains is therefore likewise a Fréchet space with the product topology, $0 \le q < \infty$. We summarize some relevant properties of these spaces in the following:

Theorem 7. *Every cochain space $C^q(\mathfrak{U}, \mathcal{S})$, $q = 0, 1, 2, \ldots$, is a Fréchet space. All coboundary maps $\partial: C^{q-1}(\mathfrak{U}, \mathcal{S}) \to C^q(\mathfrak{U}, \mathcal{S})$ are continuous, and the cocycle spaces $Z^q(\mathfrak{U}, \mathcal{S}) = \operatorname{Ker} \partial$, $q = 0, 1, \ldots$, are Fréchet subspaces of $C^q(\mathfrak{U}, \mathcal{S})$. If $\varphi: \mathcal{S}' \to \mathcal{S}$ is an $\mathcal{O}_X$-homomorphism, then every induced ($\mathbb{C}$-linear) map $C^q(\mathfrak{U}, \mathcal{S}') \to C^q(\mathfrak{U}, \mathcal{S})$ is continuous.*

Proof: Every coboundary map ∂ is a finite sum of restriction maps, and is therefore continuous by Theorem 6a). Since the space $Z^q(\mathfrak{U}, \mathcal{S})$ is the kernel of ∂, it is closed and is consequently a Fréchet subspace of $C^q(\mathfrak{U}, \mathcal{S})$. The map $\varphi: \mathcal{S}' \to \mathcal{S}$ induces continuous maps $\mathcal{S}'(U_{\iota_0 \cdots \iota_q}) \to \mathcal{S}(U_{\iota_0 \cdots \iota_q})$ (see Theorem 6b)). Thus the induced mapping $C^q(\mathfrak{U}, \mathcal{S}') \to C^q(\mathfrak{U}, \mathcal{S})$ is also continuous. □

The following is needed in Chapter VI.3.4:

Suppose that the cover $\mathfrak{U}' = \{U_i\}$ is finer than $\mathfrak{U} = \{U_i\}$ with $U_i' \subset U_i$ for all $i \in I$. Then every $\mathbb{C}$-linear restriction map $\rho: C^q(\mathfrak{U}, \mathcal{S}) \to C^q(\mathfrak{U}', \mathcal{S})$ is continuous.

Proof: By Theorem 6a), all of the restriction maps $\mathcal{S}(U_{\iota_0 \cdots \iota_q}) \to \mathcal{S}(U'_{\iota_0 \cdots \iota_q})$ are continuous. Thus the claim is immediate. □

Remark: Every coboundary space $\operatorname{Im} \partial \subset C^q(\mathfrak{U}, \mathcal{S})$ is a topological $\mathbb{C}$-vector space. Consequently every cohomology module $H^q(\mathfrak{U}, \mathcal{S})$ ($\simeq \operatorname{Ker} \partial / \operatorname{Im} \partial$) is also a topological vector space. In this way (by taking the limit) one obtains a natural topology on the space $H^q(X, \mathcal{S})$, $q = 1, 2, \ldots$. The main problem is that this is in general *not* a Fréchet topology, because as a rule the coboundary spaces $\operatorname{Im} \partial \subset C^q(\mathfrak{U}, \mathcal{S})$, $q = 1, 2, \ldots$, are not closed and are therefore not Fréchet spaces.

6. Reduced Complex Spaces and Compact Convergence. In this section X denotes a *reduced* complex space. The $\mathbb{C}$-vector space $\mathcal{O}(X)$ is thus a subspace of the

$\mathbb{C}$-vector space $\mathscr{C}(X)$ of continuous, complex valued functions. Hence, along with the canonical topology of Theorem 5, $\mathcal{O}(X)$ carries the topology of compact convergence. We let V denote $\mathcal{O}(X)$ with the canonical topology and let W be $\mathcal{O}(X)$ with the topology induced from $\mathscr{C}(X)$. Our goal is to show that $V = W$. We begin by remarking that

the identity map id: $V \to W$ *is continuous.*

Proof: Let $\{f_i\}$ be a sequence in V which converges to $f \in V$ (in the canonical topology). Let S denote the analytic set of singular points of X. Now the restriction map $\mathcal{O}(X) \to \mathcal{O}(X \setminus S)$ is continuous in the canonical topology, and on the complex manifold $X \setminus S$ the canonical topology and the topology of compact convergence are the same (see the remark at the end of Paragraph 3). Thus $f_i | X \setminus S$ converges compactly to $f | X \setminus S$. The singular set S is a proper analytic subset of X which is nowhere dense. Thus compact convergence on $X \setminus S$ implies compact convergence on X. □

The following is an immediate consequence of Banach's Theorem:

If W is a Fréchet space, then $V = W$.

By a well-known theorem of analysis, the space $\mathscr{C}(X)$ of all continuous functions on X is a Fréchet space. Hence it only remains to prove the following:

Lemma. *The space $W = \mathcal{O}(X) \subset \mathscr{C}(X)$ is closed in $\mathscr{C}(X)$.*

Proof: Let $\{f_i\} \subset \mathcal{O}(X)$ be a sequence which converges compactly to $f \in \mathscr{C}(X)$. Let S be the nowhere dense analytic set of singular points of X. Then, by Theorem 2, $f | X \setminus S \in \mathcal{O}(X \setminus S)$. Now if X is normal, then the Riemann removability theorem (Chapter A.3.8) implies that $f \in \mathcal{O}(X)$.

In the general case we take a normalization $\xi: \tilde{X} \to X$ of X. Hence the lifted sequence $\tilde{f}_i := f_i \circ \xi$ converges in the canonical topology of $\mathcal{O}(\tilde{X})$ to $\tilde{f} := f \circ \xi \in \mathcal{O}(\tilde{X})$. Since ξ is finite, Theorem 6c) implies that the section $\xi_*(\tilde{f}) \in \xi_*(\mathcal{O}_{\tilde{X}})(X)$ is the limit of the sequence $\xi_*(\tilde{f}_i) \in \xi_*(\mathcal{O}_{\tilde{X}})(X)$ in the canonical topology on the vector space $\xi_*(\mathcal{O}_X)(X)$ of global sections of the coherent $\mathcal{O}_X$-sheaf $\xi_*(\mathcal{O}_X)$. Now, since $\mathcal{O}_X$ is an $\mathcal{O}_X$-subsheaf of $\xi_*(\mathcal{O}_{\tilde{X}})$, the space $\mathcal{O}(X)$ is closed in $\xi_*(\mathcal{O}_{\tilde{X}})(X)$ (by Theorem 6a)). But $\xi_*(\tilde{f}_i) = f_i \in \mathcal{O}(X)$. Thus $\xi_*(\tilde{f}) \in \mathcal{O}(X)$. Furthermore, since ξ is biholomorphic on $X \setminus S$, it follows that $\xi_*(\tilde{f}) = f$ on $X \setminus S$. Finally we recall that S is nowhere dense in X. Hence $f = \xi_*(\tilde{f}) \in \mathcal{O}(X)$. □

In summary, we have proved the following:

Theorem 8. *If X is a reduced complex space, then the topology of compact convergence on $\mathcal{O}(X)$ yields a Fréchet space structure which coincides with the canonical topology on $\mathcal{O}(X)$.*

7. Convergent Series. In this paragraph $(X, \mathcal{O}_X)$ denotes an arbitrary complex space and $(X, \mathcal{O}_{\mathrm{red}\,X})$ its reduction. If $(f_\nu)_{\nu \geq 1}, f_\nu \in \mathcal{O}_X(X)$, is a sequence, then the notion of convergence of the associated series $\sum_{\nu=1}^{\infty} f_\nu$ in the canonical topology of $\mathcal{O}_X(X)$ to $f \in \mathcal{O}(X)$ is well-defined. If $\sum_{\nu=1}^{\infty} f_\nu$ converges to $f \in \mathcal{O}_X(X)$, then the "reduced series" $\sum_{\nu=1}^{\infty} \operatorname{red} f_\nu$, where $\operatorname{red} f_\nu \in \mathcal{O}_{\operatorname{red} X}(X)$, converges compactly to $\operatorname{red} f$ (by Theorem 6b)). In general it is not the case that the convergence of $\sum_{\nu=1}^{\infty} \operatorname{red} f_\nu$ implies the convergence of $\sum_{\nu=1}^{\infty} f_\nu$. As an example we consider the following complex subspace X of the z-plane $\mathbb{C}$:

$$X := \bigcup_{n=1}^{\infty} \{x_n\} \subset \mathbb{C} \quad \text{with} \quad x_n := n$$

and

$$\mathcal{O}_X = \bigcup_{n=1}^{\infty} \mathcal{O}_{X,x_n} \text{ with } \mathcal{O}_{X,x_n} := \mathcal{O}_{\mathbb{C},x_n}/(z - x_n)^n \mathcal{O}_{\mathbb{C},x_n}.$$

On X we consider the sequence

$$f_\nu := (f_{1\nu}, \ldots, f_{n\nu}, \ldots) \quad \text{with} \quad f_{n\nu} := \nu(z - x_n) \text{ modulo } (z - x_n)^n \mathcal{O}_{\mathbb{C},x_n}.$$

Since $\operatorname{red} f_\nu = 0$ for all ν, the series $\sum_{\nu=1}^{\infty} \operatorname{red} f_\nu = 0$ converges. Nevertheless for $m = 1, 2, \ldots$ the series $\sum_{\nu=1}^{\infty} f_\nu^m$ does not converge. However $\sum_{\nu=1}^{\infty} f_\nu^\nu$ has a non-zero limit in $\mathcal{O}_X(X)$, because $f_\nu^\nu = (0, \ldots, 0, *, \ldots)$.

The above example is indicative of the general situation:

Theorem 9 (Convergence Theorem). *Let X be an arbitrary complex space and $(f_\nu)_{\nu \geq 1}, f_\nu \in \mathcal{O}_X(X)$, be a sequence so that the reduced series $\sum_{\nu=1}^{\infty} \operatorname{red} f_\nu$ converges in $\mathcal{O}_{\operatorname{red} X}(X)$. Then there is a sequence $(m_\nu)_{\nu \geq 1}$ of natural numbers so that the series $\sum_{\nu=1}^{\infty} f_\nu^{n_\nu}$ converges whenever $n_\nu \geq m_\nu$.*

For the proof we construct a sequence of semi-norms $(|\ \ |_i)_{i \geq 1}$ on $\mathcal{O}_X(X)$, and for every i a sequence of natural numbers so that the following hold:

1) The semi-norms $|\ \ |_i$, $i \geq 1$, determine the Fréchet topology $\mathcal{O}_X(X)$.

2) The series $\sum_{\nu=1}^{\infty} |f_\nu^{k_\nu}|_i$ of real numbers converge, whenever $k_\nu \geq l_{i\nu}$

Having done this, we define

$$m_\nu := \max\{l_{1\nu}, \ldots, l_{\nu\nu}\}$$

and note that whenever $n_\nu \geq m_\nu$ the series $\sum_{\nu=1}^{\infty} f_\nu^{n_\nu}$ converges for all of these semi-norms, thus proving the theorem.

It remains to carry out the construction. Now every block (P, π) in X (in the sense of Paragraph 3) determines a semi-norm $|\ \ |_\pi$ on $\mathcal{O}_X(X)$ as follows: Since $\pi: U \to V$ is a holomorphic embedding of a non-empty open set U in X into a domain V in some $\mathbb{C}^n$, and since $P = \pi^{-1}(Q)$ is the preimage of a compact block $Q \subset V$ with $Q \neq \emptyset$, one has an $\mathcal{O}_V$-epimorphism

$$\varepsilon: \mathcal{O}_V \to \pi_*(\mathcal{O}_U)$$

which induces an epimorphism

$$\varepsilon_Q: \mathcal{O}_V(Q) \to \pi_*(\mathcal{O}_U)(Q) = \mathcal{O}_U(P)$$

at the section level. One sets

$$|f|_\pi := \inf\{|g|_Q \mid g \in \mathcal{O}(Q),\ \varepsilon_Q(g) = f \mid P\}, \qquad f \in \mathcal{O}_X(X).^5$$

A block (P, π) is called distinguished in U if it is relatively compact in X and V is a Stein domain. Since X has countable topology, there exists a sequence of distinguished blocks (P_i, π_i) in X so that the relatively compact sets $\bigcup_{i=1}^{n} P_i^0$, $n = 1, 2, \ldots$, exhaust the space X. It is clear from the definition of the canonical topology that, for every such sequence of blocks, the sequence $(|\ \ |_i)_{i \geq 1}$ with $|\ \ |_i := |\ \ |_{\pi_i}$ determines the Fréchet topology of $\mathcal{O}_X(X)$. Thus the condition 1) above can be fulfilled by a sequence of semi-norms coming from distinguished blocks.

The following lemma is now important:

Lemma. *Let $q := \frac{1}{2}$ and let (P, π) be a distinguished block in X. Then there exists $l = l(\pi) \in \mathbb{Z}$ and $t = t(\pi) > 0$ satisfying the following:*

For every $f \in \mathcal{O}_X(U)$ with $|f|_U < t$ there exists a constant $M_f > 0$ so that

$$|f^k|_\pi \leq M_f \cdot q^k \quad \textit{for all} \quad k > l.$$

Given this lemma, we can easily find the natural numbers for condition 2): One chooses $l_\nu > l$ big enough so that $M_{f_\nu} \cdot q^{l_\nu} \leq 2^{-\nu}$, $\nu \geq 1$. The convergence of the reduced series $\sum_{\nu=1}^{\infty} \operatorname{red} f_\nu$ along with the relative compactness of U implies that there exists an index $j = j(U)$ so that $|f_\nu|_U < t$ for all $\nu \geq j$. Thus $\sum_{\nu=1}^{\infty} |f_\nu^{k_\nu}|_\pi < \infty$ whenever $k_\nu \geq l_\nu$.

---

[5] The reader should note the analogy between this construction and that for "good" semi-norms in Chapter IV.4.1.

We now come to the proof of the lemma. Using the epimorphism $\varepsilon\colon \mathcal{O}_V \to \pi_*(\mathcal{O}_U)$, we identify $\pi_*(\mathcal{O}_U)$ with $\mathcal{O}_V/\mathcal{J}$ where $\mathcal{J} := \mathcal{K}\mathit{er}\ \varepsilon$. We may further identify $\pi_*(\mathcal{O}_{\mathrm{red}\,U})$ with $\mathcal{O}_V/\mathrm{rad}\,\mathcal{J}$, where $\mathrm{rad}\,\mathcal{J}$ is the nilradical of $\mathcal{J}$. Let $\rho\colon \mathcal{O}_V \to \mathcal{O}_V/\mathrm{rad}\,\mathcal{J}$. Since V is Stein, ε and ρ induce epimorphisms at the section level: $\varepsilon_V\colon \mathcal{O}_V(V) \to \pi_*(\mathcal{O}_U)(V)$ and $\rho_V\colon \mathcal{O}(V) \to \pi_*(\mathcal{O}_{\mathrm{red}\,U})(V)$. Finally, identifying $\pi_*(\mathcal{O}_U)(V)$ with $\mathcal{O}_X(U)$ and $\pi_*(\mathcal{O}_{\mathrm{red}\,U}(V)$ with $\mathcal{O}_{\mathrm{red}\,X}(U)$, one has a commutative triangle

$$\begin{array}{ccc} & \mathcal{O}_V(V) & \\ {\scriptstyle\varepsilon}\swarrow & & \searrow{\scriptstyle\rho} \\ \mathcal{O}_X(U) & \xrightarrow{\ \mathrm{red}\ } & \mathcal{O}_{\mathrm{red}\,X}(U) \end{array}$$

of *continuous*, $\mathbb{C}$*-linear epimorphisms*, where these vector spaces carry their canonical topologies. In this case red denotes the homomorphism at the section level which is induced by the reduction epimorphism $\mathcal{O}_X \to \mathcal{O}_{\mathrm{red}\,X}$. Since the domain $U \subset X$ is Stein, red is surjective. For simplicity we have written ε, ρ instead of ε_V, ρ_V.

The following fact is now useful:

Proposition. *For every $r \in \mathbb{R}$, $0 < r < 1$, there exists a $t > 0$ so that for every $f \in \mathcal{O}_X(U)$ with $|f|_U < t$ there is a $g \in \mathcal{O}_V(V)$ with*

$$\rho(g) = \mathrm{red}\, f \quad \textit{and} \quad |g|_Q \leq r.$$

Proof: By Banach's theorem ρ is open. Since $W := \{h \in \mathcal{O}_V(V) \mid |h|_Q \leq r\}$ is a neighborhood of zero in $\mathcal{O}_V(V)$, it follows that $\rho(W)$ is a neighborhood of zero in $\mathcal{O}_{\mathrm{red}\,X}(U)$. In other words there exists a compact set $K \subset U$ and $t > 0$ so that $\{v \in \mathcal{O}_{\mathrm{red}\,X}(U) \mid |v|_K < t\} \subset \rho(W)$. In particular, for every $f \in \mathcal{O}_X(U)$ with $|f|_U < t$ there exists $g \in \mathcal{O}_V(V)$ with $|g|_Q \leq r$ and $\rho(g) = \mathrm{red}\, f$. □

Continuing with the proof of the lemma, we note that since $Q \subset V$ is compact, there exists $l \geq 1$ so that $(\mathrm{rad}\,\mathcal{J})^l \,|\, Q \subset \mathcal{J} \,|\, Q$. Let $r := q/2$ and let $t > 0$ be the number guaranteed by the above proposition. Take $f \in \mathcal{O}_X(U)$ with $|f|_U < t$. There exists $h \in \mathcal{O}_V(V)$ with $\varepsilon(h) = f$, and (by the proposition) there is a $g \in \mathcal{O}_V(V)$ with $\rho(g) = \mathrm{red}\, f$ and $|g|_Q \leq r$. Define $v := h - g \in \mathcal{O}_V(V)$. Then $\rho(v) = 0$. In other words $v \in (\mathcal{K}\mathit{er}\ \rho)(V) = (\mathrm{rad}\,\mathcal{J})(V)$. Hence

$$v^l \,|\, Q \in \mathcal{J}(Q).$$

Now let

$$y_k := \sum_{\mu = l+1}^{k} \binom{k}{\mu} g^{k-\mu} v^{\mu} \in \mathcal{O}_V(V), \qquad k = l+1,\, l+2, \ldots.$$

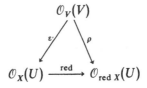

Then $y_k | Q \in \mathscr{I}(Q)$. If one sets

$$x_k := \sum_{\mu=0}^{l} \binom{k}{\mu} g^{k-\mu} v^{\mu} \in \mathcal{O}_V(V), \qquad k > l,$$

then

$$h^k = (g+v)^k = x_k + y_k, \quad \text{and} \quad f^k = \varepsilon(h^k) = \varepsilon(x_k) + \varepsilon(y_k) \quad \text{for all} \quad k > l.$$

Since $y_k | Q \in \mathscr{I}(Q)$, it follows that $\varepsilon(y_k | Q) = 0$. Consequently $f^k | P = \varepsilon(x_k | Q) = \varepsilon_Q(x_k)$ for $k > l$. Hence, by the definition of the semi-norm $|\ \ |_\pi$,

$$|f^k|_\pi \le |x_k|_Q \quad \text{for all} \quad k > l.$$

Recalling that $|g|_Q < r$, we estimate $|x_k|_Q$ as follows:

$$|x_k|_Q \le \sum_{\mu=0}^{l} \binom{k}{\mu} |g|_Q^{k-\mu} \cdot |v|_Q^{\mu} = r^k \sum_{\mu=0}^{l} \binom{k}{\mu} (|g|_Q^{-1} |v|_Q)^{\mu}.$$

Since l is fixed, the number

$$M_f := \max_{1 \le \mu \le l} \{(|g|_Q^{-1} |v|_Q)^{\mu}\} \in \mathbb{R}$$

depends only on f. Now $\sum_{\mu=0}^{l} \binom{k}{\mu} \le \sum_{\mu=0}^{k} \binom{k}{\mu} = 2^k$ and $q = 2r$. Thus

$$|x_k|_Q \le r^k M_f 2^k = M_f \cdot q^k \quad \text{for all} \quad k > l.$$

Thus we have concluded the proof of the Lemma as well as that for Theorem 9. □

As an application of the Convergence Theorem we now show that Theorem 12 in Chapter IV.2.5 holds for *arbitrary*, holomorphically convex spaces X. Recalling the proof of that theorem, given an infinite discrete set $D \subset X$ and a family $\{r_p\}_{p \in D}$ of real numbers, we constructed a sequence $(h_\nu)_{\nu \ge 0}$, $h_\nu \in \mathcal{O}(X)$, so that every series $\sum_{\nu=0}^{\infty} (\text{red } h_\nu)^{m_\nu}$, $m_\nu \ge 1$, converges compactly to a function $g \in \mathscr{C}(X)$ with $|g(p)| \ge r_p$ for all $p \in D$ (see the remark at the end of Chapter IV.2.5). By Theorem 9 (i.e. the Convergence Theorem) one can choose particular m_ν's so that the series $\sum_{\nu=0}^{\infty} h_\nu^{m_\nu}$ converges in $\mathcal{O}(X)$ to $h \in \mathcal{O}(X)$. Since red $\mathcal{O}_X(X) \to \mathcal{O}_{\text{red } X}(X)$ is continuous, red $h = g$ and thus $|h(p)| \ge r_p$ for all $p \in D$. □

A second, completely pretentious proof of Theorem 12 of Chapter IV.2.5 should be noted. One uses the following "Reduction Theorem," the proof of which requires further preparation (in particular the "big coherence theorem" [CAS]):

For every weakly holomorphically convex complex space X there is a holomorphic map $\rho\colon X \to \check{X}$ of X onto a Stein space $\check{X}$ so that the following hold:

1) *ρ is proper and all fibers $\rho^{-1}(\rho(x))$, $x \in X$ are connected.*
2) ρ induces an isomorphism $\rho^*\colon \mathcal{O}_{\check{X}}(\check{X}) \to \mathcal{O}_X(X)$.

From this, Theorem 12 for weakly holomorphically convex spaces X follows in three lines: If D is discrete and infinite in X, then the properness of ρ guarantees that $\rho(D)$ is discrete and infinite in $\check{X}$. Since $\check{X}$ is Stein, there exists an $\check{h} \in \mathcal{O}_{\check{X}}(\check{X})$ so that $|\check{h}(\rho(p))| \geq r_p$ for all $p \in D$. For $h := \rho^*(\check{h}) \in \mathcal{O}_X(X)$ it follows that $|h(p)| \geq r_p$ for all $p \in D$. □

In particular we see that

every weakly holomorphically convex space X is holomorphically convex.

This result also goes back to K.-W. Wiegmann: Structuren auf Quotienten komplexer Räume, Comm. Math. Helv. **44**, 93–116 (1969).

§ 7. Character Theory for Stein Algebras

A $\mathbb{C}$-algebra is called a *Stein algebra* if it is algebraically isomorphic to a $\mathbb{C}$-algebra $\mathcal{O}(X)$ of holomorphic functions on a Stein space X. In this section we will show by means of character theory that finite dimensional Stein spaces are *completely* determined by their Stein algebras.

We will always use X to denote a complex space and $T := \mathcal{O}(X)$ the $\mathbb{C}$-algebra of all functions holomorphic on X.

1. Characters and Character Ideals. A $\mathbb{C}$-algebra homomorphism $\chi\colon T \to \mathbb{C}$ is called a *character* (of T), and the ideal $\operatorname{Ker} \chi \subset T$ is called a *character ideal.* Character ideals are *maximal* ideals in T, and a character is uniquely determined by its character ideal.

Every point $p \in X$ determines the *point character* $\chi_p\colon T \to \mathbb{C}, f \to f(p)$, whose ideal is

$$\operatorname{Ker} \chi_p = \{f \in T \mid f_p \in \mathfrak{m}(\mathcal{O}_p)\}.$$

Every *point character χ_p is continuous* (T is assumed to be equipped with the canonical Fréchet topology), because χ_p is just the composition of the continuous restriction map $\mathcal{O}(X) \to \mathcal{O}_p$ with the continuous quotient map $\mathcal{O}_p \to \mathcal{O}_p/\mathfrak{m}(\mathcal{O}_p) = \mathbb{C}$.

Different points can determine the same point character. For example on a compact space they are all the same. On the other hand

if X is holomorphically separable, then $\chi_p \neq \chi_q$ for all $p, q \in X$ with $p \neq q$.

It is possible to give characters which are not point characters. In fact, as Theorem 5.7 shows, this is the case for every non-Stein domain in $\mathbb{C}^m$.

The following remark is quite useful:

(*) *If X is Stein and $I \neq T$ is an ideal in T, then every finite set $\{f_1, \ldots, f_l\} \subset I$ has at least one common zero $p \in X$.*

Proof: If $f_1, \ldots, f_l$ had no common zeros, then we could find $g_1, \ldots, g_l \in T$ with $1 = \sum_{i=1}^{l} g_i f_i$ (see Theorem 5.5). Since $I \neq T$, this is impossible. □

As an immediate consequence of (*) we note that

if X is Stein, then for every finitely generated, maximal ideal M in T there is a unique point $p \in X$ so that $M = \operatorname{Ker} \chi_p$.

Proof: Let $M = Tf_1 + \cdots + Tf_l$. By (*) the functions $f_1, \ldots, f_l$ have a common zero $p \in X$. Thus $M \subset \operatorname{Ker} \chi_p$. But the maximality of M implies that $M = \operatorname{Ker} \chi_p$. Since X is holomorphically separable, p is uniquely determined. □

We are now able to show the following:

Theorem 1. *Let X be a Stein space, $p \in X$, and let I be an ideal in T which is generated by finitely many functions $h_1, \ldots, h_l \in T$. Then the following statements are equivalent:*

i) *The point p is the only common zero of $h_1, \ldots, h_l$, and the germs $h_{1p}, \ldots, h_{lp} \in \mathcal{O}_p$ generate the ideal $\mathfrak{m}(\mathcal{O}_p)$.*
ii) *$I^s = \{f \in T \mid f_p \in \mathfrak{m}(\mathcal{O}_p)^s\}$ for all $s = 1, 2, \ldots$.*
iii) *$I = \operatorname{Ker} \chi_p$.*

Proof: i)⇒ii): Let $s \geq 1$ be fixed. Let $\mathcal{J}$ be the ideal sheaf generated by the finite many sections $h_1^{\nu_1} \cdot h_2^{\nu_1} \cdot \cdots \cdot h_l^{\nu_l}$, $\nu_1 + \cdots + \nu_l = s$. Then $\mathcal{J}_p = \mathfrak{m}(\mathcal{O}_p)^s$, and $\mathcal{J}_x = \mathcal{O}_x$ for $x \neq p$. Thus $\mathcal{J}(X) = \{f \in T \mid f_p \in \mathfrak{m}(\mathcal{O}_p)^s\}$. Now the products $h_1^{\nu_1} h_2^{\nu_2} \cdot \cdots \cdot h_l^{\nu_l}$, $\nu_1 + \cdots + \nu_l = s$, generate the ideal I^s. So by Theorem 5.4 it follows that $I^s = \mathcal{J}(X)$.

ii)⇒iii): This is clear, because $\operatorname{Ker} \chi = \{f \in T \mid f_p \in \mathfrak{m}(\mathcal{O}_p)\}$.

iii)⇒i): For every $q \in X \backslash p$ there exists an $f \in T$ with $f(q) = 1$ and $f(p) = 0$. Thus p is the only common zero of $h_1, \ldots, h_l$. By Theorem 4.3, v) there exist functions $g_1, \ldots, g_t \in T$ whose germs at p generate $\mathfrak{m}(\mathcal{O}_p)$. Since $g_1, \ldots, g_t \in \operatorname{Ker} \chi_p$, the germs $h_{1p}, \ldots, h_{lp}$ also generate $\mathfrak{m}(\mathcal{O}_p)$. □

2. Finiteness Lemma for Character Ideals. The goal of this paragraph is to give a proof of the following:

Theorem 2 (Finiteness Lemma). *Let X be a finite dimensional Stein space. Then every character $\chi: T \to \mathbb{C}$ is a point character χ_p with finitely generated character ideal* Ker χ.

We need a dimension theoretical remark for the proof:

Lemma. *Let A be a d-dimensional analytic set in a Stein space X, $1 \leq d < \infty$. Then there exists a holomorphic function $g \in T$ so that every set $A \cap \{x \in X \mid g(x) = c\}$, $c \in \mathbb{C}$, is at most $(d-1)$-dimensional.*

Proof: Let A_i, $i = 1, 2, \ldots$, be the d-dimensional prime components of A. Choose two points $p_i, q_i \in A_i$ which are *not* in any other prime component of A (see Chapter A.3.5-6). The set D of all such p_i and q_i is discrete in X. By the results in Paragraph 4.1 there exists $g \in T$ which has different values at the different points of D. If the set $A_c := A \cap \{x \in X \mid g(x) = c\}$ were d-dimensional, then A_c and A would have a common d-dimensional prime component A_i. This would imply that $g(p_i) = g(q_i) = c$, and this is impossible. □

We now prove Theorem 2. Let χ be an arbitrary character of T, and let $m := \dim X$. If $m \geq 1$, then let $g_1 \in T$ be the function guaranteed by the lemma for $A := X$. It follows that $h_1 := g_1 - \chi(g_1) \in \operatorname{Ker} \chi$, and the m_1-dimensional set $X_1 := \{x \in X \mid h_1(x) = 0\}$ is non-empty with $m_1 < m$. If $m_1 \geq 1$, then let g_2 be the function guaranteed by the lemma where $A := X_1$. With $h_2 := g_2 - \chi(g_2) \in \operatorname{Ker} \chi$, it follows from $(*)$ that $X_2 := \{x \in X \mid h_1(x) = h_2(x) = 0\} \neq \emptyset$. Furthermore, since $X_2 = X_1 \cap \{x \in X \mid h_2(x) = 0\}$, we have $m_2 := \dim X_2 < m_1$. Continuing on for at most m steps, we obtain a 0-dimensional, non-empty set

$$X_k = \{x \in X \mid h_1(x) = h_2(x) = \cdots = h_k(x) = 0\}, \quad \text{where} \quad h_1, \ldots, h_k \in \operatorname{Ker} \chi.$$

Since X_k is discrete in X, the results in Paragraph 4.1 show that there exists $g_0 \in \mathcal{O}(T)$ which maps X_k injectively into $\mathbb{C}$. We set $h_0 := g_0 - \chi(g_0)$ and note that the set of common zeros of $h_0, \ldots, h_k \in \operatorname{Ker} \chi$, which by $(*)$ must be non-empty, consists of one point p. Thus (again by $(*)$) $h(p) = 0$ for every $h \in \operatorname{Ker} \chi$. In other words $\operatorname{Ker} \chi \subset \operatorname{Ker} \chi_p$. By the maximality of $\operatorname{Ker} \chi$, it follows that $\operatorname{Ker} \chi = \operatorname{Ker} \chi_p$ and consequently $\chi = \chi_p$.

By Theorem 4.3 there are finitely many more functions $h_{k+1}, \ldots, h_l \in T$ whose germs generate the ideal $\mathfrak{m}(\mathcal{O}_p)$. Thus (by Theorem 1) $\operatorname{Ker} \chi_p$ is generated by h_0, $h_1, \ldots, h_l \in T$. □

Corollary. *If X is Stein and finite dimensional, then every character χ is continuous with closed ideal* $\operatorname{Ker} \chi$ *(with respect to the canonical Fréchet topology on T).*

This is clear, because χ must be a point character. □

Remark 1. The assumption in Theorem 2 that X is Stein is *indispensible.* In fact an example of a 3-dimensional manifold Y is constructed in [18] with points $p \in Y$ whose character ideals $\operatorname{Ker} \chi_p$ in $\mathcal{O}(Y)$ can *not* be finitely generated. The algebra $\mathcal{O}(Y)$ is therefore *not* a Stein algebra. There exists an analytic set $A \subset Y$ so that the difference $Y \backslash A$ is a ramified holomorphically separable domain over $\mathbb{C}^3$, and whose function algebra $\mathcal{O}(Y \backslash A)$ is isomorphic to $\mathcal{O}(Y)$. In particular this shows that there are holomorphically separable domains G over $\mathbb{C}^3$ whose algebras $\mathcal{O}(G)$ are not Stein.

Remark 2: The assumption in Theorem 2 that X is finite dimensional is dispensible. In fact Theorem 2 can be sharpened as follows:

Sharp Version of Theorem 2. *If X is a Stein space, then the following statements about a maximal ideal $M \subset T$ are equivalent:*

i) *Relative to the canonical Fréchet topology M is closed in T.*
ii) *There exists a point $p \in X$ so that $M = \operatorname{Ker} \chi_p$.*
iii) *M is finitely generated.*

We want to say a few words about the proof. The main difficulty is the implication i)$\Rightarrow$ii). For this one needs the following theorem of H. Cartan:

Let X be Stein and $I \neq T$ be a closed ideal in T. Then the ideal sheaf $\mathscr{I}$ which I generates is coherent and $I = \mathscr{I}(X)$.

This theorem is proved in [CAS]. As a corollary one gets the following sharpened version of $(*)$ in Paragraph 1:

$(\stackrel{*}{*})$ *If X is Stein, then every closed ideal $I \neq T$ has a non-empty zero set.*

If I had an empty zero set, then $\mathscr{I}_x = \mathcal{O}_x$ for all $x \in X$ (i.e. $\mathscr{I} = \mathcal{O}$). Thus, by Cartan's theorem, $I = T$.

The implication i)$\Rightarrow$ii) is immediate from $(\stackrel{*}{*})$.

In order to verify ii)$\Rightarrow$iii), given $p \in X$, one first constructs finitely many functions $h_1, \ldots, h_k \in \operatorname{Ker} \chi_p$ which have no common zeros other than p on the union X' of all prime components of X which go through p (X' is Stein and finite dimensional!). Next one can find $h_{k+1} \in \operatorname{Ker} \chi_p$ which is nowhere zero on $X \backslash X'$. This is done by letting $\tilde{X}$ be the union of the prime components of X which don't contain p and noting that $Y := \{p\} \cup \tilde{X}$ is a closed subspace of X. Thus the function $h \in \mathcal{O}_Y(Y)$ with $h(p) = 0$ and $h | \tilde{X} = 1$ is by Theorem 4.4 continuable to a function $h_{k+1} \in \mathcal{O}(X)$. If one adds to $\{h_1, \ldots, h_{k+1}\}$ functions $g_1, \ldots, g_t \in \operatorname{Ker} \chi_p$ whose germs at p generate the ideal $\mathfrak{m}(\mathcal{O}_p)$, then by Theorem 1 one has a generating system for $\operatorname{Ker} \chi_p$.

The implications iii)$\Rightarrow$ii)$\Rightarrow$i) are clear. □

Remark 3: In every Stein algebra T which belongs to a complex space X which contains a discrete infinite set D, there are maximal ideals which can not be finitely generated: The set $L := \{f \in T \mid f(x) = 0$ for *almost all* $x \in D)$ generates an ideal $I \neq T$ whose zero set is empty. By Zorn's lemma I is contained in a *maximal ideal* $M \subset T$. This ideal can *not be finitely generated*. It is *dense* in T, because if $\bar{M} \neq T$ then $\bar{M}$ would be an ideal which contains M as a proper subset which is contrary to the maximality of M.

The quotient field T/M of any Stein algebra by a maximal ideal M is in a natural way an extension field of $\mathbb{C}$ which is in fact $\mathbb{C}$ for character ideals. It is amusing to note that T/M *is (in a highly non-canonical way) always algebraically isomorphic to* $\mathbb{C}$.

3. The Homeomorphism $\Xi: X \to \mathscr{X}(T)$. For every complex space X with $\mathbb{C}$-algebra $T = \mathcal{O}(X)$ we call the set $\mathscr{X}(T)$ of all characters $\chi: T \to \mathbb{C}$ the (analytic) *spectrum* of X. One has the canonical map

$$\Xi: X \to \mathscr{X}(T), \qquad p \mapsto \chi_p.$$

The map Ξ is injective if and only if X is holomorphically separable.

We equip $\mathscr{X}(T)$ with the so-called *weak topology*: The sets

$$V(\chi; f_1, \ldots, f_n; \varepsilon) := \{\varphi \in \mathscr{X}(T) \mid |\varphi(f_1) - \chi(f_1)| < \varepsilon, \ldots, |\varphi(f_n) - \chi(f_n)| < \varepsilon\},$$

where ε ranges over all positive real numbers and $\{f_1, \ldots, f_n\}$ is an arbitrary finite subset of T, form a basis of open neighborhoods about the character χ. It is easy to see that

$$\Xi: X \to \mathscr{X}(T) \qquad \textit{is continuous.}$$

Proof: Let $p \in X$ and $V := V(\chi_p; f_1, \ldots, f_n; \varepsilon)$ be arbitrary. One can choose a neighborhood U of p in X so that for all $p \in U$ it follows that $|f_\nu(x) - f_\nu(p)| < \varepsilon$, $\nu = 1, \ldots, n$. Thus $\Xi(U) \subset V$. □

The following is proved by applying the generalized embedding theorem of Paragraph 1.1:

Theorem 3. *The following statements about a finite-dimensional complex space X are equivalent:*

i) *X is Stein.*
ii) *$\Xi: X \to \mathscr{X}(T)$ is a homeomorphism.*

Proof: i) ⇒ ii): Since Ξ is injective and (by Theorem 2) is also surjective, it is enough to show the *openness* of Ξ. For this let $p \in X$ and U be a neighborhood of p in X. In order to give a $V := V(\chi_p; f_1, \ldots, f_n; \varepsilon)$ with $\Xi(U) \supset V$, we have to find functions $f_1, \ldots, f_n \in T$ and $\varepsilon > 0$ so that

$$\{x \in X \mid |f_1(x) - f_1(p)| < \varepsilon, \ldots, |f_n(x) - f_n(p)| < \varepsilon\} \subset U.$$

Since X is finite dimensional, the generalized embedding theorem guarantees the existence of finitely many functions $f_1, \ldots, f_n \in T$ so that $F := (f_1, \ldots, f_n): X \to \mathbb{C}^n$ maps X homeomorphically onto a closed topological subspace of $\mathbb{C}^n$. Now the polycylinders

$$\{(z_1, \ldots, z_n) \in \mathbb{C}^n \mid |z_\nu - f_\nu(p)| < r, \nu = 1, \ldots, n\}$$

form a basis for the topology of $\mathbb{C}^n$ at the point $(f_1(p), \ldots, f_n(p))$. Thus the sets

$$W_r := \{x \in X \mid |f_\nu(x) - f_\nu(p)| < r, \nu = 1, \ldots, n\}, \qquad r > 0,$$

form a neighborhood basis at p in X. Consequently there exists an $\varepsilon > 0$ with $W_\varepsilon \subset U$.

ii) ⇒ i): First, since Ξ is injective, the space X is holomorphically separable. In order to show that X is holomorphically convex we let K be an arbitrary compact set in X and $\hat{K}$ its holomorphically convex hull. We wish to show that $\hat{K}$ is compact in X. Thus it is enough to prove that every ultrafilter $\mathfrak{F}$ on $\hat{K}$ has a limit $p \in X$. For every $f \in T$ we know that $f(\mathfrak{F})$ is an ultrafilter basis on the compact disk $\{z \in \mathbb{C} \mid |z| \leq |f|_K < \infty\}$ in $\mathbb{C}$. Thus $\lim f(\mathfrak{F}) \in \mathbb{C}$ exists.

The map $f \mapsto \lim f(\mathfrak{F}), f \in T$, is obviously a character. Thus, since Ξ is surjective, there is a point $p \in X$ so that for all $f \in T$ we have $f(p) = \lim f(\mathfrak{F})$. Consequently $\Xi(\mathfrak{F})$ is an ultrafilter basis in $\mathscr{X}(T)$ which converges in the (weak) topology of $\mathscr{X}(T)$ to $\chi_p = \Xi(p)$. That is, $\Xi(p) = \lim \Xi(\mathfrak{F}$. Since Ξ^{-1} is continuous, it follows that $p = \lim \mathfrak{F}$. □

In the proof of i) ⇒ ii) above we actually proved the following:

If X is a finite dimensional Stein space, then for every character $\chi \in \mathscr{X}(T)$ there exist finitely many functions $h_1, \ldots, h_n \in \operatorname{Ker} \chi$ so that the system $\{U_\delta\}_{\sigma > 0}$, $U_\delta := V(\chi; h_1, \ldots, h_n; \delta)$, forms a neighborhood basis of χ in $\mathscr{X}(T)$ (Just set $h_i := f_i - f_i(p)!$).

Remark 1: The implication ii) ⇒ i) of Theorem 3 was proved by R. Iwahashi. It should be noted that for domains X in $\mathbb{C}^m$ the theorem of Igusa (Theorem 5.7) shows that the bijectivity of Ξ implies that X is Stein. Hence for domains if Ξ is bijective, then it is a homeomorphism.

Remark 2: As in the case of Theorem 2, the finite dimensionality assumption in Theorem 3 is superfluous. In the proof of ii) ⇒ i), $\dim X < \infty$ was never used. The proof of i) ⇒ ii) can be modified as follows: First we note that Ξ is injective and (by the sharpened version of Theorem 2) is also surjective. Let X' be the union of the prime components of X which go through p. As above, since $\dim X' < \infty$, we have functions $f_1, \ldots, f_n \in T$ so that $F := (f_1, \ldots, f_n): X' \to \mathbb{C}^n$ is injective and proper. We can also find $f_{n+1} \in \operatorname{Ker} \chi_p$ with $f_{n+1} \mid (X \setminus X') \equiv 1$. Thus for sufficiently small $\varepsilon > 0$ the domain $\{x \in X \mid |f_\nu(x) - f_\nu(p)| < \varepsilon, 1 \leq \nu \leq n+1\}$ is contained in U.

Remark 3: The use of the embedding theorem in the proof of Theorem 3 seems unavoidable. In the proof of Theorem 2, the much weaker dimension theoretical lemma sufficed.

4. Complex Analytic Structure on $\mathscr{X}(T)$. Let $p \in X$ be an arbitrary point in the complex space X, and continue the notation $T := \mathcal{O}(X)$. The restriction homomorphism $\tau: T \to \mathcal{O}_p$ maps $I := \operatorname{Ker} \chi_p$ into $\mathfrak{m} := \mathfrak{m}(\mathcal{O}_p)$. Thus for every $n = 1, 2, \ldots$, we have an induced $\mathbb{C}$-algebra homomorphism $\tau_n: T/I^n \to \mathcal{O}_p/\mathfrak{m}^n$ so that the diagram below (where the vertical arrows denote the quotient homomorphisms) is commutative.

$$\begin{array}{ccc} T & \xrightarrow{\tau} & \mathcal{O}_p \\ \downarrow & & \downarrow \\ T/I^n & \xrightarrow{\tau_n} & \mathcal{O}_p/\mathfrak{m}^n. \end{array}$$

Lemma. *If X is Stein and finite dimensional, then every map τ_n: $T/I^n \to \mathcal{O}_p/\mathfrak{m}^n$ is bijective.*

Proof: Let n be fixed. We first show that τ_n is surjective. Let $\bar{f}_p \in \mathcal{O}_p/\mathfrak{m}^n$ with preimage $f_p \in \mathcal{O}_p$ be given. By Theorem 4.2 there exists $h \in T$ with $h_p - f_p \in \mathfrak{m}^n$. Thus the equivalence class $\bar{h} \in T/I^n$ satisfies $\tau_n(\bar{h}) = \bar{f}_p$.

In order to prove the injectivity of τ_n, we let $\bar{g} \in T/I^n$ with $\tau_n(\bar{g}) = 0$. Take $g \in T$ to be some preimage of $\bar{g}$. Thus $\tau(g) \in \mathfrak{m}^n$. Now Theorem 2 implies that I is finitely generated. Hence by Theorem 1, ii) it follows that $g \in I^n$, and consequently $\bar{g} = 0$. □

Given a character $\chi \in \mathscr{X}(T)$ we consider now the *completion* $\hat{T}_\chi$ *of* T *with respect to the* Ker χ*-adic topology*:

$$\hat{T}_\chi = \varprojlim T/(\operatorname{Ker} \chi)^n.$$

The completion of $\mathcal{O}_p$ with respect to the $\mathfrak{m}(\mathcal{O}_p)$-adic topology is denoted by $\hat{\mathcal{O}}_p$. That is, $\hat{\mathcal{O}}_p := \varprojlim \mathcal{O}_p/(\mathfrak{m}(\mathcal{O}_p))^n$. The following is a consequence of the above lemma:

Theorem 4. *Let X be a finite dimensional Stein space and χ: $T \to \mathbb{C}$ a character. Then the restriction $T \to \mathcal{O}_{\Xi^{-1}(\chi)}$ induces a $\mathbb{C}$-algebra isomorphism*

$$\hat{\tau}_\chi\colon \hat{T}_\chi \xrightarrow{\sim} \hat{\mathcal{O}}_{\Xi^{-1}(\chi)}.$$

Proof: This is clear, because the maps τ_n: $T/I^n \to \mathcal{O}_p/\mathfrak{m}^n$ are isomorphisms. □

We now set

$$T_\chi := \hat{\tau}_\chi^{-1}(\mathcal{O}_{\Xi^{-1}(\chi)}), \qquad \tau_\chi := \hat{\tau}_\chi | T_\chi.$$

Then τ_χ: $T_\chi \to \mathcal{O}_{\Xi^{-1}(\chi)}$ is a $\mathbb{C}$-algebra homomorphism and in particular $\hat{T}_\chi$ and T_χ are *noetherian local rings* over $\mathbb{C}$. The natural $\mathbb{C}$-homomorphism j_χ: $T \to \hat{T}_\chi$ has the ideal $\bigcap_{\nu=1}^{\infty} (\operatorname{Ker} \chi)^\nu$ as its kernel and maps Ker χ into the maximal ideal $\mathfrak{m}(T_\chi)$ of T_χ. Thus $j_\chi(\operatorname{Ker} \chi)$ is a $\hat{T}_\chi$-generating system of the maximal ideal $\mathfrak{m}(\hat{T}_\chi)$ of $\hat{T}_\chi$. We have the commutative diagram below, where the vertical arrows denote the inclusions.

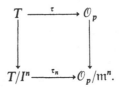

$$\begin{array}{ccc} & T & \\ & {\scriptstyle j_\chi}\swarrow \quad \searrow{\scriptstyle \tau} & \\ T_\chi & \xrightarrow[\tau_\chi]{\sim} & \mathcal{O}_{\Xi^{-1}(\chi)} \\ \downarrow & & \downarrow \\ \hat{T}_\chi & \xrightarrow[\hat{\tau}_\chi]{\sim} & \hat{\mathcal{O}}_{\Xi^{-1}(\chi)} \end{array}$$

The following invariant description of the algebras T_χ and $\hat{T}_\chi$ can now be proved:

Theorem 5. *Let T be a finite dimensional Stein algebra, take $\chi \in \mathscr{X}(T)$ and let $g_1, \ldots, g_n \in j_\chi(\operatorname{Ker} \chi)$. Then every formal power series $\sum_0^\infty a_{\nu_1 \ldots \nu_n} g_1^{\nu_1} \cdot \ldots \cdot g_n^{\nu_n}$, $a_{\nu_1 \ldots \nu_n} \in \mathbb{C}$, converges in the $\mathfrak{m}(\hat{T}_\chi)$-adic topology to a uniquely determined element $g \in \hat{T}_\chi$. Moreover $g \in T_\chi$ if and only if the series $\sum_0^\infty a_{\nu_1 \ldots \nu_n} z_1^{\nu_1} \cdot \ldots \cdot z_n^{\nu_n}$ converges compactly in some neighborhood U of $0 \in \mathbb{C}^n$. If $\{h_1, \ldots, h_l\} \subset \operatorname{Ker} \chi$ is a generating system of $\operatorname{Ker} \chi$, then every $f \in \hat{T}_\chi$ (resp. T_χ) is in fact a formal (resp. convergent) power series in $f_1 := j_\chi(h_1), \ldots, f_l := j_\chi(h_l)$.*

Proof: For every analytic local ring $\mathcal{O}_p$ with completion $\hat{\mathcal{O}}_p$ there is a $\mathbb{C}$-epimorphism of an algebra $\mathbb{C}[[Y_1, \ldots, Y_t]]$ of formal power series in finitely many indeterminants $Y_1, \ldots, Y_t$ onto $\hat{\mathcal{O}}_p$ which maps the $\mathbb{C}$-subalgebra of convergent power series onto $\mathcal{O}_p$. It follows immediately from the elementary theory of quotient algebras that every formal power series in arbitrary elements $v_1, \ldots, v_n \in \mathfrak{m}(\mathcal{O}_p)$ represents in the $\mathfrak{m}(\mathcal{O}_p)$-adic topology an element $v \in \hat{\mathcal{O}}_p$, and $v \in \mathcal{O}_p$ if and only if the series converges. If $\{w_1, \ldots, w_l\} \subset \mathfrak{m}(\mathcal{O}_p)$ is an $\hat{\mathcal{O}}_p$-generating system of $\mathfrak{m}(\hat{\mathcal{O}}_p)$, then it is clear that every $w \in \hat{\mathcal{O}}_p$ is a power series in $w_1, \ldots, w_l$.

One now observes that every T-generating system of $\operatorname{Ker} \chi$ carries over by j_χ to a $\hat{T}_\chi$-generating system of $\mathfrak{m}(\hat{T}_\chi)$. Thus all of the claims of Theorem 5 follow from the above diagram. □

It is now easy to equip the analytic spectrum $\mathscr{X}(T)$ of a Stein algebra T with a complex analytic structure sheaf $\mathscr{T}$. First we put the weak topology on $\mathscr{X}(T)$. Then let

$$\mathscr{T} := \bigcup_{\chi \in \mathscr{X}(T)} T_\chi,$$

and let $\pi\colon \mathscr{T} \to \mathscr{X}(T)$ denote the map obtained from the "stalk projections" $T_\chi \to \chi$. We introduce a topology in $\mathscr{T}$ as follows:

Let $\chi \in \mathscr{X}(T)$ be fixed and $\{h_1, \ldots, h_l\}$ be a generating system of $\operatorname{Ker} \chi$. If $f \in T_\chi$, then by Theorem 5 there is a power series $P = P(z_1, \ldots, z_l) =$

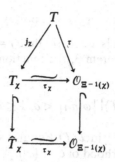

$\sum_0^\infty a_{\nu_1 \cdots \nu_l} z_1^{\nu_1} \cdot \cdots \cdot z_l^{\nu_l}$, which converges in a polycylinder $\Delta_\varepsilon := \{|z_i| < \varepsilon,\ 1 \le i \le l\}$ whose radius $\varepsilon = \varepsilon(f)$ depends on f, so that $f = P(f_1, \ldots, f_l)$ with $f_i := j_\chi(h_i)$. By the observation after Theorem 3, the functions $h_1, \ldots, h_l$ can be chosen so that the sets

$$U_\delta := \{\varphi \in \mathscr{X}(T) \mid |\varphi(h_i)| < \delta,\ 1 \le i \le l\}, \qquad \delta > 0,$$

form a neighborhood basis of χ in $\mathscr{X}(T)$. For all $\delta < \varepsilon$ and $\varphi \in U_\delta$ the power series P converges in some neighborhood of $\mathcal{O} := (\varphi(h_1), \ldots, \varphi(h_l)) \in \Delta_\varepsilon$. Thus φ is associated in a unique way to a convergent power series $P_\varphi = \sum_0^\infty a_{\nu_1 \cdots \nu_l}(\varphi) z_1^{\nu_1} \cdot \cdots \cdot z_l^{\nu_l}$, where

$$P_\varphi(z_1 - \varphi(h_1), \ldots, z_l - \varphi(h_l)) = P(z_1, \ldots, z_l)$$

for all $(z_1, \ldots, z_l)$ near 0. Since $h_i - \varphi(h_i) \in \operatorname{Ker} \varphi$, it follows from Theorem 5 that $f_\varphi := P_\varphi(f_1 - \varphi(h_1), \ldots, f_l - \varphi(h_l)) \in T_\varphi$. Obviously $f_\chi = f$. We set

$$W_\delta(f) := \{f_\varphi \mid \varphi \in U_\delta\}, \qquad \delta < \varepsilon = \varepsilon(f),$$

and let the topology in $\mathscr{T}$ be generated by such $W_\delta(f)$'s as f runs over $\mathscr{T}$. It is now routine to show that

$(\mathscr{T}, \pi, \mathscr{X}(T))$ is a sheaf of $\mathbb{C}$-algebras over $\mathscr{X}(T)$ with germs $\mathscr{T}_\chi = T_\chi$.

Further we define

$$\tau^* := \bigcup_{\chi \in \mathscr{X}(T)} \tau_\chi, \quad \text{where} \quad \tau_\chi\colon T_\chi \xrightarrow{\sim} \mathcal{O}_{\Xi^{-1}(\chi)}.$$

Then we have the commutative diagram

$$\begin{array}{ccc} \mathcal{O} & \xleftarrow{\ \tau^*\ } & \mathscr{T} \\ \downarrow & & \downarrow \pi \\ X & \xrightarrow{\ \Xi\ } & \mathscr{X}(T) \end{array},$$

and (Ξ, τ^*) is an isomorphism of ringed spaces $(X, \mathcal{O}) \to (\mathscr{X}(T), \mathscr{T})$. In summary we see that $(X, \mathcal{O})$ is completely reconstructable from T:

Theorem 6. *Let $(X, \mathcal{O})$ be a finite dimensional Stein space with Stein algebra $T = \mathcal{O}(X)$. Then the ringed space $(\mathscr{X}(T), \mathscr{T})$ is canonically isomorphic to $(X, \mathcal{O})$ (and is thus in particular Stein).*

It is easy to see that the map

$$(X, \mathcal{O}_X) \rightsquigarrow T := \mathcal{O}_X(X)$$

of finite dimensional Stein spaces to their Stein algebras is a *contravariant* functor and the map

$$T \rightsquigarrow (\mathscr{X}(T), \mathscr{T})$$

of Stein algebras to their analytic spectra is likewise a contravariant functor. Thus one quickly arrives at the following result:

The category of finite dimensional Stein spaces and the category of finite dimensional Stein algebras are anti-equivalent.

It turns out to be irrelevant whether or not one puts the additional condition of *Fréchet* on Stein algebras and whether or not the morphisms are *continuous*. In fact the following are easy to show:

Every Stein algebra T possesses a unique topology so that T is a Fréchet algebra.
Every $\mathbb{C}$*-algebra homomorphism of a finite dimensional Stein algebra into a second Stein algebra is continuous.*

In the above anti-equivalence theorem the assumption of finite dimensionality is again unnecessary. For this we refer the reader to the more detailed literature in [CAS] and [14].

Chapter VI. The Finiteness Theorem

The main purpose of this chapter is to prove the following:

Finiteness Theorem (Cartan, Serre): *Let X be a compact complex space. Then for every coherent sheaf $\mathscr{S}$ on X all of the cohomology modules $H^q(X, \mathscr{S})$, $0 \le q \le \infty$, are finite dimensional $\mathbb{C}$-vector spaces.*

It is easy to sketch the idea of the proof. If $\mathfrak{V}$, $\mathfrak{W}$ are two Stein coverings of X with $\mathfrak{V} < \mathfrak{W}$, then the maps in the bottom row of the following commutative diagram are bijective:

$$\begin{array}{ccccc}
Z^q(\mathfrak{W}, \mathscr{S}) & \longrightarrow & Z^q(\mathfrak{V}, \mathscr{S}) & & \\
\downarrow{\scriptstyle \psi} & & \downarrow{\scriptstyle \varphi} & & \\
H^q(\mathfrak{W}, \mathscr{S}) & \xrightarrow{\ \sim\ } & H^q(\mathfrak{V}, \mathscr{S}) & \xrightarrow{\ \sim\ } & H^q(X, \mathscr{S}).
\end{array}$$

Since ψ is surjective, it follows that

$$\varphi(Z^q(\mathfrak{W}, \mathscr{S}) | \mathfrak{V}) = H^q(\mathfrak{V}, \mathscr{S}).$$

Due to the compactness of X, we will be able to show that $\mathfrak{V}$ and $\mathfrak{W}$ can be chosen (dependent on $\mathscr{S}$) so that for each q there are *finitely many* cocycles $\xi_1, \ldots, \xi_d \in Z^q(\mathfrak{V}, \mathscr{S})$ with

$$Z^q(\mathfrak{W}, \mathscr{S}) | \mathfrak{V} \subset \sum_1^d \mathbb{C}\xi_i \oplus \partial C^{q-1}(\mathfrak{V}, \mathscr{S}).$$

But $\operatorname{Ker} \varphi = \partial C^{q-1}(\mathfrak{V}, \mathscr{S})$. Thus $H^q(\mathfrak{V}, \mathscr{S}) = \sum_1^d \mathbb{C}\varphi(\xi_i)$ and consequently $\dim_{\mathbb{C}} H^q(X, \mathscr{S}) \le d < \infty$.

The proof of Cartan and Serre makes use of the finiteness lemma of L. Schwartz concerning continuous linear maps between Fréchet spaces. The proof which we give here is instead dependent on the Schwarz lemma of classical function theory

We replace the sup-norm by the L^2-norm (going back to S. Bergman) on the Hilbert space of square integrable holomorphic functions. This turns out to be much easier to handle. Following the classical pattern we construct monotone orthogonal bases by means of minimal functions. The smoothing of cocycles follows in a simple way via the Banach open mapping theorem.

It should be said that this method can be further developed in order to carry out a proof of the general finiteness theorem involved in the *coherence of the image sheaf by proper holomorphic maps.* The smoothing part is however significantly more complicated.

§ 1. Square-integrable Holomorphic Functions

Let $\mathbb{C}^m$ be equipped with coordinates $z_\mu = x_\mu + iy_\mu$, $1 \le \mu \le m$. We denote its euclidean volume element by $d\lambda := dx_1\, dy_1 \cdots dx_m\, dy_m$. For every domain $B \subset \mathbb{C}^m$, the space $\mathcal{O}_h(B)$ is defined to be the *Hilbert space* of holomorphic functions on B which are square-integrable with respect to $d\lambda$. If $\mathcal{O}(B)$ is given its natural Fréchet topology, then the injection $\mathcal{O}_h(B) \to \mathcal{O}(B)$ is compact (Theorem 4). A Schwarz lemma is proved for the spaces $\mathcal{O}_h^k(B)$.

1. The Space $\mathcal{O}_h(B)$. For every function $f \in \mathcal{O}(B)$ we put

$$\|f\|_B^2 := \int_B |f(z)|^2\, d\lambda \le \infty.$$

The set of square-integrable holomorphic functions on B,

$$\mathcal{O}_h(B) := \{f \in \mathcal{O}(B) \mid \|f\|_B < \infty\},$$

is a normed $\mathbb{C}$-vector space (not a $\mathbb{C}$-algebra!) with $\|\ \|_B$ as norm. It is always the case that

(1) $$\|f\|_B \le \sqrt{\operatorname{vol} B}\, |f|_B,$$

where vol B denotes the euclidean volume of B. Thus if B is a bounded domain, all of the bounded functions in $\mathcal{O}(B)$ belong to $\mathcal{O}_h(B)$.

The following is an immediate consequence of (1):

(2) *If* vol B *is finite (e.g. B bounded) and the sequence $(f_j) \subset \mathcal{O}_h(B)$ converges uniformly to $f \in \mathcal{O}(B)$, then*

$$f \in \mathcal{O}_h(B), \qquad \lim_j \|f_j - f\|_B = 0 \quad \textit{and} \quad \lim_j \|f_j\|_B = \|f\|_B.$$

The reader should note that convergence on compact subsets does not in general imply $\|\ \|_B$-convergence. For example the sequence jz^j on the unit disk $E \subset \mathbb{C}$ has no $\|\ \|_E$-limit.

The above remark implies in particular that,

(3) *if B' is a relatively compact subdomain of a domain B, then $\mathcal{O}(B) \subset \mathcal{O}_h(B')$ and the "restriction" $\mathcal{O}(B) \to \mathcal{O}_h(B')$ is continuous (with the Fréchet topology on $\mathcal{O}(B)$ and the $\| \ \|_{B'}$-topology on $\mathcal{O}_h(B')$).*

A *positive-definite hermitian product* is defined on $\mathcal{O}_h(B)$ by

$$(f, g)_B := \int_B f\bar{g}\, d\lambda, \qquad f, g \in \mathcal{O}_h(B).$$

The associated norm is $\| \ \|_B$ and the Schwarz inequality, $|(f, g)_B| \le \|f\|_B \|g\|_B$, is valid. An explicit calculation of $(\ , \ )_B$ is rarely possible. However for monomials on polydisks one obtains the following:

Theorem 1 (Orthogonality Relations). *Let $\Delta := E_1 \times \cdots \times E_m$, where*

$$E_j := \{z_j \in \mathbb{C} \mid |z_j| < r_j\}, \qquad r_j > 0,$$

and $z^\mu := z_1^{\mu_1} \cdots z_m^{\mu_m}$, $z^\nu = z_1^{\nu_1} \cdots z_m^{\nu_m}$ be two monomials, $\mu_j \ge 0$, $\nu_j \ge 0$. Then

$$(z^\mu, z^\nu)_\Delta = 0 \quad \textit{for} \quad \mu \neq \nu \quad \textit{and} \quad \|z^\nu\|_\Delta^2 = \pi^m \cdot \frac{r_1^{2\nu_1+2}}{\nu_1 + 1} \cdot \ldots \cdot \frac{r_m^{2\nu_m+2}}{\nu_m + 1}.$$

Proof: By Fubini's Theorem, $(z^\mu, z^\nu)_\Delta = \prod_{j=1}^{m} (z_j^{\mu_j}, z_j^{\nu_j})_{E_j}$. Using $z_j = \rho e^{i\varphi}$, the claims follow from the fact that

$$(z_j^{\mu_j}, z_j^{\nu_j})_{E_j} = \int_0^{r_j} \int_0^{2\pi} \rho^{\mu_j+\nu_j+1} e^{i(\mu_j-\nu_j)\varphi}\, d\varphi\, d\rho.$$

2. The Bergman Inequality. The monotone convergence theorem from integration theory implies that

$$(4) \qquad \|f\|_B = \sup_{B' \Subset B} \|f\|_{B'}$$

for all $f \in \mathcal{O}(B)$, where B' runs through all relative compact subdomains of B. The following is an easy consequence of this:

Theorem 2 (Completeness Relations). *If Δ is a polydisk centered at $0 \in \mathbb{C}^m$, then*

$$\|f\|_\Delta^2 = \sum_0^\infty |a_\nu|^2 \|z^\nu\|_\Delta^2 \quad \textit{for all} \quad f = \sum_0^\infty a_\nu z^\nu \in \mathcal{O}(\Delta).$$

Proof: On every concentric polydisk $\Delta' \Subset \Delta$, the Taylor polynomials $\sum_0^{<\infty} a_\nu z^\nu$ converge uniformly to $f \mid \Delta'$. Then by (2) and Theorem 1

$$\|f\|_{\Delta'}^2 = \lim \left\| \sum_0^{<\infty} a_\nu z^\nu \right\|_{\Delta'} = \lim \left(\sum_0^{<\infty} |a_\nu|^2 \|z^\nu\|_{\Delta'}^2 \right) = \sum_0^{\infty} |a_\nu|^2 \|z^\nu\|_{\Delta'}^2.$$

The claim now follows from (4):

$$\|f\|_{\Delta}^2 = \sup_{\Delta' \Subset \Delta} \sum_0^{\infty} |a_\nu|^2 \|z^\nu\|_{\Delta'}^2 = \sum_0^{\infty} |a_\nu|^2 \cdot \sup_{\Delta' \Subset \Delta} \|z^\nu\|_{\Delta'}^2 = \sum_0^{\infty} |a_\nu|^2 \|z^\nu\|_{\Delta}^2. \qquad \square$$

The completeness relations, which among other things imply that yield $\mathcal{O}_h(\Delta)$ is a Hilbert space with $\{z^\nu\}$ as an orthogonal basis, also imply the "Cauchy inequalities"

$$|a_\nu| \leq \frac{\|f\|_\Delta}{\|z^\nu\|_\Delta}, \qquad \nu = (\nu_1, \ldots, \nu_m), \qquad \nu_\mu \geq 0,$$

for the $\| \quad \|_\Delta$-norm. The inequality for $\nu_1 = \cdots = \nu_m = 0$ is quite important for us. Since $f(0) = a_0$ and $\|1\|_\Delta^2 = \pi^m r_1^2 \cdots r_m^2$ (see Theorem 1), it can be written as follows:

$$(5) \qquad |f(0)| \leq \frac{1}{(\sqrt{\pi})^m r_1 \cdots r_m} \|f\|_\Delta \quad \text{for all} \quad f \in \mathcal{O}_h(\Delta).$$

This implies an analogous inequality for arbitrary domains B:

Theorem 3 (Bergman Inequality). *Let $K \subset B$ be compact and d be the euclidean distance from K to the boundary of B. Then*

$$|f|_K = \frac{(\sqrt{m})^m}{(\sqrt{\pi}\, d)^m} \|f\|_B \quad \textit{for all} \quad f \in \mathcal{O}_h(B).$$

Proof: Let $p \in K$. Since $d\lambda$ is translation invariant, we may assume that $p = 0$. The polydisk Δ centered at p with polyradius $(r_1, \ldots, r_m)$, $r_\mu = \sqrt{m}^{-1} \cdot d$ lies in B. Since $\|f\|_\Delta \leq \|f\|_B$, (5) implies that

$$|f(p)| \leq \frac{(\sqrt{m})^m}{(\sqrt{\pi}\, d)^m} \|f\|_B \quad \text{for all points} \quad p \in K.$$

3. The Hilbert Space $\mathcal{O}_h^k(B)$. For every natural number $k \geq 1$ we have the k-fold direct sum $\mathcal{O}_h^k(B) := \bigoplus_1^k \mathcal{O}_h(B)$ equipped with the inner product.

$$(f, g)_B := \sum_1^k (f_i, g_i)_B,$$

where

$$f=(f_1,\ldots,f_k),\quad g=(g_1,\ldots,g_k),\quad \text{and}\quad f_i, g_i\in\mathcal{O}_h(B).$$

Theorem 4. *For every domain B, the space $\mathcal{O}_h^k(B)$ equipped with $(\ ,\ )_B$ is a Hilbert space, $1\le k\le\infty$. The injection $\mathcal{O}_h^k(B)\to\mathcal{O}^k(B)$ is continuous and compact (i.e. every bounded set in $\mathcal{O}_h^k(B)$ is relatively compact in $\mathcal{O}^k(B)$).*

Proof: It is enough to consider the case $k=1$. From the Bergman inequality it follows that $\|\ \|_B$-convergence implies convergence on compact subsets. Thus $\mathcal{O}_h(B)\to\mathcal{O}(B)$ is continuous.

If $f_j\in\mathcal{O}_h(B)$ is a Cauchy sequence, then it is also a Cauchy sequence in $\mathcal{O}(B)$ and therefore converges compactly to $f\in\mathcal{O}(B)$. Let $\varepsilon>0$ be given and choose n_0 so that $\|f_i-f_j\|_B<\varepsilon$ for all $i,j>n_0$. Thus, for every relatively compact subdomain B' of B, (1) implies that

$$\|f_j-f\|_{B'}\le\|f_j-f_i\|_{B'}+\|f_i-f\|_{B'}\le\varepsilon+(\operatorname{vol}B')^{1/2}|f_i-f|_{B'}.$$

Letting $i\to\infty$ one sees that $\|f_j-f\|_{B'}<\varepsilon$ for $j\ge n_0$. Since this is true for all $B'\Subset B$, it follows from (4) that $\|f_j-f\|_B\le\varepsilon$ for $j\ge n_0$. In other words, $f\in\mathcal{O}_h(B)$ and $\lim\|f_j-f\|_B=0$. Hence $\mathcal{O}_h(B)$ is complete.

Every bounded set in $\mathcal{O}_h(B)$ is a family of functions holomorphic on B which by Theorem 3 is uniformly bounded on any compact $K\subset B$. But the classical theorem of Montel states that such a set is relatively compact in $\mathcal{O}(B)$. □

In addition to Theorem 4 one can say that every closed and bounded set in $\mathcal{O}^k(B)$ is compact in $\mathcal{O}_h^k(B)$. This is an immediate corollary of Theorem 4 and the following:

(6) *If $f_j\in\mathcal{O}_h^k(B)$ is a $\|\ \|_B$-bounded sequence which converges compactly to $f\in\mathcal{O}^k(B)$ then $f\in\mathcal{O}_h^k(B)$ and $\|f\|_B\le\underline{\lim}\|f_j\|_B$.*

4. Saturated Sets and the Minimum Principle. A subset $S\subset\mathcal{O}_h^k(B)$ is said to be *saturated* if there exists a Fréchet closed set $T\subset\mathcal{O}^k(B)$ so that $S=T\cap\mathcal{O}_h^k(B)$.

Theorem 5 (Minimum Principle). *Every non-empty saturated set $S\subset\mathcal{O}_h^k(B)$ contains an element g with $\|g\|_B=\inf\{\|v\|_B\,|\,v\in S\}$.*

Proof: There exists a sequence $g_j\in S$ with $\lim\limits_j\|g_j\|_B=m:=\inf\{\|v\|_B\,|\,v\in S\}$. Since this sequence is $\|\ \|_B$-bounded, there is a compactly convergent subsequence g_j^* with limit $g\in\mathcal{O}^k(B)$. From (6) it follows that $\|g\|_B\le\lim\|g_j^*\|_B=m$ Since S is saturated, $g\in S$ and therefore $\|g\|_B=m$ by the definition of m. □

5. The Schwarz Lemma. The classical Schwarz lemma can be stated as follows

Let E, E' be disks centered at the origin in the w-plane with radii $0<r'<r$. Let $a:=r'r^{-1}$. Suppose $h\in\mathcal{O}(E)$ vanishes of order e at the origin. Then

$$|h|_{E'}\le a^e|h|_E.$$

The Schwarz lemma for the sup-norm in polydisks is an easy consequence:

Let $\Delta := \{z \in \mathbb{C}^m \mid |z_\mu| < r\}$ *and* $\Delta' := \{z \in \mathbb{C}^m \mid |z_\mu| < r'\}, 0 < r' < r$, *be polydisks in* $\mathbb{C}^m$. *If the function* $f \in \mathcal{O}(\Delta)$ *vanishes of order e at the origin, then*

$$|f|_{\Delta'} \leq a^e |f|_\Delta.$$

Proof: Let $c \in \Delta', c \neq 0$. The complex line $L_c := \{z = cw \mid w \in \mathbb{C}\} \subset \mathbb{C}^n$ cuts Δ, Δ' in concentric disks $E := L_c \cap \Delta, E' := L_c \cap \Delta'$. Considering E and E' as disks in the w-plane, they have radii $r\gamma$ and $r'\gamma$ respectively where $\gamma = (\max |c_\mu|)^{-1}$. The function $f \mid E := h$ is holomorphic on E and vanishes of at least order e at the origin. Since the ratio of the radii is still a, an application of the one variable Schwarz lemma yields

$$|f(c)| = |h(c)| \leq |h|_{E'} \leq a^e |h|_E \leq a^e |f|_\Delta. \qquad \square$$

Theorem 6 (Schwarz Lemma for the Hilbert Norm). *Let*

$$\Delta := \{z \in \mathbb{C}^m \mid |z_\mu| < r\}, \qquad \Delta' := \{z \in \mathbb{C}^m \mid |z_\mu| < r'\}, \qquad 0 < r' < r,$$

be polydisks in $\mathbb{C}^m$. *Let a be any real number with* $r'r^{-1} < a < 1$. *Then for any* $k \geq 1$, *there exists a constant* $M > 0$ *(dependent on r, r' and a) with the following property:*

If $f \in \mathcal{O}_h^k(\Delta)$ *vanishes of order e at the origin, then*

$$\|f\|_{\Delta'} \leq a^e M \|f\|_\Delta.$$

Proof: It is enough to consider the case where $k = 1$. Let $\tilde{r} := a^{-1} r'$ and $\tilde{\Delta} := \{z \in C^m \mid |z_\mu| < \tilde{r}\}$. Hence $r' < \tilde{r} < r$ and $\Delta' \Subset \tilde{\Delta} \Subset \Delta$. By the Bergman inequality, there exists a constant L so that

$$|f|_{\tilde{\Delta}} \leq L \|f\|_\Delta \quad \text{for all} \quad f \in \mathcal{O}_h(\Delta).$$

Applying the Schwarz lemma for the sup-norm to the disks Δ' and $\tilde{\Delta}$, it follows that

$$|f|_{\Delta'} \leq a^e |f|_{\tilde{\Delta}}.$$

Since $\|f\|_{\Delta'} \leq \sqrt{\mathrm{vol}(\Delta')} |f|_{\Delta'}$, we see that $M := L \cdot \sqrt{\mathrm{vol}(\Delta')}$ does the job.

§ 2. Monotone Orthogonal Bases

Using classical techniques, monotone orthogonal bases are constructed for saturated Hilbert subspaces of $\mathcal{O}_h^k(B)$ by means of minimal functions.

1. Monotonicity. Let B be a domain in $\mathbb{C}^m$ and fix $p \in B$ as well as the natural number k. For short we write $F := \mathcal{O}^k(B)$. Every vector $f = (f_1, \ldots, f_k) \in F$ has a well-defined order at p:

$$o_p(f) := \min(o_p(f_1), \ldots, o_p(f_k)).$$

A sequence $(g_j) \subset F$ is said to be monotone at p if

$$o_p(g_1) \leq o_p(g_2) \leq \cdots \leq o_p(g_j) \leq \cdots \quad \text{and} \quad \lim_j o_p(g_j) = \infty.$$

For H a Hilbert subspace of $\mathcal{O}_h^k(B)$, we will look for orthogonal bases $\{g_1, g_2, \ldots\}$ of H which are monotone at p. Such bases were originally considered by S. Bergman. The monomials in the example of the polydisk Δ form (when properly indexed) a monotone orthogonal bases for $\mathcal{O}_h(\Delta)$ at the origin.

The key property of monotone orthogonal bases is expressed in the following theorem. This will be the main ingredient for a convergence argument in Section 4.

Theorem 1. *Let Δ and Δ' be polydisks centered at $0 \in \mathbb{C}^m$ with $r > r'$ and take $a \in \mathbb{R}$ such that $r'r^{-1} < a < 1$. Let $k \geq 1$ and denote by $M > 0$ the constant of the Schwarz lemma. Suppose that H is a Hilbert subspace of $\mathcal{O}_h^k(\Delta)$ equipped with an orthogonal bases $\{g_1, g_2, \ldots\}$ which is monotone at 0. Then for every $e \in \mathbb{N}$ there exists $d \in \mathbb{N}$ such that for all $v \in \mathcal{O}_h^k(\Delta)$*

$$\left\| v - \sum_{i=1}^{d} (v, g_i) g_i \right\|_{\Delta'} \leq M a^e \|v\|_\Delta.$$

Proof: The monotonicity implies that for every $e \in \mathbb{N}$ there exists $d \in \mathbb{N}$ such that for *every* $v \in H$ the vector $w := v - \sum_{i=1}^{d} (v, g_i) g_i$ vanishes of order at least e at the origin. The Schwarz lemma implies that $\|w\|_{\Delta'} \leq M a^e \|w\|_\Delta$ and the orthogonality of $\{g_j\}$ yields $\|w\|_\Delta \leq \|v\|_\Delta$.

2. The Subdegree. In order to construct monotone orthogonal bases we need a slightly more general function than o_p. For this we set $I := \{1, \ldots, l\}$, $\mathbb{N}^m := \{v = (v_1, \ldots, v_m) \mid v_\mu \geq 0\}$, $A = I \times \mathbb{N}^m$ and $|v| := v_1 + \cdots + v_m$. For two elements $\alpha = (i, v)$, $\alpha' = (i', v') \in A$ we say that $\alpha < \alpha'$ whenever anyone of the following three conditions is satisfied:

1) $|v| < |v'|$
2) $|v| = |v'|$ and there exists j, $1 \leq j \leq m$, with $v_j \leq v'_j$ and $v_k = v'_k$ for $k > j$.
3) $v = v'$ and $i < i'$.

This relation $<$ is a *linear ordering* of A with each non-empty subset of A having a uniquely determined *smallest* element. The *subdegree* at p of a non-zero $f = (f_1, \ldots, f_k) \in F$, $\omega_p(f)$, is now defined as follows: Let $f_i = \sum a_{iv}(z - z(p))^v$ be the Taylor development of f_i at p. Then

$$\omega_p(f) := \min\{(i, v) \in A \mid a_{iv} \neq 0\}.$$

If $\omega_p(f) = (i^*, v^*)$, then $o_p(f) = |v^*|$. Thus $\omega_p(f) \leq \omega_p(g)$ implies that $o_p(f) \leq o_p(g)$. Consequently

every sequence $g_1, g_2, \ldots$ with $\omega_p(g_1) < \cdots < \omega_p(g_j) < \cdots$ is monotone in p.

In that which follows we write ω instead of ω_p.

3. Construction of Monotone Orthogonal Bases by Means of Minimal Functions. Maintaining the notation of the last section, we begin with the following remark:

For every index $\alpha \in A$ the set $F(\alpha) := \{f \in F \mid \omega(f) \geq \alpha\} \cup \{0\}$ is a Fréchet subspace of F.

Proof: If $c \in \mathbb{C}^*$, then $\omega(cf) = \omega(f)$. Furthermore $\omega(f+g) \geq \min(\omega(f), \omega(g))$. Thus $F(\alpha)$ is a linear subspace of F. It is obviously closed. □

In $F(\alpha)$ we have the set

$$F(\alpha)^* := \left\{ f = \left(\sum_{\nu} a_{i\nu}(z - z(p))^\nu \right)_{i \in I} \in F \mid a_{i^*\nu^*} = 1 \right\},$$

where $(i^*, \nu^*) := \alpha$. It is clear that $F(\alpha)^*$ is closed in F. □

If H is a vector subspace of F, then the set $A_H := \{\omega(f) \mid f \in H, f \neq 0\} \subset A$ is countable (perhaps finite) and consequently $A_H = \{\alpha_1, \alpha_2, \ldots\}$ where $\alpha_1 < \alpha_2 < \cdots$. We set

$$H_j := H \cap F(\alpha_j) \quad \text{and} \quad H_j^* := H \cap F(\alpha_j)^*, \qquad j = 1, 2, \ldots.$$

It follows that $H = H_1 \supsetneqq H_2 \supsetneqq \cdots$. This sequence is strictly decreasing unless A_H is finite. Every $g_j \in H_j^*$ with $\|g_j\| = \inf\{\|g\| \mid g \in H_j^*\}$ is called a *minimal function.*

Existence Theorem. *If H is a saturated Hilbert subspace of $\mathcal{O}_h^k(B)$, then every H_j^* contains a minimal function.*

Proof: Since H is saturated, there exists a set T which is closed in F with $H = T \cap \mathcal{O}_h^k(B)$. Thus $H_j^* = (T \cap F(\alpha_j)^*) \cap \mathcal{O}_h^k(B)$. Since $F(\alpha_j)^*$ is closed in F, this implies that H_j^* is saturated. Hence the minimum principle yields a minimal function in each H_j^*. □

Lemma. *Let H be a Hilbert subspace of $\mathcal{O}_h^k(B)$ such that every H_j^* contains a minimal function g_j. Then H_j is the orthogonal complement of $((g_1, \ldots, g_{j-1}))_{\mathbb{C}}$ in H. Furthermore g_j is the unique minimal function in H_j^*.*

Proof: Let $h \in H$ with $b := \|h\|^2 > 0$. We set $a_i := (h, g_i)$, $i = 1, 2, \ldots$. Suppose $h \in H_j$. Then, for all $i < j$, the vectors $v := g_i + ch$, $c \in \mathbb{C}$, are in H_i^*. Since g_i is a minimal function,

$$\|v\|^2 = \|g_i\|^2 + ca_i + \overline{ca_i} + |c|^2 b \geq \|g_i\|^2$$

for all $c \in \mathbb{C}$. Since $b > 0$, this is impossible unless $a_i = 0$. Consequently every $h \in H_j$ is perpendicular to the linear span $((g_1, \ldots, g_{j-1}))_{\mathbb{C}}$.

On the other hand suppose $a_1 = \cdots = a_{j-1} = 0$. We set $\alpha_s := \omega(h)$ and normalize h so that $h \in H_s^*$. Thus for all $t \in \mathbb{R}$ the vector $w = (1-t)g_s + th$ lies in H_s^*. Hence

$$\|w\|^2 = (1-t)^2\|g_s\|^2 + ta_s + t\bar{a}_s + t^2 b \geq \|g_s\|^2$$

for all $t \in \mathbb{R}$. But $(1-t)^2\|g_s\|^2 + t^2 b < \|g_s\|^2$ for small $t > 0$. Thus $a_s \neq 0$ and $s \geq j$. In other words $h \in H_j$.

If $g'_j \in H_j^*$ is another minimal function, then $u := g'_j - g_j \in H_{j+1}$. From the above remarks, $(u, g_j) = (u, g'_j) = 0$. Hence $\|u\|^2 = 0$ and $g'_j = g_j$. □

The following important fact is now easy to prove.

Theorem 2. *If H is a saturated Hilbert subspace of $\mathcal{O}_h^k(B)$, then every set H_j^* contains a unique minimal function g_j. The family $\{g_1, g_2, \ldots\}$ is a monotone orthogonal system for H at $p \in B$. If in addition B is connected, then $\{g_1, g_2, \ldots\}$ is a monotone orthogonal basis for H.*

Proof: It is only necessary to prove the last statement. If $h \in H$ is a vector which is perpendicular to each of the g_j's, then $\omega(h) > \omega(g_j)$ for all j. Consequently $o_p(h) \geq \lim o_p(g_j) = \infty$. Since B is connected, this implies that $h = 0$. □

Remark: For $k = m = 1$ and $H = \mathcal{O}_h(B)$ one finds a presentation of the theory of the minimal functions in the book of H. Behnke and F. Sommer: Theorie der analytischen Funktionen einer komplexen Veranderlichen, Springer Verlag, 1962, (2. Aufl.), p. 270.

Furthermore reference should be made to the Ergebnisbericht of H. Behnke and P. Thullen: Theorie der Funktionen mehrerer komplexer Veranderlichen, Springer-Verlag, 1970, (2. Aufl.), p. 1970.

§ 3. Resolution Atlases

In this section X *always* denotes a *compact* complex space and $\mathcal{S}$ is a *coherent* analytic sheaf on X. If $\mathfrak{U} = \{U_i\}_{1 \leq i \leq i_*}$ is an open cover of X, then the vector space $C^q(\mathfrak{U}, \mathcal{S}) = \prod_{i_0, \ldots, i_q} \mathcal{S}(U_{i_0 \cdots i_q})$ of q-cochains is a finite product of Fréchet spaces and is thus a Fréchet space in a natural way. In order to specify a subset $C_h^q(\mathfrak{U}, \mathcal{S})$ which is a Hilbert space we introduce the idea of a *resolution atlas*. The purpose of this section, in particular the reason for studying these Hilbert space, is to make the necessary preparations for the proof of the finiteness theorem.

1. Existence. A triple (U, Φ, P) is called a *chart* on X whenever $U \neq \emptyset$ is open in X and $\Phi\colon U \to P$ is *closed holomorphic embedding* of U into a polycylinder

$$P = \{z \in \mathbb{C}^n \mid |z_1| < r, \ldots, |z_n| < r\}, \qquad r \in \mathbb{R}, r > 0.$$

Since X is compact, it is trivial to show that

there exist finitely many charts $(U_\iota, \Phi_\iota, P_\iota)$, $\iota = 1, \ldots, \iota_*$, *on* X *such that all of the polycylinders* P_ι *are the same, and* $X = \bigcup_\iota U_\iota$.

Every such family $(U_\iota, \Phi_\iota, P_\iota)_{1 \le \iota \le \iota_*}$ is called an *atlas of charts* on X. Let $q \ge 0$ be given. Then $P_{\iota_0 \ldots \iota_q} := P_{\iota_0} \times \cdots \times P_{\iota_q}$ is the polycylinder about 0 in $\mathbb{C}^m$, $m := n(q+1)$, with each radius r. If $U_{\iota_0 \ldots \iota_q} = U_{\iota_0} \cap \cdots \cap U_{\iota_q}$ is non-empty, then

$$\Phi_{\iota_0 \ldots \iota_q}: U_{\iota_0 \ldots \iota_q} \to P_{\iota_0 \ldots \iota_q}, \qquad x \to (\Phi_{\iota_0}(x), \ldots, \Phi_{\iota_q}(x)),$$

is likewise a closed holomorphic embedding.[1] The image sheaf

$$(\Phi_{\iota_0 \ldots \iota_q})_*(\mathscr{S} \mid U_{\iota_0 \ldots \iota_q})$$

is therefore coherent on $P_{\iota_0 \ldots \iota_q}$ and

$$\mathscr{S}(U_{\iota_0 \ldots \iota_q}) \cong (\Phi_{\iota_0 \ldots \iota_q})_*(\mathscr{S} \mid U_{\iota_0 \ldots \iota_q})(P_{\iota_0 \ldots \iota_q}).$$

Remark: Even though all of the polycylinders $P_{\iota_0 \ldots \iota_q}$ are equal to the polycylinder of radius r about $0 \in \mathbb{C}^m$, it is for cohomological reasons advantageous to keep the indices.

Definition 1 (Resolution Atlas). *A system* $\mathfrak{A} = \{U_\iota, \Phi_\iota, P_\iota, \varepsilon_{\iota_0 \ldots \iota_q}\}$ *is called a resolution atlas for* $\mathscr{S}$ *on* X *if*

1) $(U_\iota, \Phi_\iota, P_\iota)_{1 \le \iota \le \iota_*}$ *is an atlas of charts on* X

and

2) *there exists* $l \in \mathbb{N}$ *so that the maps*

$$\varepsilon_{\iota_0 \ldots \iota_q}: \mathcal{O}^l \mid P_{\iota_0 \ldots \iota_q} \to (\Phi_{\iota_0 \ldots \iota_q})_*(\mathscr{S} \mid U_{\iota_0 \ldots \iota_q})$$

are analytic sheaf epimorphisms for $q = 0, 1, \ldots, \iota_* - 1$ *and* $\iota_\nu = 1, \ldots, \iota_*$.

Note that if $\mathfrak{A}$ is a resolution atlas then $\mathfrak{U} = \{U_1, \ldots, U_{\iota_*}\}$ is a *Stein* covering of X.

Theorem 2. *For every coherent sheaf* $\mathscr{S}$ *on* X *there exists a resolution atlas.*

Proof: We begin with an arbitrary atlas of charts $(\tilde{U}_\iota, \tilde{\Phi}_\iota, \tilde{P}_\iota)_{1 \le \iota \le \iota_*}$ on X and choose concentric polycylinders $P_\iota \Subset \tilde{P}_\iota$ of radius r, $r < \tilde{r}$, so that $U_\iota := \tilde{\Phi}^{-1}(P_\iota)$, $\iota = 1, \ldots, \iota_*$, still gives a covering of X (the Shrinking Theorem shows that this is possible). The induced maps $\Phi_\iota: U_\iota \to P_\iota$, $\Phi_\iota := \tilde{\Phi}_\iota \mid U_\iota$, are likewise closed holomorphic embeddings. For all indices $\iota_0, \ldots, \iota_q$ it follows that

$$(\tilde{\Phi}_{\iota_0 \ldots \iota_q})_*(\mathscr{S} \mid \tilde{U}_{\iota_0 \ldots \iota_q}) \mid P_{\iota_0 \ldots \iota_q} = (\Phi_{\iota_0 \ldots \iota_q})_*(\mathscr{S} \mid U_{\iota_0 \ldots \iota_q}).$$

[1] We only need $\Phi_{\iota_0 \ldots \iota_q}$ to be holomorphic and *finite*. If $U_{\iota_0 \ldots \iota_q}$ is empty, we define $\Phi_{\iota_0 \ldots \iota_q}$ to be the mpty map.

Since $P_{\iota_0 \cdots \iota_q}$ is a relatively compact subset of $\tilde{P}_{\iota_0} \times \cdots \times \tilde{P}_{\iota_q}$, Theorem A guarantees the existence of sheaf epimorphisms

$$\varepsilon_{\iota_0 \cdots \iota_q}: \mathcal{O}^l | P_{\iota_0 \cdots \iota_q} \to (\Phi_{\iota_0 \cdots \iota_q})_*(\mathcal{S} | U_{\iota_0 \cdots \iota_q}).$$

Note that, since for all $n \in \mathbb{N}$ we have the epimorphism $\mathcal{O}^{n+1} \to \mathcal{O}^n$, $(f_1, \ldots, f_{n+1}) \mapsto (f_1, \ldots, f_n)$, we may choose l to be independent of $\iota_0, \ldots, \iota_q$. The system $\mathfrak{U} = \{U_\iota, \Phi_\iota, P_\iota, \varepsilon_{\iota_0 \cdots \iota_q}\}$ is the desired resolution atlas for $\mathcal{S}$. □

2. The Hilbert Space $C_h^q(\mathfrak{U}, \mathcal{S})$. Suppose that $\mathfrak{A} = \{U_\iota, \Phi_\iota, P_\iota, \varepsilon_{\iota_0 \cdots \iota_q}\}$ is a resolution atlas for $\mathcal{S}$. Then $P_{\iota_0 \cdots \iota_q}$ is always the polycylinder Δ of radius r about $0 \in \mathbb{C}^m$, where $m := n(q+1)$. Thus

$$C^q(\mathfrak{A}) := \prod_{\iota_0, \ldots, \iota_q} \mathcal{O}^l(P_{\iota_0 \cdots \iota_q}), \qquad q = 0, 1, \ldots, \iota_* - 1,$$

is canonically isomorphic to some $\mathcal{O}^k(\Delta)$, and is therefore a *Fréchet* space in a natural way.

Every sheaf epimorphism $\varepsilon_{\iota_0 \cdots \iota_q}$ determines a $\mathbb{C}$-linear, continuous surjective (by Theorem B) map

$$(\varepsilon_{\iota_0 \cdots \iota_q})_{P_{\iota_0 \cdots \iota_q}}: \mathcal{O}^l(P_{\iota_0 \cdots \iota_q}) \to (\Phi_{\iota_0 \cdots \iota_q})_*(\mathcal{S} | U_{\iota_0 \cdots \iota_q})(P_{\iota_0 \cdots \iota_q}) \cong \mathcal{S}(U_{\iota_0 \cdots \iota_q}),$$

where $\mathcal{S}(U_{\iota_0 \cdots \iota_q})$ carries the canonical Fréchet topology. The most important map for our considerations is the induced product map

$$\varepsilon: C^q(\mathfrak{A}) \to C^q(\mathfrak{U}, \mathcal{S}).$$

It is clear that

$\varepsilon: C^q(\mathfrak{A}) \to C^q(\mathfrak{U}, \mathcal{S})$ is a $\mathbb{C}$-linear, surjective, continuous map between Fréchet spaces.

In the Fréchet space $C^q(\mathfrak{A})$ we have the Hilbert space

$$C_h^q(\mathfrak{A}) := \prod_{\iota_0, \ldots, \iota_q} \mathcal{O}_h^l(P_{\iota_0 \cdots \iota_q}) \cong \mathcal{O}_h^k(\Delta).$$

Recall that the injection $C_h^q(\mathfrak{A}) \to C^q(\mathfrak{A})$ is continuous and compact (Theorem 1.4). We define now the space $C_h^q(\mathfrak{U}, \mathcal{S}) \subset C^q(\mathfrak{U}, \mathcal{S})$ of the "square-integrable q-cochains (with respect to $\mathfrak{A}$)" by

$$C_h^q(\mathfrak{U}, \mathcal{S}) := \varepsilon(C_h^q(\mathfrak{A})) \cong C_h^q(\mathfrak{A}) / \operatorname{Ker} \varepsilon \cap C_h^q(\mathfrak{A}).$$

We let $\| \quad \|_\Delta$ denote the Hilbert space norm on $C_h^q(\mathfrak{A})$ and set

$$\|\zeta\|_{\mathfrak{A}}^\bullet := \inf\{\|v\|_\Delta \mid v \in C_h^q(\mathfrak{A}), \varepsilon(v) = \zeta\}, \qquad \zeta \in C_h^q(\mathfrak{U}, \mathcal{S}).$$

The following is now easy to prove:

Theorem 3. *The space $C^q_h(\mathfrak{U}, \mathcal{S})$ is a Hilbert space with $\| \quad \|^{\bullet}_{\mathfrak{A}}$ as norm. The injection $C^q_h(\mathfrak{U}, \mathcal{S}) \to C^q(\mathfrak{U}, \mathcal{S})$ is continuous and compact.*

Proof: Since ε is continuous, Ker ε is closed in $C^q(\mathfrak{A})$. Furthermore the injection $C^q_h(\mathfrak{A}) \to C^q(\mathfrak{A})$ is continuous. Thus Ker $\varepsilon \cap C^q_h(\mathfrak{A})$ is a Hilbert subspace of $C^q_h(\mathfrak{A})$ and $C^q_h(\mathfrak{U}, \mathcal{S})$ is a quotient Hilbert space with $\| \quad \|^{\bullet}_{\mathfrak{A}}$ as its Hilbert norm.
The continuity and compactness of the injection $C^q_h(\mathfrak{U}, \mathcal{S}) \to C^q(\mathfrak{U}, \mathcal{S})$ follows immediately by considering the diagram

$$\begin{array}{ccc} C^q_h(\mathfrak{A}) & \hookrightarrow & C^q(\mathfrak{A}) \\ \downarrow{\scriptstyle \varepsilon} & & \downarrow{\scriptstyle \varepsilon} \\ C^q_h(\mathfrak{U}, \mathcal{S}) & \overset{\cdot}{\hookrightarrow} & C^q(\mathfrak{U}, \mathcal{S}). \end{array}$$

It is obviously commutative, the injection in the first row is continuous and compact, and the projections ε are open.

3. The Hilbert Space $Z^q_h(\mathfrak{U}, \mathcal{S})$. We now define the Hilbert space of "square-integrable q-cocycles (with respect to $\mathfrak{A}$)," $Z^q_h(\mathfrak{U}, \mathcal{S})$, as a Hilbert subspace of $C^q_h(\mathfrak{U}, \mathcal{S})$. If ∂ is the q-th boundary map in the cochain complex $\{C^*(\mathfrak{U}, \mathcal{S}), \partial_*\}$, then ∂ is a continuous map of Fréchet spaces and $Z^q(\mathfrak{U}, \mathcal{S}) = \text{Ker } \partial$ is a Fréchet subspace of $C^q(\mathfrak{U}, \mathcal{S})$. Thus, since ε is continuous,

$$Z^q(\mathfrak{A}) := \varepsilon^{-1}(Z^q(\mathfrak{U}, \mathcal{S}))$$

is a Fréchet subspace of $C^q(\mathfrak{A})$. Recalling that the injection $C^q_h(\mathfrak{A}) \to C^q(\mathfrak{A})$ is continuous, we therefore have the following:

The space $Z^q_h(\mathfrak{A}) := Z^q(\mathfrak{A}) \cap C^q_h(\mathfrak{A})$ is a Hilbert subspace of $C^q_h(\mathfrak{A}) \cong \mathcal{O}^k_h(\Delta)$ which is saturated in $C^q(\mathfrak{A}) \cong \mathcal{O}^k(\Delta)$. The injection $Z^q_h(\mathfrak{A}) \to Z^q(\mathfrak{A})$ is continuous.

The space $Z^q_h(\mathfrak{U}, \mathcal{S})$ along with its norm $\| \quad \|_{\mathfrak{A}}$ is now defined in a way which is analogous to that for $C^q_h(\mathfrak{U}, \mathcal{S})$:

$$Z^q_h(\mathfrak{U}, \mathcal{S}) := \varepsilon(Z^q_h(\mathfrak{A})) \cong Z^q_h(\mathfrak{A})/\text{Ker } \varepsilon \cap Z^q_h(\mathfrak{A}),$$
$$\|\zeta\|_{\mathfrak{A}} := \inf\{\|v\|_{\Delta} \mid v \in Z^q_h(\mathfrak{A}), \varepsilon(v) = \zeta\}, \qquad \zeta \in Z^q_h(\mathfrak{U}, \mathcal{S}).$$

Theorem 4. *The space $Z^q_h(\mathfrak{U}, \mathcal{S})$ is a Hilbert space with $\| \quad \|_{\mathfrak{A}}$ as norm. The injection $Z^q_h(\mathfrak{U}, \mathcal{S}) \to Z^q(\mathfrak{U}, \mathcal{S})$ is continuous. The injection $Z^q_h(\mathfrak{U}, \mathcal{S}) \to C^q_h(\mathfrak{U}, \mathcal{S})$ is an isometry.*

Proof: Since $\operatorname{Ker} \varepsilon \cap Z_h^q(\mathfrak{A})$ is a Hilbert subspace of $Z_h^q(\mathfrak{A})$ (see the proof of Theorem 3), the first statement is obvious. The second statement follows (as in the proof of Theorem 3) from the commutative diagram

$$\begin{array}{ccc} Z_h^q(\mathfrak{A}) & \hookrightarrow & Z^q(\mathfrak{A}) \\ \downarrow{\scriptstyle \varepsilon} & & \downarrow{\scriptstyle \varepsilon} \\ Z_h^q(\mathfrak{U}, \mathscr{S}) & \hookrightarrow & Z^q(\mathfrak{U}, \mathscr{S}) \end{array}$$

It suffices to note that ε is open and that the injection in the first row is continuous.

Since per definition $\|\zeta\|^{\bullet}_{\mathfrak{A}} \le \|\zeta\|_{\mathfrak{A}}$ for all $\zeta \in Z_h^q(\mathfrak{U}, \mathscr{S})$, the map $Z_h^q(\mathfrak{U}, \mathscr{S}) \to C_h^q(\mathfrak{U}, \mathscr{S})$ is a contraction. But, observing that $C_h^q(\mathfrak{A}) = Z_h^q(\mathfrak{A}) + Z_h^q(\mathfrak{A})^{\perp}$, it is clear that if $v \in C_h^q(\mathfrak{A})$ has non-trivial projection on $Z_h^q(\mathfrak{A})^{\perp}$, then $|v|_{\Delta} > |v_0|_{\Delta}$, where $v_0 \in Z_h^q(\mathfrak{A})$ is the element of minimal length which is mapped to $\zeta \in Z_h^q(\mathfrak{U}, \mathscr{S})$. Hence the injection is an isometry.

Every Hilbert subspace of a Hilbert space has an orthogonal complement. Thus every vector in the quotient space has a preimage of the same length. In the case of $Z_h^q(\mathfrak{U}, \mathscr{S})$ this says that

given a vector $\zeta \in Z_h^q(\mathfrak{U}, \mathscr{S})$ there exists an ε-preimage $v \in Z_h^q(\mathfrak{A})$ with $\|v\|_{\Delta} = \|\zeta\|_{A}$. □

4. Refinements. For every refinement $\mathfrak{U}'$ of $\mathfrak{U}$ one can define the groups $Z_h^q(\mathfrak{U}, \mathscr{S})|\mathfrak{U}'$. In the following, special refinements of $\mathfrak{U}$ are introduced. Let $\mathfrak{A} = \{U_\iota, \Phi_\iota, P_\iota, \varepsilon_{\iota_0 \cdots \iota_q}\}$ and $\mathfrak{A}' = \{U'_\iota, \Phi'_\iota, P'_\iota, \varepsilon'_{\iota_0 \cdots \iota_q}\}$ be resolution atlases for $\mathscr{S}$ on X which have the same index set $\{1, \ldots, \iota_*\}$. Furthermore assume that the polycylinders P_ι, P'_ι are both n-dimensional with $l = l'$ (see Definition 1.2).

Definition 5 (Refinement). *The resolution atlas $\mathfrak{A}'$ is called a refinement of $\mathfrak{A}$, denoted by $\mathfrak{A}' < \mathfrak{A}$, if*

1) *P'_ι is relatively compact in P_ι (i.e. $r' < r$),*
2) *$U'_\iota = \Phi_\iota^{-1}(P'_\iota)$ and $\Phi'_\iota = \Phi_\iota | U'_\iota$, $1 \le \iota \le \iota_*$,*

and

3) *$\varepsilon'_{\iota_0 \cdots \iota_q} = \varepsilon_{\iota_0 \cdots \iota_q} | (\mathcal{O}^l | P'_{\iota_0 \cdots \iota_q})$ for all $\iota_0, \ldots, \iota_q$, where $P'_{\iota_0 \cdots \iota_q} := P'_{\iota_0} \times \cdots \times P'_{\iota_q}$.*

Condition 3) above makes sense, because $P'_{\iota_0 \cdots \iota_q} \Subset P_{\iota_0 \cdots \iota_q}$ by 1) and

$$(\Phi_{\iota_0 \cdots \iota_q})_*(\mathscr{S} | U_{\iota_0 \cdots \iota_q}) | P'_{\iota_0 \cdots \iota_q} = (\Phi'_{\iota_0 \cdots \iota_q})_*(\mathscr{S} | U'_{\iota_0 \cdots \iota_q})$$

by 2). It is easy to construct such refinements.

Theorem 6. *Every resolution atlas $\mathfrak{A}$ for $\mathscr{S}$ possesses a refinement.*

Proof: One shrinks the polycylinder P_ι so that the Φ_ι-preimages still cover X. Then it is enough to restrict the maps to the smaller preimage sets. □

If $\mathfrak{A}$, $\mathfrak{A}^*$ are resolution atlases for $\mathscr{S}$ with associated covers $\mathfrak{U} = \{U_i\}$ and $\mathfrak{U}^* = \{U_i^*\}$ such that $\mathfrak{A} < \mathfrak{A}^*$, then $U_i \Subset U_i^*$ for all. Thus $\mathfrak{U}$ is a refinement of $\mathfrak{U}^*$, and one has (see V.6.5) the continuous, $\mathbb{C}$-linear restriction map $\rho\colon C^q(\mathfrak{U}^*, \mathscr{S}) \to C^q(\mathfrak{U}, \mathscr{S})$. One also has the $\mathbb{C}$-linear, continuous restriction map

$$\sigma\colon C^q(\mathfrak{A}^*) \to C^q(\mathfrak{A}), \quad v \mapsto v|\Delta,$$

which restricts component-wise the vectors in $C^q(\mathfrak{A}^*) \cong \mathcal{O}^k(\Delta^*)$ to Δ. (All objects in $\mathfrak{A}^*$ are starred. Thus $\Delta^* = P^*_{i_0 \ldots i_q}$).

It follows from the definitions of ρ and σ (see Definition 5, 3)) that the diagram

$$(\#) \qquad \begin{array}{ccc} C^q(\mathfrak{A}^*) & \xrightarrow{\ \sigma\ } & C^q(\mathfrak{A}) \\ \Big\downarrow{\scriptstyle \varepsilon^*} & & \Big\downarrow{\scriptstyle \varepsilon} \\ C^q(\mathfrak{U}^*, \mathscr{S}) & \xrightarrow{\ \rho\ } & C^q(\mathfrak{U}, \mathscr{S}) \end{array}$$

is commutative. Writing $C^q(\mathfrak{U}^*, \mathscr{S})|\mathfrak{U}$ for $\operatorname{Im}\rho$, this implies the following:

Theorem 7. *If $\mathfrak{A}$ is a refinement of the resolution atlas $\mathfrak{A}^*$, then*

$$C^q(\mathfrak{U}^*, \mathscr{S})|\mathfrak{U} \subset C^q_h(\mathfrak{U}, \mathscr{S}), \qquad Z^q(\mathfrak{U}^*, \mathscr{S})|\mathfrak{U} \subset Z^q_h(\mathfrak{U}, \mathscr{S})$$

and the induced mappings $C^q(\mathfrak{U}^*, \mathscr{S}) \to C^q_h(\mathfrak{U}, \mathscr{S})$, $Z^q(\mathfrak{U}^*, \mathscr{S}) \to Z^q_h(\mathfrak{U}, \mathscr{S})$ are continuous.

Proof: Since $\Delta \Subset \Delta^*$,

$$C^q(\mathfrak{A}^*)|\Delta := \operatorname{Im}\sigma \subset C^q_h(\mathfrak{A}), \qquad Z^q(\mathfrak{A}^*)|\Delta \subset Z^q_h(\mathfrak{A}),$$

and the induced maps $C^q(\mathfrak{A}^*) \to C^q_h(\mathfrak{A})$, $Z^q(\mathfrak{A}^*) \to Z^q_h(\mathfrak{A})$ are continuous (see Section 1.2). The commutativity of (#) implies that $\operatorname{Im}\rho = \varepsilon(\operatorname{Im}\sigma)$ and $C^q_h(\mathfrak{U}, \mathscr{S}) = \varepsilon(C^q_h(\mathfrak{A}))$ by definition. Thus $C^q(\mathfrak{U}^*, \mathscr{S})|\mathfrak{U} \subset C^q_h(\mathfrak{U}, \mathscr{S})$. The diagram (#) induces the commutative diagram below. The map in the first row is continuous. By the definition of the Hilbert quotient topology, ε^* and ε are open maps. Thus the map in the second row is continuous.

$$\begin{array}{ccc} C^q(\mathfrak{A}^*) & \longrightarrow & C^q_h(\mathfrak{A}) \\ \Big\downarrow{\scriptstyle \varepsilon^*} & & \Big\downarrow{\scriptstyle \varepsilon} \\ C^q(\mathfrak{U}^*, \mathscr{S}) & \longrightarrow & C^q_h(\mathfrak{U}, \mathscr{S}) \end{array}$$

The statements about $Z^q(\mathfrak{U}^*, \mathscr{S})$ and $Z^q(\mathfrak{U}, \mathscr{S})$ are verified analogously. □

§ 4. The Proof of the Finiteness Theorem

As before, X denotes a *compact* complex space and $\mathscr{S}$ is a coherent sheaf on X.

1. The Smoothing Lemma. If $\mathfrak{A}$, $\mathfrak{A}'$ are resolutions for $\mathscr{S}$ with $\mathfrak{A}' < \mathfrak{A}$ then the cochains in $C^q_h(\mathfrak{U}, \mathscr{S})$ look "smoother" than those in $C^q_h(\mathfrak{U}', \mathscr{S})$. An important problem is to "smooth" a cocycle $\xi' \in Z^q_h(\mathfrak{U}', \mathscr{S})$ by finding a cocycle $\xi \in Z^q(\mathfrak{U}, \mathscr{S})$ such that ξ' and ξ determine the same cohomology class in $H^q(X, \mathscr{S})$. As in the last section, if $\mathfrak{A}''$, $\mathfrak{A}'$, ... are resolution atlases, then we let $\| \ \|^{\bullet}_{\mathfrak{A}''}$, $\| \ \|^{\bullet}_{\mathfrak{A}'}$, ... and $\| \ \|_{\mathfrak{A}''}$, $\| \ \|_{\mathfrak{A}'}$, ... denote the Hilbert norms on $C^q_h(\mathfrak{A}'', \mathscr{S})$, $C^q_h(\mathfrak{A}', \mathscr{S})$, ..., and $Z^q_h(\mathfrak{A}'', \mathscr{S})$, $Z^q_h(\mathfrak{A}', \mathscr{S})$, ... respectively.

Smoothing Lemma. *Let $\mathfrak{A}$, $\mathfrak{A}'$, $\mathfrak{A}''$, $\mathfrak{A}^*$ denote resolution atlases for $\mathscr{S}$ on X with $\mathfrak{A}'' < \mathfrak{A}' < \mathfrak{A} < \mathfrak{A}^*$. Let $q \in \mathbb{N}$. Then there exists a real constant $L > 0$ so that, given $\xi' \in Z^q_h(\mathfrak{U}', \mathscr{S})$, there is a cocycle $\xi \in Z^q_h(\mathfrak{U}, \mathscr{S})$ and a cochain $\eta \in C^{q-1}_h(\mathfrak{U}'', \mathscr{S})$ with*

$$\xi' \,|\, \mathfrak{U}'' = \xi \,|\, \mathfrak{U}'' + \partial\eta, \qquad \|\xi\|_{\mathfrak{A}} \le L\|\xi'\|_{\mathfrak{A}'} \quad \textit{and} \quad \|\eta\|_{\mathfrak{A}''} \le L\|\xi\|_{\mathfrak{A}'}.$$

Proof: The map

$$\alpha\colon Z^q(\mathfrak{U}^*, \mathscr{S}) \times C^{q-1}(\mathfrak{U}', \mathscr{S}) \to Z^q(\mathfrak{U}', \mathscr{S}), \qquad (\xi^*, \eta') \to \xi^* \,|\, \mathfrak{U}' + \partial\eta',$$

is continuous. Since $\mathfrak{U}^*$, $\mathfrak{U}'$ are Stein coverings, α is surjective (see the introduction of this chapter). Consequently Banach's Theorem says that α is an open map.

The restrictions $\beta\colon Z^q(\mathfrak{U}^*, \mathscr{S}) \to Z^q_h(\mathfrak{U}, \mathscr{S})$, $\gamma\colon C^{q-1}(\mathfrak{U}', \mathscr{S}) \to C^{q-1}_h(\mathfrak{U}'', \mathscr{S})$ are continuous (Theorem 3.7). Thus there exists a neighborhood W of 0 in $Z^q(\mathfrak{U}^*, \mathscr{S}) \times C^{q-1}(\mathfrak{U}', \mathscr{S})$ so that $\beta \circ \pi_1(W)$ and $\gamma \circ \pi_2(W)$ are contained in the unit balls of $Z^q_h(\mathfrak{U}, \mathscr{S})$ and $C^{q-1}_h(\mathfrak{U}'', \mathscr{S})$ respectively. The set $\alpha(W)$ is an open neighborhood of $0 \in Z^q(\mathfrak{U}', \mathscr{S})$. Recalling that the injection $Z^q_h(\mathfrak{U}', \mathscr{S}) \to Z^q(\mathfrak{U}', \mathscr{S})$ is continuous (Theorem 3.4), it follows that $\alpha(W) \cap Z^q_h(\mathfrak{U}', \mathscr{S})$ is a neighborhood of $0 \in Z^q_h(\mathfrak{U}', \mathscr{S})$. Thus there exists $\rho > 0$ so that

$$\{\zeta' \in Z^q_h(\mathfrak{U}', \mathscr{S}) \mid \|\zeta'\|_{\mathfrak{A}'} = \rho\} \subset \alpha(W).$$

We claim that $L := \rho^{-1}$ is the desired constant. To see this let ξ' be an arbitrary ($\neq 0$) element of $Z^q_h(\mathfrak{U}', \mathscr{S})$. We find $c \in \mathbb{C}$ so that $\|c\xi'\|_{\mathfrak{A}'} = \rho$. Hence there are elements $\xi^* \in Z^q(\mathfrak{U}^*, \mathscr{S})$ and $\omega' \in C^{q-1}(\mathfrak{U}', \mathscr{S})$ with

$$c\xi' = \xi^* \,|\, \mathfrak{U}' + \partial\omega' \quad \text{and} \quad (\xi^*, \omega') \in W.$$

Now we define $\zeta := \beta(\xi^*) = \xi^* \,|\, \mathfrak{U} \in Z^q_h(\mathfrak{U}, \mathscr{S})$ and $\omega := \gamma(\omega') = \omega' \,|\, \mathfrak{U}'' \in C^{q-1}_h(\mathfrak{U}'', \mathscr{S})$. Certainly

$$c\xi' \,|\, \mathfrak{U}'' = \zeta \,|\, \mathfrak{U}'' + \partial\omega \quad \text{and} \quad \|\zeta\|_{\mathfrak{A}} \le 1, \qquad \|\omega\|^{\bullet}_{\mathfrak{A}''} \le 1.$$

For $\xi := c^{-1}\zeta \in Z_h^q(\mathfrak{U}, \mathcal{S})$ and $\eta := c^{-1}\omega \in C_h^{q-1}(\mathfrak{U}'', \mathcal{S})$ we have

$$\xi' | \mathfrak{U}'' = \xi | \mathfrak{U}'' + \partial\eta \quad \text{and} \quad \|\xi\|_{\mathfrak{A}} \le c^{-1}, \qquad \|\eta\|_{\mathfrak{A}''} \le c^{-1}.$$

Since $|c| \, \|\xi'\|_{\mathfrak{A}'} = L^{-1}$, the proof is complete. □

2. Finiteness Lemma. We continue with the notation of the smoothing lemma. Let $q \in \mathbb{N}$ be fixed. For the sake of brevity, we write $\|\cdot\|$ for the norm $\|\cdot\|_\Delta$ on $Z_h^q(\mathfrak{A})$. Since $Z_h^q(\mathfrak{A})$ is a saturated subset of $\mathcal{O}_h^k(\Delta)$, it has an orthogonal basis $\{g_1, g_2, \ldots\}$ which is *monotone* at the origin. We may assume that $\|g_j\| = 1$ for all j.

For the polydisks Δ', Δ we have $r' < r$. Let $a \in \mathbb{R}$ be such that $r'r^{-1} < a < 1$ and let $M > 0$ and $L > 0$ be the constants of the Schwarz Lemma and the Smoothing Lemma respectively. We choose $e \in \mathbb{N}$ large enough so that $t := LMa^e < 1$.

By Theorem 2.1 there exists a positive integer d so that for every vector $v \in Z_h^q(\mathfrak{A})$ the vector $w := v - \sum_{i=1}^{d} (v, g_i)g_i$ satisfies

(1) $$\|w\|_{\Delta'} \le Ma^e \|v\|.$$

Combining the projection $Z_h^q(\mathfrak{A}) \xrightarrow{\varepsilon} C_h^q(\mathfrak{U}, \mathcal{S})$ and the injection $Z_h^q(\mathfrak{U}, \mathcal{S}) \to Z_h^q(\mathfrak{U}'', \mathcal{S})$ we have a *continuous* map $Z_h^q(\mathfrak{A}) \to Z_h^q(\mathfrak{U}'', \mathcal{S})$ which we denote by $v \to \bar{v} = \varepsilon(v) | \mathfrak{U}''$. For d as above we have the following:

Finiteness Lemma.

$$Z_h^q(\mathfrak{U}, \mathcal{S}) | \mathfrak{U}'' = \sum_{i=1}^{d} \mathbb{C}\bar{g}_i + \partial C_h^{q-1}(\mathfrak{U}'', \mathcal{S}) \quad \textit{where} \quad \bar{g}_1, \ldots, \bar{g}_d \in Z_h^q(\mathfrak{U}, \mathcal{S}).$$

Proof: Let $\zeta \in Z_h^q(\mathfrak{U}, \mathcal{S})$ be given. It is enough to construct two sequences $v_0, v_1, \ldots$ and $\eta_1, \eta_2, \ldots$, where $v_j \in Z_h^q(\mathfrak{A})$ and $\eta_j \in C_h^{q-1}(\mathfrak{U}'', \mathcal{S})$, so that with w_j defined by $w_j := v_j - \sum_{i=1}^{d} (v_j, g_i)g_i$ we have

(2) $$\bar{v}_0 = \zeta | \mathfrak{U}'', \qquad \bar{w}_j = \bar{v}_{j+1} + \partial\eta_{j+1}, \qquad j \ge 0,$$

and

(3) $$\|v_j\| \le t^j \|v_0\|, \qquad \|\eta_j\|^{\bullet}_{\mathfrak{A}''} \le t^j \|v_0\|, \qquad j \ge 1.$$

Assuming we have such sequences, (2) implies

(4) $$\zeta | \mathfrak{U}'' - \bar{v}_{n+1} = \sum_{j=0}^{n} (\bar{v}_j - \bar{v}_{j+1}) = \sum_{i=1}^{d} \sum_{j=0}^{n} (v_j, g_i)\bar{g}_i + \sum_{j=0}^{n} \partial\eta_j, \qquad n \ge 0.$$

Since $t < 1$, it is immediate from (3) that the following converges:

$$\eta := \sum_0^\infty \eta_{j+1} \in C_h^{q-1}(\mathfrak{U}'', \mathscr{S}). \tag{5}$$

Now $\partial: C^{q-1}(\mathfrak{U}'', \mathscr{S}) \to C^q(\mathfrak{U}'', \mathscr{S})$ is continuous and $\sum_0^\infty \eta_{j+1}$ converges to η in the Fréchet topology (Theorem 3). Thus $\partial\eta = \sum_0^\infty \partial\eta_{j+1}$. Furthermore, using (3) and the fact that $|(v_j, g_i)| \le \|v_j\|$ (Schwarz inequality),

$$c_i := \sum_{j=0}^\infty (v_j, g_i) \in \mathbb{C}, \qquad 1 \le i \le d. \tag{6}$$

By continuity, $\lim v_{n+1} = 0$ implies $\lim \bar{v}_{n+1} = 0$. Thus, putting (4), (5) and (6) together, $\zeta | \mathfrak{U}'' = \sum_{i=1}^d c_i \bar{g}_i + d\eta$.

We now construct the sequences $v_0, v_1, \ldots$ and $\eta_1, \eta_2, \ldots$ using an inductive procedure. Suppose v_j, η_j have been constructed. Then $w_j := v_j - \sum_{i=1}^d (v_j, g_j) g_i$. Let $w' := w_j | \Delta' \in Z_h^q(\mathfrak{A}')$ and let $\varepsilon': Z_h^q(\mathfrak{A}') \to Z_h^q(\mathfrak{U}'', \mathscr{S})$ be the projection map which is in fact a contraction. From the smoothing lemma we deduce the existence of $\xi \in Z_h^q(\mathfrak{U}, \mathscr{S})$ and $\eta_{j+1} \in C_h^{q-1}(\mathfrak{U}'', \mathscr{S})$ so that

$$\varepsilon'(w') | \mathfrak{U}'' = \xi | \mathfrak{U}'' + \partial\eta_{j+1}, \quad \text{with} \tag{7}$$
$$\|\xi\| \le L \|\varepsilon'(w')\|_\Delta, \quad \|\eta_{j+1}\|_{\mathring{\mathfrak{A}}''} \le L \|\varepsilon'(w')\|_{\Delta'}.$$

Since $Z_h^q(\mathfrak{U}, \mathscr{S})$ carries the ε-quotient norm from $Z_h^q(\mathfrak{U}, \mathscr{S})$, there is a preimage vector $v_{j+1} \in Z_h^q(\mathfrak{A})$ of ξ such that $\|v_{j+1}\| = \|\xi\|$ (see Paragraph 2.2). Now $\varepsilon'(w') = \varepsilon(w_j) | \mathfrak{U}'$ (see diagram (#) in Section 3.4). Thus $\bar{v}_{j+1} = \xi | \mathfrak{U}''$ and $\bar{w}_j = \varepsilon'(w') | \mathfrak{U}''$. Since ε' is a contraction, $\|\varepsilon'(w')\|_{\Delta'} \le \|w'\|_{\Delta'}$. Furthermore $\|w'\|_{\Delta'} = \|w_j\|_{\Delta'}$. Then (7) implies (2) with the estimates

$$\|v_{j+1}\| \le L \|w_j\|_{\Delta'} \quad \text{and} \quad \|\eta_{j+1}\|_{\mathring{\mathfrak{A}}''} \le L \|w_j\|_{\Delta'}.$$

Since $\|w_j\|_{\Delta'} \le M a^e \|v_j\|$ (see (1) above) and $t = LMa^e$,

$$\|v_{j+1}\| \le t \|v_j\|, \qquad \|\eta_{j+1}\|_{\mathring{\mathfrak{A}}''} \le t \|v_j\|.$$

The induction assumption (3) now yields the desired estimates for v_{j+1}, η_{j+1}.

3. Proof of the Finiteness Theorem.

Finiteness Theorem. *If X is a compact complex space and $\mathscr{S}$ is a coherent analytic sheaf on X, then every cohomology module $H^q(X, \mathscr{S})$, $0 \le q \le \infty$, is finite dimensional.*

Proof: From 3.3 we have four resolution atlases $\mathfrak{A}'' < \mathfrak{A}' < \mathfrak{A} < \mathfrak{A}^*$ for $\mathscr{S}$. Theorem 3.7 implies that $Z^q(\mathfrak{U}^*, \mathscr{S})|\mathfrak{U} \subset Z^q_h(\mathfrak{U}, \mathscr{S})$. Now there exists a positive integer d so that we can apply the finiteness lemma: There exist cocycles $\bar{g}_1, \ldots, \bar{g}_d \in Z^q(\mathfrak{U}'', \mathscr{S})$ so that

$$Z^q(\mathfrak{U}^*, \mathscr{S})|\mathfrak{U}'' \subset Z^q_h(\mathfrak{U}, \mathscr{S})|\mathfrak{U}'' \subset \sum_1^d \mathbb{C}\bar{g}_i + \partial C^{q-1}(\mathfrak{U}'', \mathscr{S}).$$

Hence the cohomology classes determined by $\bar{g}_1, \ldots, \bar{g}_d$ generate $H^q(\mathfrak{U}'', \mathscr{S})$, and, since $\mathfrak{U}''$ is a Stein cover, $\dim_{\mathbb{C}} H^q(X, \mathscr{S}) \leq d < \infty$ (see the introduction to this chapter, where $\mathfrak{B}$ corresponds to $\mathfrak{U}''$, $\mathfrak{W}$ to $\mathfrak{U}^*$ and $\xi_i = \bar{g}_i$). □

The finiteness theorem is the foundation for the theory of compact Riemann surfaces which is presented in the next chapter. In this special situation only the groups $H^0(X, \mathscr{S})$ and $H^1(X, \mathscr{S})$ for *locally-free* sheaves play a role. If one is only interested in Riemann surfaces, then the above considerations for resolution atlases are technically simpler. A sketch of a proof of the finiteness theorem for this special case, which goes along the lines of the Cartan-Serre proof, can be found in the book of R. Gunning: Lectures on Riemann Surfaces, Princeton University Press, 1966, p. 59 ff. Instead of Fréchet spaces, the Hilbert spaces of square integrable cocycles are likewise used. Instead of monotone orthogonal bases, the lemma of L. Schwarz for Hilbert spaces (which is substantially easier than the Fréchet version) is used.

Chapter VII. Compact Riemann Surfaces

In the theory of compact Riemann surfaces it is possible to make particularly elegant applications of the finiteness theorem. For such considerations we will always let X denote a connected, compact Riemann surface with structure sheaf $\mathcal{O}$. With script letters like $\mathcal{S}$ we will denote, as before, coherent analytic sheaves over X. If the support of such a sheaf is finite then $\mathcal{T}$ will usually be written. For such a sheaf it is easy to see that $H^1(X, \mathcal{T}) = (0)$. The symbols $\mathcal{F}$, $\mathcal{G}$ are reserved for locally free $\mathcal{O}$-sheaves. The letter $\mathcal{L}$ is usual exclusively for locally free sheaves of rank 1. All tensor products are formed over $\mathcal{O}$.

Since X is 1-dimensional, every stalk $\mathcal{M}_x$ of the sheaf $\mathcal{M}$ of germs of meromorphic functions on X is a discrete valuation field with respect to the order function o_x. Recall that, for $h \in \mathcal{M}_x$ and $t \in \mathfrak{m}_x$ a local coordinate, $o_x(h) = n$, where $h = t^n e$ and $e \in \mathcal{O}_x$ is a unit. The order of the identically zero germ is defined to be infinity. The valuation ring associated to o_x is $\mathcal{O}_x$ with $\mathfrak{m}_x$ as maximal ideal.

The goal of this chapter is the derivation of the Riemann-Roch theorem along with the Serre duality theorem. Further a criterion for the splitting of locally free sheaves is proved. An immediate corollary of this is the classification of the locally free sheaves over the Riemann sphere.

§ 1. Divisors and Locally Free Sheaves

Every locally free sheaf $\mathcal{F}$ is a subsheaf of the $\mathcal{O}$-sheaf,

$$\mathcal{F}^{\infty} := \mathcal{F} \otimes \mathcal{M},$$

of germs of meromorphic sections of $\mathcal{F}$. We identify every stalk $\mathcal{F}_x^{\infty}$ with $\mathcal{M}_x \mathcal{F}_x = \bigcup_{n \in \mathbb{Z}} t^n \mathcal{F}_x$, where $t \in \mathfrak{m}_x$ is a local coordinate at x. The sheaf $\mathcal{F}^{\infty}$ is not coherent. However it contains important locally free subsheaves which are related to $\mathcal{F}$. These will be introduced in this section. We write $\mathcal{F}^{\infty}(X)^*$ (resp. $\mathcal{F}(X)^*$) for $\mathcal{F}^{\infty}(X)$ minus the identically zero section (resp. $\mathcal{F}(X)$ minus the identically zero section).

0. Divisors. In Chapter V.2.2, we considered the exact sequence

$$0 \to \mathcal{O}^* \to \mathcal{M}^* \to \mathcal{D} \to 0$$

for arbitrary complex spaces. The sheaf $\mathcal{D} := \mathcal{M}^*/\mathcal{O}^*$ is called the sheaf of germs of divisors. In the case of a Riemann surface, every stalk $\mathcal{D}_x$ is isomorphic to $\mathbb{Z}$, and every non-trivial $s \in \mathcal{D}(U)$ over an open set U, has discrete support $|s|$ in U. This is the so-called skyscraper property of $\mathcal{D}$.

For a compact Riemann surface X, the *divisor group*,

$$\operatorname{Div} X := \mathcal{D}(X),$$

is canonically isomorphic to the free abelian group generated by the points $x \in X$. Consequently every divisor D is of the form

$$D = \sum_{x \in X} n_x x,$$

where $n_x \in \mathbb{Z}$ and $n_x = 0$ for almost all x. Throughout we write $o_x(D)$ instead of n_x. The integer

$$\deg D := \sum_{x \in X} o_x(D)$$

is called the *degree of* D. The mapping $\operatorname{Div} X \to \mathbb{Z}$, which associates $\deg D$ to D, is a group epimorphism.

If $o_x(D) \geq 0$ for every $x \in X$ then the divisor D is called *positive*. For divisors $D_1, D_2 \in \operatorname{Div} X$, we write $D_1 \leq D_2$ when $D_2 - D_1$ is positive. The group of divisors is directed with respect to this relation. In other words, given two divisors D_1, $D_2 \in \operatorname{Div} X$, there exists $D_3 \in \operatorname{Div} X$ so that $D_1 \leq D_3$ and $D_2 \leq D_3$.

The set $|D| := \{x \in X: o_x(D) \neq 0\}$, called the support of D, is always finite.

1. Divisors of Meromorphic Sections. Let $\mathcal{F}$ be a locally free sheaf. Given a local coordinate $t \in \mathfrak{m}_x$, every germ $s_x \in \mathcal{F}_x^\infty$ can be uniquely written in the form

$$s_x = t^m \hat{s}_x,$$

where $\hat{s}_x \in \mathcal{F}_x \setminus \mathfrak{m}_x \mathcal{F}_x$. The exponent m is uniquely determined by s_x. We define

$$o(s_x) := m$$

to be the order of s_x with respect to $\mathcal{F}$. If $o(s_x) > 0$ (resp. $o(s_x) < 0$) then we call x a zero (resp. a pole) of s_x. The situation $o(s_x) = \infty$ only occurs for the identically zero germ. It follows that $\mathcal{F}_x = \{s_x \in \mathcal{F}_x^\infty: o(s_x) \geq 0\}$ and $\mathfrak{m}_x \mathcal{F}_x = \{s_x \in \mathcal{F}_x^\infty: o(s_x) > 0\}$.

Each section $s \in \mathcal{F}^\infty(X)$, $s \not\equiv 0$, is called a global meromorphic section of $\mathcal{F}$. If $s \not\equiv 0$ then it has only finitely many zeros and poles. Thus the following definition makes sense.

Definition 1. (*The divisor and the degree of meromorphic sections*). *Given* $s \in \mathcal{F}^\infty(X)$,

$$(s) := \sum_{x \in X} o(s_x) \cdot x \in \operatorname{Div} X$$

is called the divisor (*with respect to* $\mathcal{F}$) *of s. The integer* $\deg(s)$ *is called the degree* (*with respect to* $\mathcal{F}$) *of s.*

It follows that (s) is positive if and only if s has no poles, or equivalently, if $s \in \mathcal{F}(X)$.

Warning: The order functions o and the divisors (s) of sections $s \in \mathcal{F}^\infty(X)$ depend heavily on the sheaf with which one starts out. For example, with respect to $\mathcal{O}$, the zero divisor is associated to the section $1 \in \mathcal{O}^\infty(X)$. On the other hand, if $\mathcal{O}$ is replaced by $\mathcal{O}(D)$ (see Paragraph 3) for some $D \in \operatorname{Div} X$ then $(1) = D$. In the following it will always be completely clear with respect to which sheaf we are forming divisors.

From the above remarks it follows that every meromorphic function $h \in \mathcal{O}^\infty(X)^*$ has an associated divisor (h). Such divisors are called principal divisors. The map $\mathcal{M}(X)^* \to \operatorname{Div} X$, which sends h to its divisor (h), is a group homomorphism. For all $h \in \mathcal{O}^\infty(X)^*$ and $s \in \mathcal{F}^\infty(X)^*$, it follows easily that $(hs) = (h) + (s)$. The image group,

$$P(X) := \operatorname{Im}(\mathcal{M}(X)^* \to \operatorname{Div} X) \subset \operatorname{Div} X,$$

is called the *group of principal divisors* and the quotient group

$$\operatorname{Div} X/P(X)$$

is called the *group of divisor classes on X*. Following the classical language of algebraic geometry, one says that two divisors D and D' are *linearly equivalent* if each is a representative of the same class of divisors. In other words D and D' are linearly equivalent if $D - D' \in P(X)$.

2. The Sheaves $\mathcal{F}(D)$. Given a locally free sheaf $\mathcal{F}$ and a divisor D, we define an *analytic subsheaf* $\mathcal{F}(D)$ of $\mathcal{F}^\infty$ by

$$\mathcal{F}(D)_x := \{s_x \in \mathcal{F}_x^\infty : o(s_x) \geq -o_x(D)\} \quad \text{and} \quad \mathcal{F}(D) := \bigcup_{x \in X} \mathcal{F}(D)_x \subset \mathcal{F}^\infty.$$

Except for points in $|D|$, $\mathcal{F}$ and $\mathcal{F}(D)$ agree. For $x \in |D|$, $\mathcal{F}_x$ is made smaller (resp. larger) when $o_x(D) < 0$ (resp. $o_x(D) > 0$). More precisely, it follows that,

$t \in \mathfrak{m}_x$ is a local coordinate at x, then

$$\mathscr{F}(D)_x = t^{-o_x(D)}\mathscr{F}_x.$$

One obtains an $\mathcal{O}$-isomorphism of $\mathscr{F}(D)$ onto $\mathscr{F}$ at x by multiplication by $t^{o_x(D)}$. In particular $\mathscr{F}(D)$ is a locally free sheaf with the same rank as $\mathscr{F}$ and it is always true that $\mathscr{F}^\infty(D) = \mathscr{F}^\infty$.

In the following lemma we summarize the laws which follow immediately from the general sheaf calculus:

Lemma. Let $\mathscr{F}$, $\mathscr{F}_1$, and $\mathscr{F}_2$ be locally free sheaves and D, D_1, D_2 divisors on a compact Riemann surface X. Then

1) Every exact $\mathcal{O}$-sequence $0 \to \mathscr{F}_1 \to \mathscr{F} \to \mathscr{F}_2 \to 0$ determines in a natural way an exact $\mathcal{O}$-sequence $0 \to \mathscr{F}_1(D) \to \mathscr{F}(D) \to \mathscr{F}_2(D) \to 0$.
2) If $\mathscr{F} = \mathscr{F}_1 + \mathscr{F}_2$ then $\mathscr{F}(D) = \mathscr{F}_1(D) + \mathscr{F}_2(D)$.
3) There is a natural $\mathcal{O}$-isomorphism $\mathscr{F}(D_1)(D_2) \cong \mathscr{F}(D_1 + D_2)$.
4) If $D_1 \leq D_2$ then $\mathscr{F}(D_1)$ is an analytic subsheaf of $\mathscr{F}(D_2)$.

The reader can easily carry out the proof. We note here that property 4) will play an important role in the next section.

3. The Sheaves $\mathcal{O}(D)$. The above considerations are in particular valid for $\mathscr{F} = \mathcal{O}$. All sheaves $\mathcal{O}(D)$, $D \in \operatorname{Div} X$, are locally free of rank 1. *Two sheaves, $\mathcal{O}(D_1)$ and $\mathcal{O}(D_2)$, are analytically isomorphic if and only if D_1 and D_2 are linearly equivalent.*

Proof: The sheaves $\mathcal{O}(D_1)$ and $\mathcal{O}(D_2)$ are analytically isomorphic if and only if $\mathcal{O}(D_1 - D_2) \cong \mathcal{O}$. Let $D := D_1 - D_2$. Assuming that D_1 and D_2 are linearly equivalent, $D = (h)$ with $h \in \mathcal{M}(X)^*$. In this case we obtain an $\mathcal{O}$-isomorphism $\mathcal{O} \to \mathcal{O}(D)$ by $f_x \mapsto f_x h_x$. Conversely, given an $\mathcal{O}$-isomorphism $\mathcal{O} \to \mathcal{O}(D)$, the image of $1 \in \mathcal{O}(X)$ in $\mathcal{O}(D)(X)$ is a meromorphic function $h \in \mathcal{M}(X)^*$ with $D = (h)$. □

By tensoring $\mathscr{F}$ with $\mathcal{O}(D)$, one gets all sheaves $\mathscr{F}(D)$. This is seen from the natural $\mathcal{O}$-isomorphism $\mathscr{F} \otimes \mathcal{O}(D) \to \mathscr{F}(D)$, defined by $s_x \otimes h_x \mapsto h_x s_x$.

Remark: If $\mathscr{S}$ is a coherent sheaf and $D \in \operatorname{Div} X$ then $\mathscr{S}(D) := \mathscr{S} \otimes \mathcal{O}(D)$ is coherent. The reader should note that the statements 1)–3) in the lemma of Paragraph 2 are in fact valid for coherent sheaves as well as locally free sheaves. The tensor product of two sheaves of the type $\mathcal{O}(D)$ is again a sheaf of that type.

Given two divisors D_1 and D_2, there is a natural isomorphism,

$$\mathcal{O}(D_1) \otimes \mathcal{O}(D_2) \to \mathcal{O}(D_1 + D_2),$$

defined stalkwise by $s_{1x} \otimes s_{2x} \mapsto s_{1x} \cdot s_{2x}$.

One usually identifies the group $H^1(X, \mathcal{O}^*)$ with the group of analytic isomorphism classes of locally free sheaves of rank 1 over X. In this way, the group operation in $H^1(X, \mathcal{O}^*)$ corresponds to the tensor product of sheaves. The homomorphism δ in the long exact cohomology sequence,

$$0 \longrightarrow \mathbb{C}^* \longrightarrow \mathcal{M}^*(X) \longrightarrow \operatorname{Div} X \xrightarrow{\delta} H^1(X, \mathcal{O}^*),$$

related to the short exact sequence of sheaves, $0 \to \mathcal{O}^* \to \mathcal{M}^* \to \mathcal{D} \to 0$, is in fact defined by $D \mapsto \mathcal{O}(D)$. The kernel of δ is just the group $P(X)$ of principal divisors and therefore one has a natural injection,

$$\operatorname{Div} X/P(X) \hookrightarrow H^1(X, \mathcal{O}^*),$$

of the group of divisor classes on X into $H^1(X, \mathcal{O}^*)$.

§ 2. The Existence of Global Meromorphic Sections

We will show here that every locally free sheaf (not identically zero) on a compact Riemann surface has "many" global meromorphic sections. This follows from a "characteristic theorem" which will give rise in the next section to a preliminary version of the Riemann-Roch theorem. In particular it is proved that for every $p \in X$ there is a non-constant holomorphic function on $X \backslash p$ which has a pole at p. This shows that $X \backslash p$ is Stein, from which it follows that $H^q(X, \mathcal{S}) = 0$, $q \geq 2$, for any coherent sheaf $\mathcal{S}$ on X.

1. The Sequence $0 \to \mathcal{F}(D) \to \mathcal{F}(D') \to \mathcal{T} \to 0$. Let D, D' be divisors with $D \leq D'$ and let $\mathcal{F}$ be a locally free sheaf of rank r. Then there is a natural exact sequence

$$(*) \qquad 0 \to \mathcal{F}(D) \to \mathcal{F}(D') \to \mathcal{T} \to 0,$$

where $\mathcal{T} := \mathcal{F}(D')/\mathcal{F}(D)$. Since this sequence plays such an important role in our considerations, we will now write down its basic properties: From the definitions it follows that for every $x \in X$

$$(1) \qquad \mathcal{T}_x = \mathbb{C}^{rn_x}, \qquad \text{with} \quad n_x := o_x(D') - o_x(D).$$

The support of $\mathcal{T}$ is therefore the support of $D' - D$ and

$$(2) \qquad \dim_{\mathbb{C}} \mathcal{T}(X) = r \deg(D' - D).$$

Since $\mathcal{T}$ has *finite* support,

$$H^1(X, \mathcal{T}) = (0).$$

Consequently the first part of the cohomology sequence associated to $(*)$ is th exact sequence

$$(3) \qquad 0 \to \mathcal{F}(D)(X) \to \mathcal{F}(D')(X) \to \mathcal{T}(X) \to H^1(X, \mathcal{F}(D)) \to H^1(X, \mathcal{F}(D')) \to$$

Thus

(4) $$\text{if } D \leq D' \text{ then } \dim_{\mathbb{C}} H^1(X, \mathscr{F}(D)) \geq \dim_{\mathbb{C}} H^1(X, \mathscr{F}(D')).$$

2. The Characteristic Theorem and an Existence Theorem. Let $\mathscr{S}$ be a coherent sheaf on X. The *characteristic* of $\mathscr{S}$, $\chi_0(\mathscr{S})$, is defined by

$$\chi_0(\mathscr{S}) := \dim_{\mathbb{C}} H^0(X, \mathscr{S}) - \dim_{\mathbb{C}} H^1(X, \mathscr{S}).$$

It will be shown (Paragraph 4) that $\chi_0(\mathscr{S})$ is the Euler-Poincaré characteristic.

In particular it is proved that for every $p \in X$ there is a non-constant holomorphic function on $X \backslash p$ which has a pole at p. This shows that $X \backslash p$ is Stein, from which it follows that $H^q(X, \mathscr{S}) = 0$, $q \geq 2$, for any coherent sheaf $\mathscr{S}$ on X.

Lemma 1 (The characteristic theorem). Let $\mathscr{F}$ be a locally free sheaf of rank r and D a divisor on a compact Riemann surface X. Then

$$\chi_0(\mathscr{F}(D)) = r \deg D + \chi_0(\mathscr{F}).$$

Proof: We will show that, for arbitrary divisors D and D',

($\circ$) $$\chi_0(\mathscr{F}(D)) - r \deg D = \chi_0(\mathscr{F}(D')) - r \deg D'$$

The claim of the lemma will then follow with $D' := 0$. First we suppose that $D \leq D'$. The alternating sum of the dimensions of the vector spaces in the exact cohomology sequence (3) is therefore zero. In other words

$$0 = \chi_0(\mathscr{F}(D)) - \chi_0(\mathscr{F}(D')) + \dim_{\mathbb{C}} \mathscr{T}(X).$$

If one substitutes $r \deg(D' - D)$ for $\dim_{\mathbb{C}} \mathscr{T}(X)$ (see (2) above) then ($\circ$) follows immediately.

If D' is arbitrary, then one chooses $D'' \in \operatorname{Div} X$ with $D \leq D''$ and $D' \leq D''$. Then it follows from the above that

$$\chi_0(\mathscr{F}(D)) - r \deg D = \chi_0(\mathscr{F}(D'')) - r \deg D'' = \chi_0(\mathscr{F}(D')) - r \deg D'. \quad \square$$

Theorem 2 (*Existence theorem*). *Let $\mathscr{F}$ be a locally free sheaf of* rank *r and D a divisor on a compact Riemann surface X. Then*

$$\dim_{\mathbb{C}} \mathscr{F}(D)(X) \geq r \deg D + \chi_0(\mathscr{F}).$$

n particular if $\mathscr{F} \neq 0$ and $\deg D > 0$, *then*

$$\lim_{n \to \infty} \dim_{\mathbb{C}} \mathscr{F}(nD)(X) = \infty.$$

We have therefore established that *every locally free sheaf* $\mathscr{F} \neq 0$ *has non-identically zero meromorphic sections*. This is clear from Theorem 2, since $\mathscr{F}(D)(X)$ is always contained in $\mathscr{F}^{\infty}(X)$.

3. The Vanishing Theorem. By Theorem 2,

$$\dim_{\mathbb{C}} \mathcal{O}(np)(X) \geq n + \chi_0(\mathcal{O})$$

for any $p \in X$, and all $n \in \mathbb{Z}$. Hence there exists $n_0 \in \mathbb{N}$ so that $\mathcal{O}(np)(X)$ contains a non-constant function h for all $n > n_0$. Since $(h) + np \geq 0$, every such function is non-constant, and holomorphic on $X \backslash p$, and has a pole of order at most n at p. This shows the following:

For every $p \in X$ there is a non-constant meromorphic function on X which is holomorphic on $X \backslash p$.[1] *Moreover, $X \backslash p$ is Stein. In particular every compact Riemann surface can be covered by two Stein domains.*

Proof: Let h be as above. Then $h: X \backslash p \to \mathbb{C}$ is a finite holomorphic map. Thus $X \backslash p$ is Stein. If $p_1, p_2 \in X$ are different points, then $\{X \backslash p_1, X \backslash p_2\}$ is a Stein cover of X. □

The following is now a consequence of the general theory.

Theorem 3 (Vanishing theorem). *Let $\mathscr{S}$ be a coherent analytic sheaf on X. Then*

$$H^q(X, \mathscr{S}) = 0, \qquad q \geq 2.$$

For every *compact* complex space X and every coherent sheaf $\mathscr{S}$ on X, *almost all* of the groups $H^q(X, \mathscr{S})$ vanish. Thus the Finiteness Theorem allows us to define the *Euler-Poincaré Characteristic*,

$$\chi(\mathscr{S}) = \sum_{i=0}^{\infty} (-1)^i \dim_{\mathbb{C}} H^i(X, \mathscr{S}) \in \mathbb{Z}.$$

Hence Theorem 3 shows that for compact Riemann surfaces

$$\chi(\mathscr{S}) = \chi_0(\mathscr{S}).$$

4. The Degree Equation. An amusing consequence of the characteristic formula, $\chi_0(\mathcal{O}(D)) = \deg D + \chi_0(\mathcal{O})$, is the degree equation: *For linearly equivalent divisors D and D'*, $\deg D = \deg D'$. *In particular* $\deg D = 0$ *for all principal divisors*

[1] With a bit more effort one can show at this point that for every $p \in X$ there exists $n_0 \in \mathbb{N}$ so that for every $n > n_0$ there is $h \in \mathscr{M}(X)$ which is holomorphic on $X \backslash p$, and which has a pole of order n at (a forerunner of the Weierstrass gap Theorem).

Proof: If D and D' are linearly equivalent then, from Paragraph 1.3, $\mathcal{O}(D) \cong \mathcal{O}(D')$ and consequently $\chi_0(\mathcal{O}(D)) = \chi_0(\mathcal{O}(D'))$. The characteristic formula yields

$$\deg D + \chi_0(\mathcal{O}) = \deg D' + \chi_0(\mathcal{O}). \qquad \square$$

The reader should note that if $X \neq \mathbb{P}_1$, then not every divisor D with $\deg D = 0$ is a principal divisor.

Remark: The degree equation can also be interpreted mapping theoretically, and proved in this way as well. For this, note that every $h \in \mathcal{M}(X)$ defines a branched covering $h: X \to \mathbb{P}_1$ (the case of h identically constant is trivial). Certainly, if every point of $h^{-1}(0)$ and $h^{-1}(\infty)$ is counted with its branching multiplicity, then $(h) = h^{-1}(0) - h^{-1}(\infty)$. For every $p \in \mathbb{P}_1$, the sum of the multiplicities of the points of $h^{-1}(p)$ is the sheet number s of the covering $h: X \to \mathbb{P}_1$. It follows that $\deg h = s - s = 0$.

§ 3. The Riemann–Roch Theorem (Preliminary Version)

The classical problem of Riemann–Roch consists of determining the dimension of the $\mathbb{C}$-vector space, $H^0(X, \mathcal{O}(D))$, of all global sections of the sheaf $\mathcal{O}(D)$. The characteristic theorem gives a preliminary solution of this problem.

1. The Genus Theorem of Riemann–Roch. The following notation is standard:

$$l(D) := \dim_{\mathbb{C}} H^0(X, \mathcal{O}(D)) \quad \text{and} \quad i(D) := \dim_{\mathbb{C}} H^1(X, \mathcal{O}(D)).$$

For linearly equivalent divisors D and D', $l(D) = l(D')$ and $i(D) = i(D')$. We further note that *the dimension $l(D) > 0$ if and only if there is a positive divisor D' which is linearly equivalent to D. In particular $l(D) = 0$ for every D with* $\deg D < 0$.

Proof: The first statement is clear, since the divisors which are linearly equivalent to D are of the form $D + (h)$ for $h \in \mathcal{M}^*(X)$. The second statement follows from the first, since linearly equivalent divisors have the same degree. $\square$

For the zero divisor $\mathcal{O}(X) \cong \mathbb{C}$ and thus $l(0) = 1$. The natural number

$$g := i(0) = \dim_{\mathbb{C}} H^1(X, \mathcal{O})$$

is called the *genus* of X. From this definition it only follows that g is a *complex analytic invariant* of X. However in Paragraph 7.1 we will show that g is in fact the *topological* genus of X (i.e. $H^1(X, \mathbb{C}) \cong \mathbb{C}^{2g}$).

For every divisor D it follows that

$$\chi_0(\mathcal{O}(D)) = l(D) - i(D),$$

and in particular

$$\chi_0(\mathcal{O}) = 1 - g.$$

Thus the characteristic formula $\chi_0(\mathcal{O}(D)) = \deg D + \chi_0(\mathcal{O})$ can be restated as follows:

Theorem 1 (Riemann–Roch, preliminary version). *If D is a divisor on a compact Riemann surface X of genus g, then*

$$l(D) - i(D) = \deg D + 1 - g.$$

Remark: The Riemann–Roch problem (i.e. the determination of the number $l(D)$) is not satisfactorily solved by Theorem 1, because the term $i(D)$ appears as a dimension of a first cohomology group. However, by means of Serre duality it can be interpreted as the dimension of a 0th cohomology group (see Paragraph 6). The final solution of the Riemann–Roch problem is given by Theorem 7.2.

2. Applications. The following *Riemann Inequality* is a special case of Theorem 1:

$$l(D) \geq \deg D + 1 - g$$

This inequality yields the first classical existence theorems. For example, since $\mathcal{O}(D)(X)$ contains a non-constant meromorphic function whenever $l(D) \geq 2$, we have the following:

For every divisor D with $\deg D \geq g + 1$ *there exists a non-constant meromorphic function h with $(h) + D \geq 0$.*

In particular, given $p \in X$, there always exist non-constant functions which have poles of order at most $g + 1$ at p, and which are holomorphic on $X \backslash p$. One can state this as a theorem about coverings of $\mathbb{P}_1$: *Every compact Riemann surface X of genus g is realizable as a branched cover with at most $g + 1$ sheets of the Riemann sphere $\mathbb{P}_1$. In particular, if $g = 0$ then $X = \mathbb{P}_1$.*

Since $l(D) = 0$ when $\deg D < 0$, it follows from Theorem 1 that, for every $D \in \operatorname{Div} X$ with $\deg D < 0$,

$$i(D) = g - 1 - \deg D.$$

One sees that, with the exception of the case $g = 0$ and $\deg D = -1$, $H^1(X, \mathcal{O}(D)) \neq (0)$ for all divisors of negative degree. Furthermore,

$$\lim_{\deg D \to -\infty} \dim_{\mathbb{C}} H^1(X, \mathcal{O}(D)) = \infty.$$

Remark: The existence of meromorphic functions which have poles (perhaps of high order) only at a prescribed point is guaranteed by the results in Paragraph 2. The improvement here is that the minimum order which can be prescribed can be estimated independent of the point p by the genus.

§ 4. The Structure of Locally Free Sheaves

We will show that every locally free sheaf, which is not the zero sheaf, contains locally free subsheaves of the form $\mathcal{O}(D)$. This theorem is the most important aid in the study of general locally free sheaves (see, for example, the supplement of this section as well as Section 8).

1. Locally Free Subsheaves. The considerations of this section are formal in nature. The following language is useful:

Definition 1 (Locally free subsheaves). An analytic subsheaf $\mathcal{F}'$ of a locally free sheaf $\mathcal{F}$ is called a *locally free subsheaf* of $\mathcal{F}$ if

0) $\mathcal{F}'$ is itself locally free.
1) The quotient sheaf $\mathcal{F}/\mathcal{F}'$ is locally free.

The rank equation for a locally free subsheaf $\mathcal{F}'$ of a locally free sheaf $\mathcal{F}$,

$$\operatorname{rk} \mathcal{F} = \operatorname{rk} \mathcal{F}' + \operatorname{rk} \mathcal{F}/\mathcal{F}',$$

is an immediate consequence of the definition. Thus every locally free sheaf $\mathcal{L}$ of rank 1 contains only 0 and $\mathcal{L}$ as locally free subsheaves.

The requirement 1) in the above definition is quite restrictive. For example a germ $t_x \in \mathcal{F}_x$ always generates a free submodule $t_x \mathcal{O}_x$ in $\mathcal{F}_x$, but the quotient module $\mathcal{F}_x/t_x\mathcal{O}_x$ is in general *not* free. For an explicit example, take $\mathcal{F} = \mathcal{O}$ and t_x a non-unit. On the other hand if t_x is a unit then there is no problem: *If $t_x \in \mathcal{F}_x$ and $o(t_x) = 0$ then $\mathcal{F}_x/t_x\mathcal{O}_x$ is a free $\mathcal{O}_x$-module.*

Proof: Let $\mathcal{F}_x = \mathcal{O}_x^r$ and $t_x = (t_1, \ldots, t_r)$, $t_i \in \mathcal{O}_x$. Since $o(t_x) = 0$, some t_i, say t_1, is a unit. Let $e := t_1^{-1}$ and define $\sigma\colon \mathcal{O}_x^r \to \mathcal{O}_x^{r-1}$ by

$$\sigma(f_1, \ldots, f_r) = (f_2 - ef_1 t_2, \ldots, f_r - ef_1 t_r).$$

Thus σ is an epimorphism with kernel $\mathcal{O}_x t_x$. □

It is easy to find locally free subsheaves in $\mathcal{F}^\infty$:

Theorem 2. *Let $\mathcal{F}$ be locally free and D the divisor of a meromorphic section $s \in \mathcal{F}^\infty(X)$. Assume $s \not\equiv 0$. Then the sheaf $\mathcal{L}$, defined by $\mathcal{L} := \mathcal{O}(D)s \cong \mathcal{O}(D)$ is a locally free subsheaf of $\mathcal{F}$. The section s is always in $\mathcal{L}^\infty(X)$ and, when $s \in \mathcal{F}(X)$, $s \in \mathcal{L}(X)$.*

Proof: Since $o(h_x s_x) = o(h_x) + o_x(D) \geq 0$ for every $h_x \in \mathcal{O}(D)_x$, $\mathcal{O}(D)_x s_x \subset \mathcal{F}_x$. Thus $\mathcal{L}$ is an analytic subsheaf of $\mathcal{F}$. Due to the fact that $s \not\equiv 0$, $\mathcal{L}$ is isomorphic to $\mathcal{O}(D)$. Furthermore $s \in \mathcal{L}(-D)(X) \subset \mathcal{L}^\infty(X)$ and, when $D \geq 0$, $s \in \mathcal{L}(X)$. Finally $\mathcal{O}(D)_x = \mathcal{O}_x g_x$, where $o(g_x) = -o_x(D)$. Defining $t_x := g_x s_x$, it follows that $\mathcal{L}_x = \mathcal{O}_x t_x$ with $o(t_x) = 0$ for all $x \in X$. Therefore, by the above remark, $\mathcal{F}/\mathcal{L}$ has everywhere free stalks. □

Remark: The sheaf $\mathscr{L} = \mathcal{O}(D)s$ *is the only locally free subsheaf of* $\mathscr{F}$ *of* rank 1 *with* $s \in \mathscr{L}^\infty(X)$. To see this, let $\hat{\mathscr{L}}$ be such a sheaf. Then $\hat{\mathscr{L}}_x = \mathcal{O}_x v_x$ for some $v_x \in \mathscr{F}_x$, where $s_x = m_x v_x$ and $m_x \in \mathscr{M}_x$. Let $h_x := m_x^{-1}$. Then $v_x = h_x s_x$ and, since $o(v_x) \geq 0$, $h_x \in \mathcal{O}(D)_x$. Consequently $v_x \in \mathcal{O}(D)_x s_x$ or $\hat{\mathscr{L}}_x \subset \mathscr{L}_x$. The $\mathcal{O}_x$-module $\mathscr{F}_x/\hat{\mathscr{L}}_x$ contains therefore a submodule which is isomorphic to $\mathscr{L}_x/\hat{\mathscr{L}}_x$. Since $\mathscr{F}_x/\mathscr{L}_x$ is free, and since $\mathscr{L}/\hat{\mathscr{L}}_x$ is in any case finite, $\mathscr{L}_x/\hat{\mathscr{L}}_x = 0$. Hence $\hat{\mathscr{L}}_x = \mathscr{L}_x$ for all x and $\hat{\mathscr{L}} = \mathscr{L}$. □

2. The Existence of Locally Free Subsheaves. The foundation for the study of the structure of locally free sheaves is the following theorem.

Theorem 3 (Subsheaf theorem). *Every locally free sheaf* $\mathscr{F} \neq 0$ *contains a locally free subsheaf which is isomorphic to* $\mathcal{O}(D)$ *for some* $D \in \operatorname{Div} X$. *One can choose* $D = (s)$, *where* $s \in \mathscr{F}^\infty(X)^*$.

Proof: From Paragraph 2.2 it follows that $\mathscr{F}$ has a non-trivial global meromorphic section. Thus the theorem is an immediate consequence of Theorem 2. □

The following is an immediate corollary of Theorem 3.

Theorem 4 (Structure theorem for locally free sheaves of rank 1). *Every locally free sheaf* $\mathscr{L}$ *of* rank 1 *is isomorphic to a sheaf* $\mathcal{O}(D)$ *with* $D \in \operatorname{Div} X$. *Furthermore one can choose* $D = (s)$ *with* $s \in \mathscr{L}^\infty(X)^*$.

This theorem says that in the cohomology sequence which is associated to the short exact sequence of sheaves $0 \to \mathcal{O}^* \to \mathscr{M}^* \to \mathscr{D} \to 0$,

$$\cdots \operatorname{Div} X \xrightarrow{\delta} H^1(X, \mathcal{O}^*) \longrightarrow H^1(X, \mathscr{M}^*) \longrightarrow H^1(X, \mathscr{D}) \longrightarrow \cdots,$$

the homomorphism δ is surjective (see Paragraph 1.3). Thus one has a natural group isomorphism,

$$\operatorname{Div} X/P(X) \xrightarrow{\sim} H^1(X, \mathcal{O}^*),$$

of the group of divisor classes on X onto $H^1(X, \mathcal{O}^*)$.

Remark: Since δ is surjective, the map $H^1(X, \mathscr{M}^*) \to H^1(X, \mathscr{D})$ is injective. Clearly $\mathscr{D}$ is a soft sheaf and thus $H^1(X, \mathscr{D}) = (0)$. Thus Theorem 4 is equivalent to the equation

$$H^1(X, \mathscr{M}^*) = (0).$$

3. The Canonical Divisors. The sheaf of germs of holomorphic 1-forms over X is a locally free sheaf of rank 1. Thus one can apply Theorem 4:

Theorem 5. *There is a unique divisor class on* X *so that for every divisor* K *in this class,* $\Omega^1 \cong \mathcal{O}(K)$.

One calls K a *canonical divisor* and its class the *canonical divisor class* on X. The significance of canonical divisors appears in Section 6.

If $X = \mathbb{P}_1$ then every divisor $-2x_0$, $x_0 \in X$, is canonical. This follows from the fact that, if z is a coordinate on $X \backslash x_0$, then dz is a differential form on X which is holomorphic and nowhere vanishing on $X \backslash x_0$ and has a pole of order 2 at x_0. Of course one must use the fact that on $\mathbb{P}_1$ the degree of a divisor determines its class. In the case of elliptic curves the zero divisor is canonical.

Supplement to Section 4: The Riemann–Roch Theorem for Locally Free Sheaves

The generalized Riemann–Roch problem consists of determining the dimension of $H^0(X, \mathscr{F}(D))$ for every locally free sheaf $\mathscr{F}$ and every divisor D. In order to do this one needs to carry the idea of degree over to the case of locally free sheaves.

We use the fact that χ_0 coincides with the Euler–Poincaré characteristic χ. Moreover the additivity of χ (i.e. for every exact sequence of coherent sheaves $0 \to \mathscr{S}' \to \mathscr{S} \to \mathscr{S}'' \to 0$, one has $\chi(\mathscr{S}) = \chi(\mathscr{S}') + \chi(\mathscr{S}'')$) plays an important role.

1. The Chern Function. We denote with $LF(X)$ the set of analytic isomorphism classes of locally free sheaves over X. A function $c\colon LF(X) \to \mathbb{Z}$ is called a *Chern function* if

1) For $D \in \operatorname{Div} X$, $c(\mathscr{O}(D)) = \deg D$.
2) For every exact sequence $0 \to \mathscr{F}' \to \mathscr{F} \to \mathscr{F}'' \to 0$ of locally free sheaves, one has $c(\mathscr{F}) = c(\mathscr{F}') + c(\mathscr{F}'')$.

The following is straight forward:

Theorem 1. *The function $c\colon LF(X) \to \mathbb{Z}$, defined by*

3) $$c(\mathscr{F}) := \chi(\mathscr{F}) - \operatorname{rank} \mathscr{F} \cdot \chi(\mathscr{O})$$

is a Chern function.

Proof: From Lemma 2.1 we have

$$c(\mathscr{O}(D)) = \chi(\mathscr{O}(D)) - \chi(\mathscr{O}) = \deg D.$$

The additivity follows from the additivity of both χ and rank. □

Remark: The results in Paragraph 4.2 imply that *this c is the only Chern function.* To see this note first that if γ is any such function then, since the locally free sheaves of rank 1 are just the divisor sheaves, $\gamma(\mathscr{L}) = c(\mathscr{L})$ for every locally free sheaf $\mathscr{L}$ of rank 1. Suppose now that $\gamma = c$ for all locally free sheaves of rank less than r and let $\mathscr{F}$ of rank $r \geq 2$ be given. Then there is an exact sequence of locally free sheaves

$$0 \to \mathscr{L} \to \mathscr{F} \to \mathscr{G} \to 0,$$

where $\mathscr{L}$ and $\mathscr{G}$ have rank 1 and $r-1$ respectively. The induction hypothesis and the additivity imply that $\gamma(\mathscr{F}) = c(\mathscr{F})$.

2. Properties of the Chern Function. The Chern function behaves nicely with respect to tensor products:

4) $\qquad c(\mathscr{F} \otimes \mathscr{L}) = \operatorname{rank} \mathscr{F} \cdot c(\mathscr{L}) + c(\mathscr{F})$, when $\operatorname{rank} \mathscr{L} = 1$.

Proof: Let $\mathscr{L} = \mathcal{O}(D)$. Then $\mathscr{F} \otimes \mathscr{L} = \mathscr{F}(D)$ and $c(\mathscr{L}) = \deg D$. Thus

$$\begin{aligned} c(\mathscr{F} \otimes \mathscr{L}) &= \chi(\mathscr{F}(D)) - \operatorname{rank} \mathscr{F}(D) \cdot \chi(\mathcal{O}) \\ &= \operatorname{rank} \mathscr{F} \cdot \deg D + \chi(\mathscr{F}) - \operatorname{rank} \mathscr{F} \cdot \chi(\mathcal{O}) \\ &= \operatorname{rank} \mathscr{F} \cdot c(\mathscr{L}) + c(\mathscr{F}). \end{aligned}$$ □

When $\mathscr{L}_1$ and $\mathscr{L}_2$ are locally free sheaves of rank 1, it follows from 4) that $c(\mathscr{L}_1 \otimes \mathscr{L}_2) = c(\mathscr{L}_1) + c(\mathscr{L}_2)$. Thus the map $H^1(X, \mathcal{O}^*) \to \mathbb{Z}$, defined by $\mathscr{L} \to c(\mathscr{L})$, is a group homomorphism from the analytic isomorphism classes of locally free sheaves of rank 1 to $\mathbb{Z}$. The reader can check that the map γ in the exact cohomology sequence

$$\cdots \longrightarrow H^1(X, \mathcal{O}) \longrightarrow H^1(X, \mathcal{O}^*) \xrightarrow{\gamma} H^2(X, \mathbb{Z}) \longrightarrow \cdots$$

is the Chern function, provided $H^2(X, \mathbb{Z})$ is identified with $\mathbb{Z}$ in the natural way. □

We remark without proof that, if $\mathscr{F}$ is locally free of rank r greater than 1, $c(\mathscr{F}) = c(\det \mathscr{F})$, where $\det \mathscr{F} := \bigwedge_1^r \mathscr{F}$ is the locally free determinant sheaf of rank 1. Thus, letting F be the vector bundle associated to $\mathscr{F}$, $c(\mathscr{F})$ is just the first Chern class of F, $c_1(F) \in H^2(X, \mathbb{Z}) = \mathbb{Z}$.

3. The Riemann–Roch Theorem. Equation 3) in Proposition 1 above can be rewritten as a Riemann–Roch theorem:

Theorem 2 (The Riemann–Roch theorem for locally free sheaves). *Let $\mathscr{F}$ be a locally free sheaf of rank r and D a divisor on a compact Riemann surface X of genus g. Then*

$$\dim_{\mathbb{C}} H^0(X, \mathscr{F}(D)) - \dim_{\mathbb{C}} H^1(X, \mathscr{F}(D)) = r(\deg D + 1 - g) + c(\mathscr{F}).$$

Proof: On the left hand side we have $\chi(\mathscr{F}(D))$ which, by the characteristic theorem, is the same as $r \deg D + \chi(\mathscr{F})$. If one writes $r \cdot \chi(\mathcal{O}) + c(\mathscr{F})$ for $\chi(\mathscr{F})$ then, since $\chi(\mathcal{O}) = 1 - g$, the claim follows immediately. □

§ 5. The Equation $H^1(X, \mathcal{M}) = 0$

The inequality from Paragraph 3.2, $l(np) \geq n + 1 - g$, is already strong enough to show that for every divisor D on X and every point $p \in X$ the cohomology groups $H^1(X, \mathcal{O}(D + np))$ vanish for all $n \gg 0$ (we already know that the dimension of these vector spaces is constant for $n \gg 0$ and is unbounded for $n < 0$). This "Theorem B for compact Riemann surfaces," which will be made even more precise in Section 7, has an immediate consequence the fact that $H^1(X, \mathcal{M}) = 0$.

1. The $\mathbb{C}$-homomorphism $\mathcal{O}(np)(X) \to \operatorname{Hom}(H^1(X, \mathcal{O}(D)), H^1(X, \mathcal{O}(D + np)))$. Let D be a divisor in Div X, p a point in X, and $n \geq 0$ a fixed integer. Every function $f \in \mathcal{O}(np)(X)$ determines the $\mathcal{O}$-homomorphism $\varphi_f\colon \mathcal{O}(D) \to \mathcal{O}(D + np)$, $h_x \mapsto f_x h_x$. If $f \not\equiv 0$, then φ_f is *injective* and the sheaf $\mathcal{I}m\ \varphi_f$ in the exact sequence

$$0 \longrightarrow \mathcal{O}(D) \xrightarrow{\varphi_f} \mathcal{O}(D + np) \longrightarrow \mathcal{I}m\,\varphi_f \longrightarrow 0, \quad \mathcal{I}m\,\varphi_f \cong \mathcal{O}(D + np)/f \cdot \mathcal{O}(D),$$

has *finite* support $((\mathcal{I}m\,\varphi_f)_x = 0$ for every $x \notin |D| \cup \{p\}$, where f_x is a unit in $\mathcal{O}_X)$. Thus $H^1(X, \mathcal{I}m\ \varphi_f) = 0$ and

every function $f \not\equiv 0$ in $\mathcal{O}(np)(X)$ induces a $\mathbb{C}$-vector space epimorphism

$$\psi_f\colon H^1(X, \mathcal{O}(D)) \to H^1(X, \mathcal{O}(D + np)).$$

If $f = 0$, then we define φ_f to be the zero mapping.

Lemma 1. *The map*

$$\psi\colon \mathcal{O}(np)(X) \to \operatorname{Hom}_{\mathbb{C}}(H^1(X, \mathcal{O}(D)), H^1(X, \mathcal{O}(D + np))), \qquad f \to \varphi_f,$$

is $\mathbb{C}$-linear.

Proof: The claim becomes clear when one looks at how φ_f is defined via the Čech complex. Let $\mathfrak{U} = \{U_i\}$, $i \in I$, be a cover of X with cochain groups

$$C^1(\mathfrak{U}, \mathcal{O}(D)) = \prod_{i_0, i_1} \mathcal{O}(D)(U_{i_0 i_1})$$

and

$$C^1(\mathfrak{U}, \mathcal{O}(D + np)) = \prod_{i_0, i_1} \mathcal{O}(D + np)(U_{i_0 i_1}).$$

Associated to every $\mathcal{O}$-homomorphism φ_f, $f \in \mathcal{O}(np)$, are the following homomorphisms at the section level:

*) $\varphi_f(i_0, i_1)\colon \mathcal{O}(D)(U_{i_0 i_1}) \to \mathcal{O}(D + np)(U_{i_0 i_1})$, $\quad h \mapsto (f \,|\, U_{i_0 i_1})h, \quad i_0, i_1 \in I.$

The collection $\Psi_f := \{\varphi_f(i_0, i_1)\}$ is a homomorphism

$$\Psi_f\colon C^1(\mathfrak{U}, \mathcal{O}(D)) \to C^1(\mathfrak{U}, \mathcal{O}(D + np))$$

which induces the homomorphism $\psi_f\colon H^1(X, \mathcal{O}(D)) \to H^1(X, \mathcal{O}(D + np))$. From the definition of $\varphi_f(i_0, i_1)$ in (*) it is clear that

$$\varphi_{af+bg}(i_0, i_1) = a\varphi_f(i_0, i_1) + b\varphi_g(i_0, i_1),\ a, b \in \mathbb{C};\ f, g \in \mathcal{O}(np)(X);\ i_0, i_1 \in I.$$

Consequently $\Psi_{af+bg} = a\Psi_f + b\Psi_g$ and thus $\psi_{af+bg} = a\psi_f + b\psi_g$. □

2. The Equation $H^1(X, \mathcal{O}(D + np)) = 0$. The following is an easy corollary to the Riemann inequality and Lemma 1:

Theorem 2. *Let* $D \in \operatorname{Div} X$ *and* $p \in X$. *Then*

$$H^1(X, \mathcal{O}(D + np)) = 0 \quad \textit{for all} \quad n \geq n_0 := (\dim_{\mathbb{C}} H^1(X, \mathcal{O}(D)))^2 + g$$

Proof: Let $d := \dim_{\mathbb{C}} H^1(X, \mathcal{O}(D))$. Then, by 2.2(4), it follows that $\dim_{\mathbb{C}} H^1(X, \mathcal{O}(D + np)) \leq d$ for all $n \geq 0$. Hence the map ψ (in Paragraph 1 above) maps $\mathcal{O}(np)(X)$ into a $\mathbb{C}$-vector space having dimension at most d^2. But d does not depend on n and, by 3.2, $\dim_{\mathbb{C}} \mathcal{O}(np)(X) \geq n + 1 - g$. Therefore, whenever $n \geq n_0 = d^2 + g$, the $\mathbb{C}$-*linear* map ψ is not injective. Thus for such an n there exists $f \neq 0$ in $\mathcal{O}(np)(X)$ with $\varphi_f = 0$. Since φ_f must be an isomorphism, $H^1(X, \mathcal{O}(D + np)) = 0$. □

3. The Equation $H^1(X, \mathcal{M}) = 0$. We can now prove the following fundamental theorem:

Theorem 3. $H^1(X, \mathcal{M}) = 0$.

Proof: Let $\xi \in H^1(X, \mathcal{M})$ be an arbitrary cohomology class. We choose a finite cover $\mathfrak{U} = \{U_i\}$ of X, a shrinking of $\mathfrak{U}$, $\mathfrak{V} = \{V_i\}$, with $V_i \Subset U_i$, and a cocycle $\xi \in Z^1(\mathfrak{U}, \mathcal{M})$ which is a representative of ξ. We let $\xi = (\xi_{i_0 i_1})$, where $\xi_{i_0 i_1} \in \mathcal{M}(U_{i_0 i_1})$. Then

$$\xi' := (\xi_{i_0 i_1} | V_{i_0 i_1}) = \xi | \mathfrak{V} \in Z^1(\mathfrak{V}, \mathcal{M})$$

is likewise a representative of ξ. Since $V_i \Subset U_i$ for all i, each function $\xi_{i_0 i_1}$ has finitely many zeros and poles on $V_{i_0 i_1}$. Thus one can find a divisor $D \in \operatorname{Div} X$ so that $\xi' \in Z^1(\mathfrak{V}, \mathcal{O}(D))$. Let $p \in X \setminus |D|$. Then, since $\mathcal{O}(D) \subset \mathcal{O}(D + np)$ for all $n \geq 0$, it follows that $\xi' \subset Z^1(\mathfrak{V}, \mathcal{O}(D + np))$ for all such n. Since $\mathcal{O}(D + np) \subset \mathcal{M}$ and $H^1(X, \mathcal{O}(D + np)) = 0$ for $n \geq n_0$, it follows that ξ' is cohomologous to zero as an element of $Z^1(\mathfrak{V}, \mathcal{M})$. Thus $\xi = 0$, and consequently $H^1(X, \mathcal{M}) = 0$. □

The most important applications of the equation $H^1(X, \mathcal{M}) = 0$ are found in the next section. In closing this section we note in passing a rather simple consequence. Let Ω^∞ denote the sheaf of germs of meromorphic 1-forms. The differential $d: M \to \Omega^\infty$, which is defined in local coordinates by $dh_x := dh_x/dz\, dz$, is $\mathbb{C}$-linear. The $\mathbb{C}$-sheaf $d\mathcal{M} \subset \Omega^\infty$ (it is *not* an $\mathcal{O}$-sheaf!) consists of *all residue free* germs of meromorphic 1-forms (i.e. germs of abelian differentials of the second kind; for the idea of a residue see Paragraph 6.5). Since $H^1(X, \mathcal{M}) = 0$, the short exact $\mathbb{C}$-sequence $0 \to \mathbb{C} \to \mathcal{M} \to d\mathcal{M} \to 0$ induces the following exact cohomology sequence:

$$0 \longrightarrow \mathbb{C} \longrightarrow H^0(X, \mathcal{M}) \xrightarrow{d} H^0(X, d\mathcal{M}) \longrightarrow H^1(X, \mathbb{C}) \longrightarrow 0.$$

This means that

$$H^1(X, \mathbb{C}) \cong H^0(X, d\mathcal{M})/dH^0(X, \mathcal{M}).$$

In words,

the cohomology group $H^1(X, \mathbb{C})$ is isomorphic to the quotient space of global abelian differentials of the second kind modulo differentials of global meromorphic functions.

§ 6. The Duality Theorem of Serre

In this section we will establish a natural isomorphism between the space of differential forms $H^0(X, \Omega(D))$ and the dual space of the first cohomology group $H^1(X, \mathcal{O}(-D))$, where D is a given divisor on X:

$$H^0(X, \Omega(D)) \xrightarrow{\sim} H^1(X, \mathcal{O}(-D))^*.$$

Given $\omega \in \Omega(D)(X)$, the associated linear form $\Theta(\omega): H^1(X, \mathcal{O}(-D)) \to \mathbb{C}$ will be obtained as a residue map by applying the residue theorem. We reproduce here the algebraic proof of Serre ([GACC], Chapter II), making use of the equation $H^1(X, \mathcal{M}) = (0)$.

1. The Principal Part Distributions with Respect to a Divisor. Defining $\mathcal{H} = \mathcal{M}/\mathcal{O}$ as the sheaf of germs of principal parts, we considered in Chapter V.2.1 the exact sequence $0 \to \mathcal{O} \to \mathcal{M} \to \mathcal{H} \to 0$. In the following we generalize this slightly. Let $D \in \operatorname{Div} X$ and define $\mathcal{H}(D) := \mathcal{M}/\mathcal{O}(D)$ as "the sheaf of germs of principal parts with respect to D." Then we have the exact sequence

(1) $$0 \to \mathcal{O}(D) \to \mathcal{M} \to \mathcal{H}(D) \to 0.$$

It follows immediately that $\mathcal{H}(D)$, like $\mathcal{D}$, is soft. In fact, *every section over an open set U has discrete support in U.* Since every stalk $\mathcal{H}(D)_x$ is the quotient module $\mathcal{M}_x/\mathcal{O}(D)_x$, this implies that *the $\mathbb{C}$-vector space $\mathcal{H}(D)(X)$ of global distributions of principal parts is canonically isomorphic to the direct sum* $\bigoplus_{x \in X} \mathcal{M}_x/\mathcal{O}(D)_x$:

(2) $$\mathcal{H}(D)(X) \cong \bigoplus_{x \in X} \mathcal{M}_x/\mathcal{O}(D)_x \cong \bigoplus_{x \in X} \mathcal{M}_x / \bigoplus_{x \in X} \mathcal{O}(D)_x.$$

2. The Equation $H^1(X, \mathcal{O}(D)) = I(D)$. Let R be the set of all maps $F = (f_x)$ which assign to every $x \in X$ a germ $f_x \in \mathcal{M}_x$ so that *almost all* f_x are holomorphic. Clearly R is a $\mathbb{C}$-vector space. Further, given $D \in \operatorname{Div} X$,

$$R(D) := \{F \in R: f_x \in \mathcal{O}(D)_x\}$$

is a subspace of R. Every element of the direct sum $\bigoplus_{x \in X} \mathcal{M}_x$ (resp. $\bigoplus_{x \in X} \mathcal{O}(D)_x$) is a family $\{f_x\}$ $x \in X$ with $f_x \in \mathcal{M}_x$ (resp. $f_x \in \mathcal{O}(D)_x$), where *almost all* f_x vanish. Thus we have the natural $\mathbb{C}$-linear injection $\bigoplus_{x \in X} \mathcal{M}_x \to R$ which maps $\bigoplus_{x \in X} \mathcal{O}(D)_x$ into $R(D)$, and which induces a $\mathbb{C}$-isomorphism

$$\mathcal{H}(D)(X) \cong \bigoplus_{x \in X} \mathcal{M}_x / \bigoplus_{x \in X} \mathcal{O}(D)_x \cong R/R(D). \tag{3}$$

Every meromorphic function h determines an element $(h_x) \in R$. Thus, identifying $\mathcal{M}(X)$ with its image in R, $\mathcal{M}(X) \cap R(D) = \mathcal{O}(D)(X)$. We now set

$$I(D) := R/(R(D) + \mathcal{M}(X)).$$

The standard isomorphism theorems of linear algebra yield

$$I(D) \cong (R/R(D))/(\mathcal{M}(X)/\mathcal{O}(D)(X)). \tag{4}$$

The following is an easy consequence of the definitions.

Theorem 1. *For every $D \in \operatorname{Div} X$ there is a natural $\mathbb{C}$-isomorphism*

$$H^1(X, \mathcal{O}(D)) \cong I(D).$$

Proof: Associated to the short exact sequence (1) we have the exact cohomology sequence

$$0 \longrightarrow \mathcal{O}(D)(X) \longrightarrow \mathcal{M}(X) \xrightarrow{\varepsilon} \mathcal{H}(D)(X)$$
$$\longrightarrow H^1(X, \mathcal{O}(D)) \longrightarrow H^1(X, \mathcal{M}) \longrightarrow \cdots$$

By Theorem 5.3, $H^1(X, \mathcal{M}) = (0)$. Thus

$$H^1(X, \mathcal{O}(D)) \cong \mathcal{H}(D)(X)/\operatorname{Im} \varepsilon,$$

where $\operatorname{Im} \varepsilon \cong \mathcal{M}(X)/\mathcal{O}(D)(X)$. The claim now follows from (3) and (4) above. □

The reader should note that the spaces R, $R(D)$ and $\mathcal{M}(X)$ have very large infinite dimensions, but that the finiteness theorem implies that $I(D)$ is finite dimensional.

3. Linear Forms. For maps $F = (f_x)$ and $G = (g_x) \in R$, we define the product $FG := (f_x g_x)$. Equipped with this product, R is of course a $\mathbb{C}$-algebra, but it is more importantly also an algebra over the field $\mathcal{M}(X) \subset R$. Let $\alpha: R \to \mathbb{C}$ be a $\mathbb{C}$-linear form and $h \in \mathcal{M}(X)$. Then one defines the $\mathbb{C}$-linear form $h\alpha: R \to \mathbb{C}$ by $h\alpha(F) = \alpha(hF)$. Thus $\operatorname{Hom}_{\mathbb{C}}(R, \mathbb{C})$ *becomes an* $\mathcal{M}(X)$*-vector space.* We explicitly state two simple, but important, properties:

a) If $\mathcal{M}(X) \subset \ker \alpha$ then $\mathcal{M}(X) \subset \ker h\alpha$.
b) If $R(D) \subset \ker \alpha$ then $R(D + (h)) \subset \ker h\alpha$.

Proof: The statement a) is trivial. If $F = (f_x) \in R(D + (h))$ then $f_x \in \mathcal{O}(D + (h))_x$ and therefore $o(h_x f_x) \geq -o_x(D)$. Hence $hF \in R(D)$. This proves b). □

We denote by $J(D)$ the dual space of $I(D)$. It follows from a) and b) that *every meromorphic function* $h \in \mathcal{M}(X)^*$ *determines a natural* $\mathbb{C}$*-linear mapping,*

$$J(D) \to J(D + h),$$

defined by $\lambda \mapsto h\lambda$.

Proof: Every $\lambda \in J(D)$ is a $\mathbb{C}$-linear form $\lambda: R/(R(D) + \mathcal{M}(X)) \to \mathbb{C}$ and is therefore liftable to a $\mathbb{C}$-linear form $\alpha: R \to \mathbb{C}$ such that α vanishes on $R(D) + \mathcal{M}(X)$. From a) and b) it follows that $h\alpha: R \to \mathbb{C}$ vanishes on $R(D + (h)) + \mathcal{M}(X)$. Clearly $h\alpha$ induces a $\mathbb{C}$-linear form $h\lambda \in J(D + (h))$, which is uniquely determined by λ and h. It is obvious that the map $\lambda \to h\lambda$ is $\mathbb{C}$-linear. □

Since $h\lambda$ is no longer in $J(D)$, we go over to the space

$$J := \bigcup_D J(D),$$

the union of all $J(D)$'s for $D \in \operatorname{Div} X$. For two divisors $D_1 \leq D_2$, we have $R(D_1) \leq R(D_2)$. Thus $J(D_1) \geq J(D_2)$ when $D_1 \leq D_2$. This immediately implies that *every finite subset of* J *is contained in some* $J(D)$.

The set J is thus a $\mathbb{C}$-vector space which is filtered by the subspaces $J(D)$, $D \in \operatorname{Div} X$. The above defined map, $J(D) \to J(D + (h))$, gives us a mapping $\mathcal{M}(X) \times J \to J$. Since $\operatorname{Hom}_{\mathbb{C}}(R, \mathbb{C})$ is an $\mathcal{M}(X)$-vector space, the following remark is obvious.

Theorem 2. *The set* J *is, with respect to the operation* $\mathcal{M}(X) \times J \to J$*, a vector space over* $\mathcal{M}(X)$.

4. The Inequality $\operatorname{Dim}_{\mathcal{M}(X)} J \leq 1$. The critical point of the proof of the duality theorem is the following surprising dimension estimate. It is obtained by taking a limit of the preliminary form of the Riemann–Roch formula.

Theorem 3. *The* $\mathcal{M}(X)$*-vector space* J *is at most* 1*-dimensional.*

Proof: Let $\lambda, \mu \in J$. We choose $D \in \operatorname{Div} X$ with $\lambda, \mu \in J(D)$. Let $p \in X$ be fixed. For every $f \in \mathcal{O}(np)(X)$ it follows that, since $D - np \le D + (f), f\lambda \in J(D + (f)) \subset J(D - np)$. Similarly $g\mu \in J(D - np)$ for all $g \in \mathcal{O}(np)(X)$. The $\mathbb{C}$-linear map

$$(\circ) \qquad \mathcal{M}(X) \oplus \mathcal{M}(X) \to J, \qquad (f, g) \mapsto f\lambda + g\mu,$$

therefore induces by restriction a $\mathbb{C}$-linear mapping

$$(\overset{\circ}{\circ}) \qquad \mathcal{O}(np)(X) \oplus \mathcal{O}(np)(X) \to J(D - np)$$

for all $n \in \mathbb{Z}$. If λ and μ were linearly independent then both maps $(\circ)$ and $(\overset{\circ}{\circ})$ would be injective. This would mean that

$$(+) \qquad 2 \dim_{\mathbb{C}} \mathcal{O}(np)(X) \le \dim_{\mathbb{C}} J(D - np)$$

for all $n \in \mathbb{Z}$. From the Riemann inequality (Paragraph 3.2) it follows that

$$(++) \qquad \dim_{\mathbb{C}} \mathcal{O}(np)(X) = l(np) \ge \deg(np) + 1 - g = n + 1 - g.$$

Furthermore (Paragraph 3.2), as soon as $\deg(D - np) = \deg D - n$ is negative,

$$\dim_{\mathbb{C}} J(D - np) = \dim_{\mathbb{C}} I(D - np) = i(D - np) = g - 1 - \deg(D - np).$$

Thus, for large n,

$$(\overset{+}{++}) \qquad \dim_{\mathbb{C}} J(D - np) = g - 1 - \deg D + n.$$

From $(++)$ and $(\overset{+}{++})$ one infers that, for n large enough,

$$2 \dim_{\mathbb{C}} \mathcal{O}(np)(X) > \dim_{\mathbb{C}} J(D - np).$$

This is contrary to $(+)$ and therefore λ and μ must be linearly dependent over $\mathcal{M}(X)$. □

Remark: There are divisors D such that $H^1(X, \mathcal{O}(D)) \neq (0)$. In other words, $J(D) \neq (0)$. Thus J is a *1-dimensional* $\mathcal{M}(X)$-vector space.

5. The Residue Calculus. We write Ω for the sheaf Ω^1 of germs of holomorphic 1-forms on X. Note that, since $\dim_{\mathbb{C}} X = 1, \Omega^i = 0$ for $i > 1$. The sheaf Ω is *locally free of rank* 1. Thus, given a local coordinate $t \in \mathfrak{m}_x$ at x, every germ $\omega_x \in \Omega_x^\infty$ is uniquely written as $\omega_x = h_x\, dt$ with $h_x \in \mathfrak{m}_x$. The *residue*, $\operatorname{Res}_x \omega_x$, *of* ω_x *at* x is invariantly defined as the coefficient of t^{-1} in the Laurent development of h_x with respect to t. In other words

$$\operatorname{Res}_x \omega_x = \frac{1}{2\pi i} \int_{\partial H} h\, dt,$$

where H is a small disk about x, and $h \in \mathcal{O}(\bar{H})$ is a representative of h_x. I $\omega_x \in \Omega_x$ then it is clear that $\operatorname{Res}_x \omega_x = 0$.

Now let $\omega \in \Omega^{\infty}(X)$ be a global meromorphic differential form and $F = (f_x) \in R$. Then, for almost all $x \in X, f_x\omega_x \in \Omega_x$. Thus the sum

$$\langle \omega, F\rangle := \sum_{x\in X} \operatorname{Res}_x(f_x\omega_x) \in \mathbb{C}$$

is finite. We summarize some properties of this pairing in the following:

Theorem 4. *The map*

$$\langle\ ,\ \rangle : \Omega^{\infty}(X) \times R \to \mathbb{C},$$

defined by $(\omega, F) \mapsto \langle \omega, F\rangle$ *is a* $\mathbb{C}$*-bilinear form. Furthermore*

0) *for all* $h \in \mathscr{M}(X)$, $\langle h\omega, F\rangle = \langle \omega, hF\rangle$

and

1) *if* $\omega \in \Omega(D)(X)$ *and* $F \in R(-D)$ *then* $\langle \omega, F\rangle = 0.$

Proof: The $\mathbb{C}$-bilinearity of $\langle\ ,\ \rangle$, as well as 0), is clear by definition. Let $\omega \in \Omega(D)(X)$ and $F = (f_x) \in R(-D)$. Then

$$o(f_x\omega_x) = o(f_x) + o(\omega_x) \geq o_x(D) - o_x(D) = 0.$$

That is, for all $x \in X, f_x\omega \in \Omega_x$, and $\operatorname{Res}_x(f_x\omega) = 0$. This proves 1). □

The following theorem is essential for our further considerations.

Theorem 5 (Residue theorem). *If* $\omega \in \Omega^{\infty}(X)$ *and* $h \in \mathscr{M}(X)$ *then* $\langle \omega, h\rangle = 0.$

Proof: Since $h\omega \in \Omega^{\infty}(X)$, it is enough to give a proof for $h \equiv 1$. It must be shown, therefore, that $\sum_{x\in X} \operatorname{Res}_x \omega_x = 0$. Let $x_1, \ldots, x_n \in X$ be the poles of ω and let $H_1, \ldots, H_n$ be pairwise disjoint "closed disks" about the x_ν's. Applying Stokes' theorem, we have

$$\sum_{x\in X} \operatorname{Res}_x \omega_x = \sum_{\nu=1}^{n} \frac{1}{2\pi i}\int_{\partial H_\nu} \omega = -\frac{1}{2\pi i}\int_{\partial(X\setminus\cup H_\nu)} \omega = -\frac{1}{2\pi i}\int_{X\setminus\cup H_\nu} d\omega = 0,$$

since ω is holomorphic outside of $\bigcup H_\nu$, and thus $d\omega$ vanishes identically on this set. □

6. The Duality Theorem. Every differential form $\omega \in \Omega^{\infty}(X)$ determines a $\mathbb{C}$-linear form,

$$\omega^* : R \to \mathbb{C},$$

defined by $F \mapsto \langle \omega, F \rangle$. The $\mathbb{C}$-vector spaces $\Omega^\infty(X)$ and $\mathrm{Hom}_\mathbb{C}(R, \mathbb{C})$ are $\mathcal{M}(X)$-vector spaces as well. In fact, *the mapping* $\Omega^\infty(X) \to \mathrm{Hom}_\mathbb{C}(R, \mathbb{C})$, *defined by* $\omega \mapsto \omega^*$ *is* $\mathcal{M}(X)$*-linear.*

Proof: This mapping is obviously additive. Let $h \in \mathcal{M}(X)$, $F \in R$ and $\omega \in \Omega^\infty(X)$ be given. Then by 0) in Theorem 4 and by the definition of $h\omega^*$,

$$(h\omega)^*(F) = \langle h\omega, F \rangle = \langle \omega, hF \rangle = \omega^*(hF) = h\omega^*(F). \qquad \square$$

From the residue theorem we see that $\mathcal{M}(X) \subset \ker \omega^*$. Moreover, if $\omega \in \Omega(D)(X)$ then $R(-D) \subset \ker \omega^*$ (Theorem 4, 1)). Thus, if $\omega \in \Omega(D)(X)$, ω^* induces a $\mathbb{C}$-linear form,

$$\Theta_D(\omega)\colon R/(R(-D) + \mathcal{M}(X)) \to \mathbb{C}.$$

In other words, if $\omega \in \Omega(D)(X)$ then $\Theta_D(\omega) \in I(-D)^* = J(-D)$.

Thus, for every divisor $D \in \mathrm{Div}\, X$, we obtain a $\mathbb{C}$-linear map

$$\Theta_D\colon \Omega(D)(X) \to J(-D).$$

This mapping extends in a unique way to a $\mathbb{C}$-linear map

$$\Theta\colon \Omega^\infty(X) \to J.$$

The reader should note that, if $\omega \in \Omega(D) \cap \Omega(D')(X)$, the elements $\Theta_D(\omega)$ and $\Theta_{D'}(\omega)$ agree in J.

The following lemma is the preparatory step in the duality theorem:

Lemma 6. The mapping Θ is $\mathcal{M}(X)$-linear. If $\Theta(\omega) \in J(-D)$, then $\omega \in \Omega(D)(X)$.

Proof: The $\mathcal{M}(X)$-linearity of Θ follows from the definition of the $\mathcal{M}(X)$-vector space structure on J (see Theorem 2) and the $\mathcal{M}(X)$-linearity of $\omega \mapsto \omega^*$ (see the above remarks). It remains to prove the last claim.

Let $p \in X$ and $n := o_p(\omega) + 1$. Define $F_0 := (f_x) \in R$ by $f_x := 0$ for $x \neq p$ and $f_p = t^{-n}$, where $t \in \mathfrak{m}_p$ is a local coordinate at p. Clearly $o(f_p\omega_p) = -1$ and thus $\omega^*(F_0) = \mathrm{Res}_p(f_p\omega_p) \neq 0$. Now, since $\Theta(\omega) \in J(-D)$, ω^* vanishes on $R(-D)$. Thus $F_0 \notin R(-D)$. Equivalently $f_p \notin \mathcal{O}(-D)_p$ or, in other words, $o(f_p) + o_p(-D) < 0$. This is the same as saying $-n - o_p(D) < 0$. In other words $o_p(\omega) + {}_p o\,(D) \geq 0$ for all p. Thus $\omega \in \Omega(D)(X)$. $\square$

The following is now immediate.

Theorem 7. *The maps* $\Theta\colon \Omega^\infty(X) \to J$ *and* $\Theta_D\colon \Omega(D)(X) \to J(-D)$ *are bijective*

Proof: If $\Theta(\omega) = 0$ then, for every $D \in \mathrm{Div}\, X$, $\Theta(\omega) \in J(-D)$. Thus, b Lemma 6, $\omega \in \Omega(D)(X)$ for all such D. Since $\bigcap_D \Omega(D)(X) = (0)$, $\omega = 0$ and Θ i therefore injective.

By Theorem 3, $\dim_{\mathcal{M}(X)} J \leq 1$. Since $\Omega^\infty(X) \neq (0)$, the $\mathcal{M}(X)$-monomorphism Θ is automatically surjective.

Let $D \in \operatorname{Div} X$ be given. Then, as the restriction of Θ to $\Omega(D)(X)$, Θ_D is injective. Every $\lambda \in J(-D)$ has a pre-image under Θ, $\omega \in \Omega^\infty(X)$. Again by Lemma 6, $\omega \in \Omega(D)(X)$. Thus $\Theta_D: \Omega(D)(X) \to J(-D)$ is bijective. □

Finally the duality theorem is an easy consequence of Theorems 1 and 7.

Theorem 8 (The duality theorem). *Let X be a compact Riemann surface and $D \in \operatorname{Div} X$. Then there exists a natural $\mathbb{C}$-isomorphism*

$$H^0(X, \Omega(D)) \simeq H^1(X, \mathcal{O}(-D))^*.$$

Proof: By Theorem 1 there is a natural $\mathbb{C}$-isomorphism between $H^1(X, \mathcal{O}(-D))$ and $I(-D)$. This induces a $\mathbb{C}$-isomorphism, $J(-D) \to H^1(X, \mathcal{O}(-D))^*$, of the dual spaces. Composition with Θ_D gives us our isomorphism:

$$H^0(X, \Omega(D)) \simeq J(-D) \to H^1(X, \mathcal{O}(-D))^*. \qquad \square$$

In this form the duality theorem is a classical theorem in the subject of algebraic curves. The more general form (for complex manifolds of higher dimension which are not necessarily compact) was first formulated and proved by J.-P. Serre in 1954 (Un théorème de dualité, Comm. Math. Helv. **29**, 9–26 (1955)).

§ 7. The Riemann–Roch Theorem (Final Version)

The results of this section are consequences of Theorem 3.1 and the duality theorem. The strength of the Riemann–Roch theorem will be demonstrated by some selected (classical) applications.

We will always use K to denote the canonical divisor. The notation $l(D)$ and $i(D)$, which was introduced in Section 3, will be consistently applied.

1. The Equation $i(D) = l(K - D)$. Since every finite dimensional vector space has the same dimension as its dual space, the duality theorem implies the following:

$$i(D) = \dim_{\mathbb{C}} H^1(X, \mathcal{O}(D)) = \dim_{\mathbb{C}} H^0(X, \Omega(-D)).$$

Thus $i(D)$ *is the number of linearly independent meromorphic differential forms* ω *such that* $(\omega) \geq D$. In particular $i(0) = g$: *On a Riemann surface X with genus g there are exactly g linearly independent global holomorphic 1-forms*:

$$\dim_{\mathbb{C}} H^0(X, \Omega) = g.$$

For every positive D, $H^0(X, \Omega(-D)) \subset H^0(X, \Omega)$. Thus using Theorem 3.1, we can estimate $i(D)$ and $l(D)$ from above: *If the divisor class of D contains a positive divisor then $i(D) \leq g$ and $l(D) \leq$* deg $D + 1$.

Since $\Omega \cong \mathcal{O}(K)$ (see Theorem 4.5), it follows that $\Omega(-D) \cong \mathcal{O}(K-D)$ for all $D \in \operatorname{Div} X$. Thus the above dimension equation can be written in the form

$$i(D) = l(K - D). \tag{1}$$

In the case of $D = 0$ this leads to $l(K) = g$ and similarly, in the case of $D = K$, we have $i(K) = l(0) = 1$. In other words $H^1(X, \Omega) \cong \mathbb{C}$. Hence it turns out that

$$\chi(\Omega) = l(K) - i(K) = g - 1 = -\chi(\mathcal{O}). \tag{2}$$

It follows easily now that g is only a topological invariant:

Theorem 1. *The genus g of a compact Riemann surface X is a topological invariant. In fact*

$$\dim_{\mathbb{C}} H^1(X, \mathbb{C}) = 2g.$$

Proof: We consider the exact sequence of sheaves

$$0 \longrightarrow \mathbb{C} \longrightarrow \mathcal{O} \xrightarrow{d=\partial} \Omega \longrightarrow 0.$$

Due to (2) above,

$$\chi(X, \mathbb{C}) = \chi(\mathcal{O}) - \chi(\Omega) = 2 - 2g.$$

Since $H^0(X, \mathbb{C})$ and $H^2(X, \mathbb{C})$ are both isomorphic to $\mathbb{C}$,

$$2 - 2g = \chi(X, \mathbb{C}) = 1 - \dim_{\mathbb{C}} H^1(X, \mathbb{C}) + 1. \qquad \square$$

We will make further remarks about the structure of $H^1(X, \mathbb{C})$ in Section 7.

2. The Formula of Riemann–Roch. It follows from Theorem 3.1 that, for all $D \in \operatorname{Div} X$, $l(D) - i(D) = \deg D + 1 - g$. If one writes $l(K - D)$ instead of $i(D)$ then one obtains the formula of Riemann–Roch.

Theorem 2 (Riemann–Roch, final version). *For every divisor D on a compact Riemann surface X of genus g,*

$$l(D) - l(K - D) = \deg D + 1 - g.$$

Since $l(K) = g$ and $l(0) = 1$, we find by setting $D = K$ that $g - 1 = \deg K + 1 - g$. Thus we have the degree equation for differential forms: For every differential form $\omega \in \Omega^\infty(X)^*$,

$$\deg K = \deg \omega = 2g - 2.$$

This equation contains for example the fact that $H^1(\mathbb{P}_1, \mathbb{C}) = (0)$. One sees this by noting that the differential dz, where z is an inhomogeneous coordinate, has degree -2. Thus, since $2g - 2 = -2$, $\mathbb{P}_1$ has genus 0.

From the above it follows that every non-trivial $\omega \in \Omega^\infty(X)$ has degree $-\chi(X)$, where $\chi(X)$ is the topological Euler–Poincaré characteristic of X. Consequently, if $\alpha: X \to X'$ is an s-sheeted ramified covering map between compact Riemann surfaces X and X', and $W \in \operatorname{Div} X$ is the ramification divisor of α, then

$$\chi(X) + \deg W = s \cdot \chi(X')$$

Proof: Let $\omega' \in \Omega^\infty(X')$, and define $\omega = \alpha^*(\omega')$. A direct calculation shows that $\deg(\omega) = s \cdot \deg(\omega') + \deg W$. Since $\deg(\omega) = -\chi(X)$ and $\deg(\omega') = -\chi(X)$, the claim follows immediately. □

In particular this shows that $\deg W$ is always even, and, when $X' = \mathbb{P}_1$ with $\chi(X') = 2$, it follows that

$$\deg W = 2(s + g - 1)$$

3. Theorem B for Sheaves $\mathcal{O}(D)$. The following important application of the Riemann–Roch formula uses the simple fact that, when $\deg D < 0$, $l(D)$ vanishes.

Theorem 3 (Theorem B). *Let $D \in \operatorname{Div} X$ with $\deg D \geq 2g - 1$. Then*

a) $H^1(X, \mathcal{O}(D)) = (0)$
b) $l(D) = \deg D + 1 - g.$

Proof: a) Since $\deg K = 2g - 2$, $\deg(K - D) < 0$ and therefore $i(D) = l(K - D) = 0$. b) This follows immediately from a) and the Riemann–Roch formula. □

Theorem 3 is the optimal form of Theorem B in the sense that if the degree of a divisor D is less than $2g - 1$, then the cohomology group $H^1(X_1, \mathcal{O}(D))$ may not vanish. For example, by Serre duality, $H^1(X, \mathcal{O}(K)) \cong H^0(X, \mathcal{O}) \cong \mathbb{C}$ and $\deg K = 2g - 2$.

4. Theorem A for Sheaves $\mathcal{O}(D)$. Let $\mathcal{L}$ be a locally free sheaf of rank 1 over X, and let $x \in X$. Then the following are equivalent:

i) *The module of sections $\mathcal{L}(X)$ generates the stalk $\mathcal{L}_x$ as an $\mathcal{O}_x$-module.*
ii) *There is a section $s \in \mathcal{L}(X)$ with $o(s_x) = 0$.*
iii) *There is a section $s \in \mathcal{L}(X)$ with $\mathcal{L}_x = \mathcal{O}_x s_x$.*
iv) $\dim_{\mathbb{C}} H^1(X, \mathcal{L}(-x)) \leq \dim_{\mathbb{C}} H^1(X, \mathcal{L})$.

Proof: i)⇒ii): Let $t_x \in \mathcal{L}_x$ be such that $\mathcal{L}_x = \mathcal{O}_x \cdot t_x$. By i) there exist sections $s_\mu \in \mathcal{L}(X)$ and germs $f_{\mu x} \in \mathcal{O}_x$, $1 \leq \mu \leq m$, so that $t_x = \sum_1^m f_{\mu x} s_{\mu x}$. Thus some $f_{\mu x}(x) \in \mathbb{C}$ must differ from 0.
Hence for that section $\mathcal{O}(s_{\mu x}) = \mathcal{O}$.

ii)⇒iii): Every germ $t_x \in \mathcal{L}_x$ with $o(t_x) = 0$ generates $\mathcal{L}_x$ as an $\mathcal{O}_x$-module.

iii) $\Rightarrow$ i): Trivial.

ii) $\Leftrightarrow$ iv): The sheaf $\mathcal{T}$ which is defined by $0 \to \mathcal{L}(-x) \to \mathcal{L} \to \mathcal{T} \to 0$ has support only at x where its stalk is $\mathbb{C}$. Thus we have the following cohomology sequence:

$$0 \longrightarrow \mathcal{L}(-x)(X) \longrightarrow \mathcal{L}(X) \xrightarrow{\varepsilon} \mathbb{C} \longrightarrow H^1(X, \mathcal{L}(-x)) \longrightarrow H^1(X, \mathcal{L}) \longrightarrow 0$$

Therefore there is a section $s \in \mathcal{L}(X)$ with $\varepsilon(s_x) = 0$ if and only if $\mathcal{L}(-x)(x) \subsetneqq \mathcal{L}(X)$, This is equivalent to $H^1(x, \mathcal{L}(-x)) \to H^1(x, \mathcal{L})$ being injective. □

Theorem 4 (Theorem A). *Let $D \in \operatorname{Div} X$ with $\deg D \geq 2g$. Then for every $x < X$ there exists a section $s \in \mathcal{O}(D)(X)$ with $\mathcal{O}(D)_x = \mathcal{O}_x s_x$.*

Proof: Let $\mathcal{L} := \mathcal{O}(D)$. If $H^1(X, \mathcal{L}(-x)) = 0$, then the above equivalences guarantee the existence of such a section. Now $\mathcal{L}(-x) = \mathcal{O}(D - x)$. But by assumption, $\deg(D - x) \geq 2g - 1$. Hence Theorem B implies that $H^1(X, \mathcal{O}(D - x)) = 0$. □

Theorem 4 is the optimal form a Theorem A in the sense that for $\deg D < 2g$ the sheaf $\mathcal{O}(D)$ may not be generated by its global sections. For example, let $D = K + (p)$. Then $\mathcal{O}(D) = \Omega(p)$. A section in $\mathcal{O}(D)(X)$ is just a differential form ω which is holomorphic on $X \backslash p$, and, if it would generate $\mathcal{O}(D)_p$, would have a pole of order 1 at p. This would contradict the residue theorem.

There are divisors with $\deg D < 2g$, and for which Theorem A, however, *is valid*, namely K: *If $g \neq 0$ then Theorem A holds for the sheaf $\Omega \cong \mathcal{O}(K)$. In other words, given $x \in X$ there exists a holomorphic differential form $\omega \in \Omega(X)$ which does not vanish at x.*

Proof: By the above equivalences, we only need to show that $\dim_{\mathbb{C}} H^1(X, \Omega(-x)) \leq \dim_{\mathbb{C}} H^1(X, \Omega)$. Since $\dim_{\mathbb{C}} H^1(X, \Omega) = 1$, it is enough then to prove that $i(K - x) \leq 1$. Suppose $l(x) = i(K - x) \geq 2$. Then there exists $h \in \mathcal{M}(X) \cap \mathcal{O}(X \backslash x)$ with a pole of order 1 at x. The map $h\colon X \to \mathbb{P}_1$ would be biholomorphic, and, since $g \neq 0$, we have reached a contradiction. Thus $i(K - x) \leq 1$. □

5. The Existence of Meromorphic Differential Forms. The following existence theorems are immediate consequences of Theorem A for differential forms: *Let $D \in \operatorname{Div} X$ with $\deg D \geq 2$. Then every stalk of $\Omega(D)$ is generated by a global meromorphic differential form $\omega \in \Omega(D)(X)$.*

Proof: Since $\Omega(D) \cong \mathcal{O}(K + D)$ and $\deg(K + D) \geq 2g$, the result follows from Theorem 4. □

In particular if $D := mp$, $m \geq 2$, then we have the following:

Let $p \in X$ and $m \geq 2$ be given. Then there exists a meromorphic differential form *on X which is holomorphic on $X\backslash p$ and has a pole of order m at p.* Furthermore, *for every p_1 and $p_2 \in X$ with $p_1 \neq p_2$, there exists a meromorphic form on X which is holomorphic on $X\backslash\{p_1, p_2\}$ and has poles of order 1 at p_1 and p_2.*

Proof: Define $D := p_1 + p_2$. Then there exists $\omega \in \Omega(D)(X)$ which generates the stalk $\Omega(D)_{p_1}$. This form must be holomorphic on $X\backslash\{p_1, p_2\}$, have a pole of order 1 at p_1 and have a pole of *at most* order 1 at p_2. But the residue theorem requires that it has a pole of exactly order 1 at p_2. □

6. The Gap Theorem. A natural number $w \geq 1$ is called a *gap value* at $p \in X$ if there is no holomorphic function on $X\backslash p$ which has a pole of order w at p. If $X = \mathbb{P}_1$ then there are no gap values. But if $X \neq \mathbb{P}_1$ then $g \neq 0$ and $w = 1$ is always a gap value.

We write $l_v := l(vp)$ for each $v \geq 0$ and note that

$l_v \leq l_{v+1} \leq l_v + 1$ *and w is a gap value if and only if* $l_w = l_{w-1}$.

Proof: In the exact sequence

$$0 \to H^0(X, \mathcal{O}(vp)) \to H^0(X, \mathcal{O}((v+1)p)) \to \mathcal{T}(X) \to \cdots,$$

the space $\mathcal{T}(X) = \mathcal{T}_p = \mathcal{O}((v+1)p)_p/\mathcal{O}(vp)_p$ is 1-dimensional. This implies that $l_v \leq l_{v+1} \leq l_v + 1$. By definition w is a gap value if and only if every $h \in \mathcal{O}(wp)(X)$ is already in $\mathcal{O}((w-1)p)(X)$. That is, $\mathcal{O}(wp)(X) = \mathcal{O}((w-1)p)(X)$ or equivalently $l_w = l_{w-1}$. □

If $v \geq 2g - 1$ it follows from Theorem 3 that $l_v = v + 1 - g$. Thus for $v \geq 2g$, $l_v = l_{v-1} + 1$ and there are no gap values greater than $2g - 1$. One can improve this remark:

Theorem 5 (Weierstrass gap theorem). *Let X be a compact Riemann surface with genus $g > 0$. Then, for every $p \in X$, there exist exactly g gap values, $w_1, \ldots, w_g$, which we order so that*

$$1 = w_1 < w_2 < \cdots < w_g \leq 2g - 1.$$

Proof: By Theorem 3, $l_{2g-1} = g$. Hence $1 \leq l_0 \leq l_1 \leq \cdots < l_{2g-1} = g$. Since $l_{i+1} - l_i \leq 1$, there must be exactly $g - 1$ indices i such that $l_{i+1} - l_i = 1$. Thus, since $l_0 = 1$ and $l_{2g-1} = g$, there must be g indices i such that $l_{i+1} = l_i$. That is, there are exactly g gap values. □

7. Theorems A and B for Locally Free Sheaves. We call a locally free sheaf Stein" if Theorems A and B are valid:

A) $H^0(X, \mathcal{F})$ generates every stalk $\mathcal{F}_x$ as an $\mathcal{O}_x$-module.
B) $H^1(X, \mathcal{F}) = (0)$.

Using this terminology we have the following:

Lemma 6. Let $0 \to \mathscr{L} \to \mathscr{F} \to \mathscr{G} \to 0$ be a sequence of locally free sheaves. Assume further that $\mathscr{L}$ and $\mathscr{G}$ are Stein. Then $\mathscr{F}$ is also Stein.

Proof: Since $H^1(X, \mathscr{L}) = H^1(X, \mathscr{G}) = (0)$, it follows immediately from the exact cohomology sequence that $H^1(X, \mathscr{F}) = (0)$. Further we consider the commutative diagram

$$\begin{array}{ccccccccc}
0 & \longrightarrow & H^0(X, \mathscr{L}) & \longrightarrow & H^0(X, \mathscr{F}) & \longrightarrow & H^0(X, \mathscr{G}) & \longrightarrow & 0 \\
 & & \Big\downarrow{\scriptstyle \lambda_x} & & \Big\downarrow{\scriptstyle \varphi_x} & & \Big\downarrow{\scriptstyle \gamma_x} & & \\
0 & \longrightarrow & \mathscr{L}_x & \longrightarrow & \mathscr{F}_x & \longrightarrow & \mathscr{G}_x & \longrightarrow & 0
\end{array}$$

where the rows are exact and the maps in the columns associate to a section its germ at x. By assumption, the images of λ_x and γ_x generate (as $\mathcal{O}_x$-modules) $\mathscr{L}_x$ and $\mathscr{G}_x$ respectively. It follows therefore that the image of $H^0(X, \mathscr{F})$ under φ_x generates $\mathscr{F}_x$ as an $\mathcal{O}_x$-module. □

Theorem 7. *Given a locally free sheaf $\mathscr{F}$ over a Riemann surface X, there exists a natural number n^+ so that, for every $D \in \operatorname{Div} X$ with $\deg D \geq n^+$, the sheaf $\mathscr{F}(D)$ is Stein.*

Proof: (by induction on the rank r of $\mathscr{F}$). Since every locally free sheaf of rank 1 is isomorphic to a sheaf of the type $\mathcal{O}(D)$, the case of $r = 1$ handled by Theorem 3 and 4. Now we assume that $r > 1$ and that the statement holds for sheaves of rank at most $r - 1$. By the "subsheaf theorem" (Theorem 4.3), there exist locally free sheaves $\mathscr{L}$ and $\mathscr{G}$ of rank 1 and $r - 1$ so that the following sequence is exact for every $D \in \operatorname{Div} X$:

$$0 \to \mathscr{L}(D) \to \mathscr{F}(D) \to \mathscr{G}(D) \to 0.$$

By the induction assumption and Lemma 6 above, there exists $n^+ \in \mathbb{Z}$ so that, for $\deg D \geq n^+$, $\mathscr{F}(D)$ is Stein. □

The following is analogous to the fact that if $\deg D < 0$, then $H^0(X, \mathcal{O}(D)) = (0)$.

Theorem 8. *Let $\mathscr{F}$ be a locally free sheaf over a compact Riemann surface X Then there exists $n^- \in \mathbb{Z}$ so that for every $D \in \operatorname{Div} X$ with $\deg D < n^-$ $H^0(X, \mathscr{F}(D)) = (0)$.*

Proof: (by induction on the rank r of $\mathscr{F}$). If the rank of $\mathscr{F}$ is 1 then $\mathscr{F} = \mathcal{O}(D'$ for some $D' \in \operatorname{Div} X$ and $n^- := -\deg D'$ has the desired property. If rank $\mathscr{F} >$ then, as in the proof of Theorem 7, we choose locally free sheaves $\mathscr{L}$ and $\mathscr{G}$ of ran

1 and $r-1$ respectively so that, for every $D \in \operatorname{Div} X$, we have the exact sequence

$$0 \to \mathscr{L}(D) \to \mathscr{F}(D) \to \mathscr{G}(D) \to 0.$$

The claim follows immediately from the induction hypothesis and the exact cohomology sequence. □

8. The Hodge Decomposition of $H^1(X, \mathbb{C})$. We want to develop a better understanding of the vector space $H^1(X, \mathbb{C}) \cong \mathbb{C}^{2g}$. As earlier, we begin with the resolution of the constant sheaf $\mathbb{C}$:

$$0 \longrightarrow \mathbb{C} \longrightarrow \mathcal{O} \xrightarrow{d=\partial} \Omega \longrightarrow 0.$$

Since $H^0(X, \mathbb{C}) = H^0(X, \mathcal{O}) = \mathbb{C}$ and $H^2(X, \mathcal{O}) = (0)$, the associated cohomology sequence yields

$$0 \longrightarrow \Omega(X) \xrightarrow{\alpha} H^1(X, \mathbb{C}) \longrightarrow H^1(X, \mathcal{O}) \longrightarrow H^1(X, \Omega) \xrightarrow{\beta} H^2(X, \mathbb{C}) \longrightarrow 0.$$

But $H^2(X, \mathbb{C}) = H^1(X, \Omega) = \mathbb{C}$. Hence β is bijective and we therefore have the exact sequence

$$0 \longrightarrow \Omega(X) \xrightarrow{\alpha} H^1(X, \mathbb{C}) \longrightarrow H^1(X, \mathcal{O}) \longrightarrow 0.$$

The map α can be explicitely described in the following way: Let $\omega \in \Omega(X)$ and $\mathfrak{U} = \{U_i\}$ a covering of X by contractable neighborhoods. Then there exists $f_i \in \mathcal{O}(U_i)$ so that $df_i = \omega|_{U_i}$.

The function $\alpha_{ij} := f_j - f_i$ is therefore constant on U_{ij} (we take these intersections to be connected). The family $\{\alpha_{ij}(\omega)\}$ forms a 1-cocycle in $Z^1(\mathfrak{U}, \mathbb{C})$. This cocycle represents the cohomology class $\alpha(\omega) \in H^1(X, \mathbb{C})$.

Since $\mathbb{R} \subset \mathbb{C}$, $H^1(X, \mathbb{R})$ is an $\mathbb{R}$-vector subspace of $H^1(X, \mathbb{C})$. We now show

Lemma 9. $\operatorname{Im}(\alpha) \cap H^1(X, \mathbb{R}) = (0)$.

Proof: Suppose that $\alpha(\omega) \in H^1(X, \mathbb{R})$. From the above description of α it follows that there is a covering $\{U_i\}$ of X and functions $f_i \in \mathcal{O}(U_i)$ such that $df_i = \omega|_{U_i}$ and every function $f_j - f_i$ is constant and *real* on U_{ij}. Let $g_i := \exp(2\pi\sqrt{-1}f_i)$. Then $|g_i|^2 = |g_j|^2$ on U_{ij} and therefore $\{|g_i|^2\}$ determines a real valued continuous function, g, on X. By the maximum principle g is identically constant and thus $\omega = 0$. □

There are of course many $\mathbb{C}$-vector spaces V in $H^1(X, \mathbb{C})$ so that $V \oplus \operatorname{Im} \alpha = H^1(X, \mathbb{C})$. However there is one particular V that is quite natural. For this we consider the "conjugate" resolution of $\mathbb{C}$,

$$0 \longrightarrow \mathbb{C} \longrightarrow \bar{\mathcal{O}} \xrightarrow{d=\bar{\partial}} \bar{\Omega} \longrightarrow 0.$$

Here $\bar{\Omega} := \overline{\Omega^1}$ (see II.2.3, in particular the diagram). For reasons similar to those above, the associated cohomology sequence is

$$0 \to \bar{\Omega}(X) \to H^1(X, \mathbb{C}) \to H^1(X, \bar{\mathcal{O}}) \to 0.$$

Using this we prove

Theorem 10 (The Hodge decomposition). *The* $\mathbb{C}$*-vector space* $H^1(X, \mathbb{C})$ *is the direct sum of the spaces* Im α *and* Im $\hat{\alpha}$, *where* $\hat{\alpha}: \bar{\Omega}(X) \to H^1(X, \mathbb{C}) = \alpha(\Omega(X)) \oplus \hat{\alpha}(\bar{\Omega}(X))$.

Proof: Since $H^1(X, \mathbb{C}) \cong \mathbb{C}^{2g}$ and Im $\alpha \cong$ Im $\hat{\alpha} \cong \mathbb{C}^g$, it is enough to show that Im $\alpha \cap$ Im $\hat{\alpha} = (0)$. The conjugation $\mathbb{C} \to \mathbb{C}$, defined by $c \mapsto \bar{c}$, determines an $\mathbb{R}$-linear involution $\sigma: H^1(X, \mathbb{C}) \to H^1(X, \mathbb{C})$ which has $H^1(X, \mathbb{R})$ as a fixed space. Obviously σ leaves every element of Im $\alpha \cap$ Im $\hat{\alpha}$ fixed. Thus Im $\alpha \cap$ Im $\hat{\alpha} \subset$ Im $\alpha \cap H^1(X, \mathbb{R})$. Hence, by Lemma 9,

$$\text{Im } \alpha \cap \text{Im } \hat{\alpha} = (0). \qquad \square$$

§ 8. The Splitting of Locally Free Sheaves

By means of a *formal* splitting criterion we give a sufficient condition for a locally free sheaf $\mathcal{F}$ over a compact Riemann surface to contain a locally free subsheaf of rank 1 which is a *direct summand* of $\mathcal{F}$ (Theorem 4). Every such locally free sheaf contains *maximal* subsheaves of rank 1. On the Riemann sphere $\mathbb{P}_1$, the maximal subsheaves are direct summands (Splitting Lemma). As a corollary, it follows that every locally free sheaf $\mathcal{F}$ of rank r over $\mathbb{P}_1$ is isomorphic to a sheaf $\bigoplus_1^r \mathcal{O}(n_i p)$, where $n_1, \ldots, n_r \in \mathbb{Z}$ are uniquely determined up to order by $\mathcal{F}$ (a theorem of Grothendieck). The presentation here follows along the lines of [21].

1. The Number $\mu(\mathcal{F})$. Let $\mathcal{L}$ be a locally free sheaf of rank 1 over a compact Riemann surface X. Then, for every $s \in \mathcal{L}^\infty(X)^*$, deg s only depends on $\mathcal{L}$ (see Paragraph 4.2). For locally free sheaves of rank $r > 1$ this is not in general true For example, consider $\mathcal{F} = \mathcal{O}(n_1 p) \oplus \cdots \oplus \mathcal{O}(n_r p)$ for some $p \in X$. Then $\mathcal{F}$ has sections of degree $n_1, n_2, \ldots, n_r$.

In order to define an integer which can be used in place of the "degree", we make the following observations:

Every homomorphism $\pi: \mathcal{F} \to \mathcal{G}$ *between locally free sheaves induces a homomorphism* $\pi: \mathcal{F}^\infty(X) \to \mathcal{G}^\infty(X)$. *If*

$$0 \longrightarrow \mathcal{F}' \xrightarrow{\ i\ } \mathcal{F} \xrightarrow{\ \pi\ } \mathcal{G} \longrightarrow 0$$

is an exact sequence of non-zero locally free sheaves, then, for every section $s \in \mathscr{F}^\infty(X)^*$, *either*

a) $\pi(s) = 0$, *in which case* $s = i(s')$ *with* $s' \in \mathscr{F}'^\infty(X)^*$ *and* $\deg s' = \deg s$,

or

b) $\pi(s) \neq 0$, *in which case* $\deg(s) \leq \deg(\pi(s))$.

Proof: If $\mathscr{F}''_x$ is complementary to $i(\mathscr{F}'_x)$ in $\mathscr{F}_x$, then for every germ $t_x \in \mathscr{F}^\infty_x$ there exist uniquely determined germs $t'_x \in \mathscr{F}'^\infty_x$, $t''_x \in \mathscr{F}''^\infty_x$ so that

$$t_x = i(t'_x) + t''_x, \qquad o(t_x) = \min\{o(t'_x), o(t''_x)\}, \qquad x \in X.$$

If $\pi(s) = 0$, then $s \in i(\mathscr{F}'(X))$ (i.e. $s = i(s')$ with $s' \in \mathscr{F}'^\infty(X)^*$). But $o(s_x) = o(s'_x)$ for all $x \in X$. Hence $\deg s = \deg s'$.

Now suppose that $\pi(s) \neq 0$, and let $s_x = i(t'_x) + t''_x$. Since π_x maps $\mathscr{F}''_x$ isomorphically onto $\mathscr{G}_x$, it follows that $o(\pi(s)_x) = o(t''_x)$. Hence $o(s_x) \leq o(\pi(s)_x)$ for all $x \in X$, and consequently $\deg(s) \leq \deg(\pi(s))$. □

For every locally free sheaf $\mathscr{F}$ over X, we define $\mu(\mathscr{F})$ as follows:

$$\mu(\mathscr{F}) := \sup\{\deg(s) \mid s \in \mathscr{F}^\infty(X)^*\}.$$

The following direct consequence of the above remarks is useful in showing that $\mu(\mathscr{F}) < \infty$:

Let $0 \to \mathscr{F}' \to \mathscr{F} \to \mathscr{G} \to 0$ *be an exact sequence of non-zero locally free sheaves. Then*

$$\mu(\mathscr{F}') \leq \mu(\mathscr{F}) \leq \max\{\mu(\mathscr{F}'), \mu(\mathscr{G})\}.$$

Now it is easy to prove that $\mu(\mathscr{F})$ is bounded.

Theorem 1. *Let* $\mathscr{F}$ *be a locally free sheaf on a compact Riemann surface* X. *Then the degree function,* $\mu(\mathscr{F})$, *is finite.*

Proof: (by induction on the rank r of $\mathscr{F}$). For $r = 1$ the statement is trivially true. There exists an exact sequence of locally free sheaves $0 \to \mathscr{L} \to \mathscr{F} \to \mathscr{G} \to 0$, where $\mathscr{L}$ and $\mathscr{G}$ have rank 1 and $r - 1$ respectively. The proof follows immediately by the induction hypothesis and the above estimates. □

2. Maximal Subsheaves. The number $\mu(\mathscr{F})$ can be characterized by the Chern numbers of the locally free sheaves of rank 1:

2) $\quad \mu(\mathscr{F}) = \max\{c(\mathscr{L}) : \mathscr{L}$ is a locally free subsheaf of $\mathscr{F}$ of rank 1$\}$.

This follows from the facts that every section $s \in \mathscr{F}^\infty(X)^*$ determines a locally free subsheaf $\mathscr{L}$ of rank 1 with $c(\mathscr{L}) = \deg s$, and conversely that a locally free subsheaf determines such a section (see Section 4). We call a locally free subsheaf $\mathscr{L}$ of rank 1 in $\mathscr{F}$ *maximal* if $c(\mathscr{L}) = \mu(\mathscr{F})$. Such sheaves are generated by *maximal sections* (i.e. sections $s \in \mathscr{F}^\infty(X)^*$ with $\deg s = \mu(\mathscr{F})$). We have seen that

every non-zero locally free sheaf $\mathscr{F}$ *possesses maximal subsheaves.*

Theorem 2. *Let* $\mathscr{F}$ *be a locally free sheaf on a compact Riemann surface* X. *Let* $\mathscr{L}$ *be a maximal subsheaf in* $\mathscr{F}$ *and* $D \in \operatorname{Div} X$. *Then* $\mathscr{L}(D)$ *is a maximal subsheaf in* $\mathscr{F}(D)$. *In general,*

$$\mu(\mathscr{F}) + \deg D \leq \mu(\mathscr{F}(D)).$$

Proof: Since $\mathscr{L}(D)$ is a locally free subsheaf of $\mathscr{F}(D)$, $c(\mathscr{L}(D)) \leq \mu(\mathscr{F}(D))$. Of course $c(\mathscr{L}(D)) = c(\mathscr{L}) + \deg D$. Thus, since $c(\mathscr{L}) = \mu(\mathscr{F})$,

$$(*) \qquad \mu(\mathscr{F}) + \deg D \leq \mu(\mathscr{F}(D)).$$

We now apply $(*)$ with $\mathscr{F}$ replaced by $\mathscr{F}(D)$ and D by $-D$. Thus

$$\mu(\mathscr{F}(D)) + \deg(-D) \leq \mu(\mathscr{F}).$$

Hence there is equality in $(*)$ and, in particular, $c(\mathscr{L}(D)) = \mu(\mathscr{F}(D))$. □

3. The Inequality $\mu(\mathscr{G}) \leq \mu(\mathscr{F}) + 2g$. We begin with a consequence of the Riemann inequality:

Lemma. Let $\mathscr{F}$ be locally free of rank 2 and $\mathscr{L}$ a maximal subsheaf which is isomorphic to $\mathscr{O}$. Then, defining $\mathscr{G} := \mathscr{F}/\mathscr{L}$, $c(\mathscr{G}) \leq 2g$.

Proof: Since $H^1(X, \mathscr{O}) \cong \mathbb{C}^g$, the cohomology sequence associated to $0 \to \mathscr{O} \to \mathscr{F} \to \mathscr{G} \to 0$ starts out like

$$0 \longrightarrow \mathbb{C} \longrightarrow \mathscr{F}(X) \longrightarrow \mathscr{G}(X) \xrightarrow{\varphi} \mathbb{C}^g \longrightarrow \cdots.$$

Now the sheaf $\mathscr{G}$ is locally free of rank 1. Thus if $c(\mathscr{G}) > 2g$, then, by the Riemann inequality,

$$\dim_{\mathbb{C}} \mathscr{G}(X) \geq c(\mathscr{G}) + 1 - g \geq g + 2.$$

Hence the kernel of φ would be at least 2-dimensional, and $\mathscr{F}(X)$ is at least 3-dimensional. However, since the rank of $\mathscr{F}$ is 2, every $\mathbb{C}$-vector space $\mathscr{F}_x/\mathfrak{m}_x\mathscr{F}_x$ is 2-dimensional. Thus the restriction map $\mathscr{F}(X) \to \mathscr{F}_x/\mathfrak{m}_x\mathscr{F}_x$ has a non-trivial kernel. In other words, there would be a section $s \not\equiv 0$ in $\mathscr{F}(X)$ which vanishes at x. This implies that $\deg s \geq 1$. But that is contrary to the assumption that

$$\mu(\mathscr{F}) = c(\mathscr{L}) = c(\mathscr{O}) = 0.$$

□

We now show

Theorem 3. *Let $\mathscr{F}$ be a locally free sheaf of rank $r \geq 2$ and let $\mathscr{L}$ be a maximal subsheaf of $\mathscr{F}$. Then, defining $\mathscr{G} := \mathscr{F}/\mathscr{L}$,*

$$\mu(\mathscr{G}) \leq \mu(\mathscr{F}) + 2g$$

Proof: As usual we start with the exact sequence

$$0 \longrightarrow \mathscr{L} \longrightarrow \mathscr{F} \xrightarrow{\pi} \mathscr{G} \longrightarrow 0.$$

At first we consider the case $r = 2$. There exists a divisor $D \in \operatorname{Div} X$ with $\mathscr{L} \cong \mathcal{O}(D)$ and $\deg D = c(\mathscr{L}) = \mu(\mathscr{F})$. Applying the above Lemma to the sequence $0 \to \mathcal{O} \to \mathscr{F}(-D) \to \mathscr{G}(-D) \to 0$, we find that $c(\mathscr{G}(-D)) \leq 2g$. Since $c(\mathscr{G}(-D)) = c(\mathscr{G}) - \deg D$, the claim follows by noting that $\mu(\mathscr{G}) = c(\mathscr{G})$ and $\deg D = \mu(\mathscr{F})$.

Now consider the case of $r > 2$. Let $\mathscr{L}'$ be a maximal subsheaf of $\mathscr{G}$. We have an induced exact sequence $0 \to \mathscr{L} \to \mathscr{F}' \to \mathscr{L}' \to 0$, where $\mathscr{F}' := \pi^{-1}(\mathscr{L}')$. For all $x \in X$,

$$\mathscr{F}_x/\mathscr{F}'_x \cong (\mathscr{F}_x/\mathscr{L}_x)/(\mathscr{F}'_x/\mathscr{L}_x) \cong \mathscr{G}_x/\mathscr{L}'_x.$$

Thus $\mathscr{F}'$ is a locally free sheaf of rank 2. Clearly $\mathscr{L}$ is a maximal subsheaf of $\mathscr{F}'$, and therefore

$$\mu(\mathscr{F}) = c(\mathscr{L}) = \mu(\mathscr{F}').$$

Consequently

$$\mu(\mathscr{G}) \leq \mu(\mathscr{F}') + 2g = \mu(\mathscr{F}) + 2g. \qquad \square$$

4. The Splitting Criterion. One says that the short exact $\mathcal{O}$-sequence,

$$0 \longrightarrow \mathscr{S}_1 \xrightarrow{i} \mathscr{S} \longrightarrow \mathscr{S}_2 \longrightarrow 0$$

splits if there exists an $\mathcal{O}$-homomorphism $\xi: \mathscr{S}_2 \to \mathscr{S}$ with $\pi \circ \xi = \text{id}$. Thus, in this case, $\mathscr{S} = i(\mathscr{S}_1) \oplus \xi(\mathscr{S}_2) \cong \mathscr{S}_1 \oplus \mathscr{S}_2$. There is a simple formal *splitting criterion:*

An exact $\mathcal{O}$-sequence of locally free sheaves $0 \to \mathscr{S}_1 \to \mathscr{S} \to \mathscr{G} \to 0$ splits when

$$H^1(X, \mathscr{H}om(\mathscr{G}, \mathscr{S}_1)) = (0).$$

Proof: Since $\mathscr{G}$ is locally free, the above short exact sequence induces the exact $\mathcal{O}$-sequence of sheaves,

$$0 \to \mathscr{H}om(\mathscr{G}, \mathscr{S}_1) \to \mathscr{H}om(\mathscr{G}, \mathscr{S}) \to \mathscr{H}om(\mathscr{G}, \mathscr{G}) \to 0,$$

which in turn induces an exact cohomology sequence

$$\cdots \longrightarrow H^0(X, \mathscr{H}om(\mathscr{G}, \mathscr{S})) \xrightarrow{\pi_*} H^0(X, \mathscr{H}om(\mathscr{G}, \mathscr{G})) \longrightarrow$$
$$H^1(X, \mathscr{H}om(\mathscr{G}, \mathscr{S}_1)) \longrightarrow \cdots.$$

If π_* is surjective, then there exists a global section $\xi \in \mathrm{Hom}(\mathscr{G}, \mathscr{S})$ with $\pi_*(\xi) = \pi \circ \xi = \mathrm{id}$. □

Remark: The splitting criterion is valid for *any* complex space. In the case of Stein spaces the relevant cohomology group is always zero. Thus we have the following observation.

Every locally free subsheaf of a locally free sheaf over a Stein space is a direct summand.

For 1-dimensional Stein manifolds (i.e. non-compact Riemann surfaces), it follows that every locally free sheaf of rank r is isomorphic to the sheaf $\mathcal{O}^r$. (Recall that $H^1(X, \mathcal{O}) = H^2(X, \mathbb{Z}) = 0$. Thus $H^1(X, \mathcal{O}^*) = 0$, and thus every locally free sheaf of rank 1 is free.)

The application of the splitting criterion which is relevant to us is the following:

Theorem 4. *Let $\mathscr{F}$ be a locally free sheaf of rank r over a compact Riemann surface of genus g. Suppose that $\mathscr{L}$ is a locally free subsheaf of rank 1 in $\mathscr{F}$ having the following properties:*

1) *The quotient sheaf $\mathscr{G} := \mathscr{F}/\mathscr{L}$ is a direct sum $\mathscr{L}_2 \oplus \cdots \oplus \mathscr{L}_r$ of locally free sheaves of rank 1.*
2) *For $i = 2, \ldots, r$, $c(\mathscr{L}) - c(\mathscr{L}_i) \geq 2g - 1$.*

Then $\mathscr{L}$ is a direct summand of $\mathscr{F}$.

Proof: We first note that $\mathscr{H}om(\mathscr{G}, \mathscr{L}) = \bigoplus_{i=2}^{r} \mathscr{H}om(\mathscr{L}_i, \mathscr{L})$. Let $\mathscr{L} \cong \mathcal{O}(D)$ and $\mathscr{L}_i \cong \mathcal{O}(D_i)$, $2 \leq i \leq r$. Then $\mathscr{H}om(\mathscr{L}_i, \mathscr{L}) \cong \mathcal{O}(D - D_i)$. Hence, since

$$\deg(D - D_i) = c(\mathscr{L}) - c(\mathscr{L}_i) \geq 2g - 1$$

for $i = 2, \ldots, r$, it follows (by Theorem B) that, for all such i,

$$H^1(X, \mathscr{H}om(\mathscr{L}_i, \mathscr{L})) = (0).$$

Thus

$$H^1(X, \mathscr{H}om(\mathscr{G}, \mathscr{L})) \cong \bigoplus_{i=2}^{r} H^1(X, \mathscr{H}om(\mathscr{L}_i, \mathscr{L})) = (0)$$

and, by the splitting criterion, $\mathscr{F}$ is a direct summand. □

In the above proof we used the following fact:

For all divisors $D, D' \in$ Div X there is a natural $\mathcal{O}$-isomorphism

$$\mathcal{O}(D' - D) \xrightarrow{\sim} \mathscr{H}om_{\mathcal{O}}(\mathcal{O}(D), \mathcal{O}(D')).$$

Proof: Let $t \in \mathfrak{m}_x$ be a local coordinate at x. Then $\mathcal{O}(D)_x = t^{-o_x(D)} \cdot \mathcal{O}_x$ and $\mathcal{O}(D')_x = t^{-o_x(D')} \cdot \mathcal{O}_x$. Thus every germ $h_x \in \mathcal{O}(D' - D)_x = t^{-o_x(D'-D)} \cdot \mathcal{O}_x$ determines, by multiplication, an $\mathcal{O}_x$-homomorphism (homothety),

$$\mathcal{O}(D)_x \to \mathcal{O}(D')_x,$$

defined by $g_x \mapsto h_x g_x$. (Observe that $o(h_x g_x) = o(h_x) + o(g_x) = -o_x(D' - D) - o_x(D) = -o_x(D')$). Since $\mathcal{O}(D)_x$ and $\mathcal{O}(D')_x$ are free $\mathcal{O}_x$-modules of rank 1, every homomorphism $\mathcal{O}(D)_x \to \mathcal{O}(D')_x$ is such a "homothety". Thus the map $\mathcal{O}(D' - D) \to \mathcal{H}om_{\mathcal{O}}(\mathcal{O}(D), \mathcal{O}(D'))$, defined by associating to each germ the homothety defined above, is surjective. The injectivity of this map is trivial and thus we have established an isomorphism. □

5. Grothendieck's Theorem. We fix a point p "at infinity" in $\mathbb{P}_1$ and set

$$\mathcal{O}(n) := \mathcal{O}(np)$$

for every $n \in \mathbb{Z}$. Then n is the Chern number of $\mathcal{O}(n)$ and $\mathcal{O}(n) \cong \mathcal{O}(m)$ if and only if $n = m$. Furthermore, *if $\mathcal{L}$ is a locally free sheaf of rank 1 over $\mathbb{P}_1$ then $\mathcal{L} \cong \mathcal{O}(n)$, where $n = c(\mathcal{L})$.*

Proof: Certainly $\mathcal{L} \cong \mathcal{O}(D)$ for some $D \in \operatorname{Div} \mathbb{P}_1$. On $\mathbb{P}_1$, however, two divisors are linearly equivalent if and only if they have the same degree. Hence $\mathcal{O}(D)$ is isomorphic to $\mathcal{O}(\deg D)$ and, since $\deg D = c(\mathcal{L})$, we have the desired result. □

Every sheaf $\mathcal{O}(n_1) \oplus \mathcal{O}(n_2) \oplus \cdots \oplus \mathcal{O}(n_r)$, $n_1, \ldots, n_r \in \mathbb{Z}$, is locally free of rank r. The splitting theorem of Grothendieck says that one obtains *all* locally free sheaves over $\mathbb{P}_1$ in this way:

Theorem 5. (Grothendieck). *Let $\mathcal{F}$ be a locally free sheaf of rank $r \geq 1$ over $\mathbb{P}_1$. Then there exist integers $n_1, \ldots, n_r$ (uniquely determined up to a permutation) such that $\mathcal{F} \cong \mathcal{O}(n_1) \oplus \cdots \oplus \mathcal{O}(n_r)$.*

Remark: The essense of this theorem can be formulated with matrices:

Let $t > 1$, $U_1 := \{z \in \mathbb{P}_1 \mid |z| < t\}$, and $U_2 := \{z \in \mathbb{P}_1 \mid |z| > 1\}$.

Let $A \in GL(r, \mathcal{O}(U_{12}))$ be a holomorphic, invertible $r \times r$ – matrix. Then there exist holomorphic, invertible matrices $P \in GL(r, \mathcal{O}(U_1))$, $Q \in GL(r, \mathcal{O}(U_2))$ so that the matrix $D := PAQ$ is a diagonal matrix with $z^{n_1}, \ldots, z^{n_r}$, $n_i \in \mathbb{Z}$, as diagonal terms.

6. Existence of the Splitting. Using the following lemma, both the existence and uniqueness are proved by successively "splitting off" maximal subsheaves.

Splitting Lemma 6. *Let $\mathcal{F}$ be a non-zero locally free sheaf over $\mathbb{P}_1$, and let $\mathcal{L}$ be a locally free subsheaf of rank 1 with $\mu(\mathcal{L}) \geq \mu(\mathcal{F}/\mathcal{L}) - 1$. Then $\mathcal{L}$ is a direct summand of $\mathcal{F}$.*

In particular every maximal subsheaf of $\mathcal{F}$ is a direct summand.

Proof: We proceed by induction on the rank r of $\mathscr{F}$, where the induction hypothesis is the existence part of Theorem 5. If $r = 1$, then the lemma is trivially true, because $\mathscr{L} = \mathscr{F}$. If $r > 1$, then the sheaf $\mathscr{F}/\mathscr{L}$ is locally free of rank $r-1$, and by assumption is therefore a direct sum $\mathscr{L}_2 \oplus \cdots \oplus \mathscr{L}_r$ of locally free sheaves L_i of rank 1. Now,

$$\max\{c(\mathscr{L}_2), \ldots, c(\mathscr{L}_r)\} \leq \mu(\mathscr{F}/\mathscr{L}) \leq c(\mathscr{L}) + 1.$$

In particular, $c(\mathscr{L}) - c(\mathscr{L}_i) \geq -1$, $2 \leq i \leq r$. By Theorem 4 it follows that $\mathscr{L}$ is a direct sum of $\mathscr{F}$.

If $\mathscr{L}$ is a maximal subsheaf of $\mathscr{F}$, then from Theorem 3 it follows that $c(\mathscr{L}) = \mu(\mathscr{F}) \geq \mu(\mathscr{F}/\mathscr{L})$. (The genus is zero!). □

7. Uniqueness of the Splitting. For every locally free sheaf $\mathscr{F}$ on $\mathbb{P}_1$ we let $m := \mu(\mathscr{F})$. Then

$$\mathscr{F}(-m)(\mathbb{P}_1) = \{s \in \mathscr{F}^\infty(\mathbb{P}_1) \mid \deg(s) = m\} \cup \{0\} \neq 0,$$

so

$$d := \dim_{\mathbb{C}} \mathscr{F}(-m)(\mathbb{P}_1) \geq 1.$$

The sections $\mathscr{F}$ in $\mathscr{F}(-m)(\mathbb{P}_1)$ generate a non-zero $\mathcal{O}$-subsheaf $\mathscr{T}$ of $\mathscr{F}(-m)$. Thus

$$\hat{\mathscr{F}} := \mathscr{T}(m) \neq 0$$

is an *invariantly (by $\mathscr{F}$ alone) determined $\mathcal{O}$-subsheaf of $\mathscr{F}$*. With this language we now prove the uniqueness part of Theorem 5:

Uniqueness Lemma 7. *Let $\mathscr{F} = \mathscr{L}_1 \oplus \cdots \oplus \mathscr{L}_r = \mathscr{L}'_1 \oplus \cdots \oplus \mathscr{L}'_r$ be two splittings of $\mathscr{F}$ with $\mathscr{L}_i \cong \mathcal{O}(n_i)$, $\mathscr{L}'_1 \cong \mathcal{O}(n_i)$, $1 \leq i \leq r$. Assume that $n_1 \geq n_2 \geq \cdots \geq n_r$, and $n'_1 \geq n'_2 \geq \cdots \geq n'_r$. Then, with $d := \dim_{\mathbb{C}} \mathscr{F}(-m)(\mathbb{P}^1) \geq 1$,*

1) $\mathscr{L}_1 \oplus \cdots \oplus \mathscr{L}_d = \mathscr{L}'_1 \oplus \cdots \oplus \mathscr{L}'_d = \hat{\mathscr{F}}$; $\mathscr{L}_i \cong \mathcal{O}(m) \cong \mathscr{L}'_i$, *for* $i = 1, \ldots, d$,

and

2) $n_i = \cdots = n_d = n_i = \cdots = n'_d = m$; $n_i = n'_i$ *for* $i = d+1, \ldots, r$.

Proof: Since $\mathscr{F} = \mathscr{L}_1 \oplus \cdots \oplus \mathscr{L}_r$, it follows that $\mathscr{F}(-m)(\mathbb{P}_1) = \bigoplus_{i=1}^{r} \mathscr{L}_i(-m)(\mathbb{P}_1)$, where $\mathscr{L}_i(-m) \cong \mathcal{O}(l_i)$ with $l_i := n_i - m \leq 0$ (by the definition of m). Now

$$\mathcal{O}(l)(\mathbb{P}_1) = 0 \quad \text{for} \quad l < 0, \quad \text{and} \quad \mathcal{O}(\mathbb{P}_1) = \mathbb{C}.$$

Thus there must be d equations $n_i = m$. Since the n_i's are monotonically decreasing, $n_1 = \cdots = n_d = m > n_{d+1}$ and $\mathscr{L}_i(-m) \cong 0$, $1 \leq i \leq d$.

For every $i \leq d$, $\mathscr{L}_i(-m)(\mathbb{P}_1)$ generates the sheaf $\mathscr{L}_i(-m)$. Th

$\bigoplus_{i=1}^{d} \mathscr{L}_i(-m)(\mathbb{P}_1)$ generates the sheaf $\bigoplus_{i=1}^{d} \mathscr{L}_i(-m)$. Since $\mathscr{F}(-m)(\mathbb{P}_1) = \bigoplus_{i=1}^{d} \mathscr{L}_i(-m)(\mathbb{P}_1)$, it follows that $\mathscr{T} = \bigoplus_{i=1}^{d} \mathscr{L}_i(-m)$. Hence $\hat{\mathscr{F}} = \mathscr{T}(m) = \bigoplus_{i=1}^{d} \mathscr{L}_i$.

The above can obviously be repeated for the splitting $\mathscr{F} = \mathscr{L}'_1 \oplus \cdots \oplus \mathscr{L}'_r$. Thus 1) has been proved as well as $n_i = \cdots = n'_d = m$. It now follows that

$$\mathcal{O}(n_{d+1}) \oplus \cdots \oplus \mathcal{O}(n_r) \cong \mathscr{F}/\hat{\mathscr{F}} \cong \mathcal{O}(n'_{d+1}) \oplus \cdots \oplus \mathcal{O}(n'_r).$$

Since $\mathscr{F}/\hat{\mathscr{F}}$ has rank $r - d < r$, and since n_i, n_i are monotonically decreasing, it follows by induction that $n_i = n'_i$ for $i = d + 1, \ldots, r$. □

Corollary. *Let $\mathscr{F} = \mathscr{G} \oplus \mathscr{H}$, where $\mathscr{G}$, $\mathscr{H}$ are non-zero locally free sheaves over $\mathbb{P}_1$. Then*

$$\mu(\mathscr{F}) = \max\{\mu(\mathscr{G}), \mu(\mathscr{H})\}.$$

Proof: Since $\mathscr{F} \cong \mathcal{O}(n_1) \oplus \cdots \oplus \mathcal{O}(n_r)$, it follows that $\mu(\mathscr{F}) = \max\{n_1, \ldots, n_r\}$. □

We see now that every locally free subsheaf $\mathscr{L}$ of rank 1 in $\mathscr{F}$ with $c(\mathscr{L}) \geq \mu(\mathscr{F}/\mathscr{L})$ is *necessarily* maximal, because the Splitting Lemma implies that $\mathscr{F} = \mathscr{G} \oplus \mathscr{L}$, and the above corollary shows that $\mu(\mathscr{F}) = \max(\mu(\mathscr{G}), \mu(\mathscr{L})) = c(\mathscr{L})$.

The following remark is also quite easy to see.

Let $\mathscr{F} = \mathscr{G} \oplus \mathscr{H}$, where $\mathscr{G}$, $\mathscr{H}$ are non-zero locally free sheaves with $\mu(\mathscr{F}) > \mu(\mathscr{G})$. Let $\mathscr{L}$ be a locally free subsheaf of rank 1 in $\mathscr{F}$ so that $c(\mathscr{L}) > \mu(\mathscr{G})$. Then $\mathscr{L}$ is a locally free subsheaf of $\mathscr{H}$.

Proof: Let $\mathscr{L} = \mathcal{O}s$ with $s \in \mathscr{F}^\infty(\mathbb{P}_1)^*$, and $\deg(s) = c(\mathscr{L})$. Let $\pi\colon \mathscr{F} \to \mathscr{G}$ denote the natural sheaf projection. Then by the remark in Paragraph 1, either $\pi(s) = 0$ or $\deg(s) \leq \deg(\pi(s))$. The latter is not possible, because $\deg(\pi(s)) \leq \mu(\mathscr{G})$ and $c(\mathscr{L}) > \mu(\mathscr{G})$. Thus $s \in (\mathscr{K}er\ \pi)(\mathbb{P}_1) = \mathscr{H}(\mathbb{P}_1)$. Hence $\mathscr{L} = \mathcal{O}s \subset \mathscr{H}$. □

In particular we have the following:

A sheaf $\mathscr{F} \cong \mathcal{O}(n_1) \oplus \cdots \oplus \mathcal{O}(n_r)$ with $n_1 > \max\{n_2, \ldots, n_r\}$ has only one locally free subsheaf $\mathscr{L}$ which is isomorphic to $\mathcal{O}(n_1)$, and there are no such with $\mathscr{L} \cong \mathcal{O}(n)$, $n_1 > n > \max\{n_2, \ldots, n_r\}$.

However, for every $n \leq \max\{n_2, \ldots, n_r\}$, there are in the above setting locally free subsheaves $\mathscr{L} \cong \mathcal{O}(n)$ in $\mathscr{F}$. For this, see Prop. 2.4 in "On holomorphic fields of complex line elements with isolated singularities", Ann. Inst. Fourier **14**, 99-130 (1964), by A van de Van.

Bibliography

Monographs

[BT] Behnke, H., Thullen, P.: Theorie der Funktionen mehrerer komplexer Veränderlichen. 2. Aufl. Erg. Math. 51. Heidelberg: Springer-Verlag 1970.

[ENS_1] Cartan, H.: Séminaire: Théorie des fonctions de plusieurs variables. Paris 1951/52.

[ENS_2] Cartan, H.: Séminaire: Théorie des fonctions de plusieurs variables. Paris 1953/54.

[CAG] Fischer, G.: Complex Analytic Geometry. Lecture Notes Math. 538. Heidelberg: Springer-Verlag 1976.

[TF] Godement, R.: Théorie des faisceaux. Act. sci. ind. 1252. Paris: Hermann 1958.

[AS] Grauert, H., Remmert, R.: Analytische Stellenalgebren. Grundl. Math. Wiss. 176. Heidelberg: Springer-Verlag 1971.

[CAS] Grauert, H., Remmert, R.: Coherent Analytic Sheaves. In Preparation.

[CA] Hörmander, L.: Complex Analysis in Several Variables. 2. Aufl. Amsterdam-London: North Holland Publ. Comp. 1973.

[ARC] Narasimhan, R.: Analysis on Real and Complex Manifolds. Amsterdam: North Holland Publ. Comp. 1968.

[FAC] Serre, J.-P.: Faisceaux Algébriques Cohérents. Ann. Math. **61,** 197–278 (1955).

[GACC] Serre, J.-P.: Groupes Algébriques et Corps de Classes. Act. sci. ind. 1264. Paris: Hermann 1959.

[TAG] Hirzebruch, F.: Topological Methods in Algebraic Geometry, Grund. der Math. Wiss. 131, Springer-Verlag, 1966.

[SCV] Grauert, H. and Fritzsche, K.: Several Complex Variables, Grad. Texts in Math. 38, Springer-Verlag, 1976.

[CAR] Cartan, H.: Collected Works (Oeuvres Scientifique), Springer-Verlag, 1979.

[DF] Cartan, H.: Differential Forms, Boston: H. Mifflin Co., 1970.

Articles

[1] Andreotti, A., Frankel, T.: The Lefschetz theorem on hyperplane sections. Ann. Math. **69,** 713–717 (1959).

[2] Le Barz, P.: A propos des revêtements ramifiés d'espaces de Stein. Math. Ann. **222,** 63–69 (1976).

[3] Behnke, H., Stein, K.: Analytische Funktionen mehrerer Veränderlichen zu vorgegebenen Null- und Polstellenflächen. Jahr. DMV **47,** 177–192 (1937).

[4] Behnke, H., Stein, K.: Konvergente Folgen von Regularitätsbereichen und die Meromor phiekonvexität. Math. Ann. **116,** 204–216 (1938).

[5] Behnke, H., Stein, K.: Entwicklung analytischer Funktionen auf Riemannschen Flächen. Mat Ann. **120,** 430–461 (1948).

[6] Behnke, H., Stein, K.: Elementarfunktionen auf Riemannschen Flächen. Canad. Journ. Math. 152–165 (1950).

[7] Cartan, H.: Les problèmes de Poincaré et de Cousin pour les fonctions de plusieurs variabl complexes. C. R. Acad. Sci. Paris **199,** 1284–1287 (1934).

[8] Cartan, H.: Sur le premier problème de Cousin. C. R. Acad. Sci. Paris **207**, 558–560 (1937).
[9] Cartan, H.: Variétés analytiques complexes et cohomologie. Coll. Plus. Var., Bruxelles 1953, 42–55.
[10] Cartan, H., Serre, J.-P.: Un théorème de finitude concernant les variétés analytiques compactes. C. R. Acad. Sci. Paris **237**, 128–130 (1953).
[11] Fornaess, J. E.: An increasing sequence of Stein manifolds whose limit is not Stein. Math. Ann. **223**, 275–277 (1976).
[12] Fornaess, J. E.: 2 dimensional counterexamples to generalizations of the Levi problem. Math. Ann. **230**, 169–173 (1977).
[13] Fornaess, J. E., Stout, E. L.: Polydiscs in Complex Manifolds. Math. Ann. **227**, 145–153 (1977).
[14] Forster, O.: Zur Theorie der Steinschen Algebren und Moduln. Math. Z. **97**, 376–405 (1967).
[15] Forster, O.: Topologische Methoden in der Theorie Steinscher Räume. Act. Congr. Int. Math. 1970, Bd. 2, 613–618.
[16] Grauert, H.: Charakterisierung der holomorph-vollständigen Räume. Math. Ann. **129**, 233–259 (1955).
[17] Grauert, H.: On Levi's problem and the imbedding of real-analytic manifolds. Ann. Math. **68**, 460–472 (1958).
[18] Grauert, H.: Bemerkenswerte pseudokonvexe Mannigfaltigkeiten. Math. Z. **81**, 377–391 (1963).
[19] Grauert, H., Remmert, R.: Konvexität in der komplexen Analysis: Nicht holomorph-konvexe Holomorphiegebiete und Anwendungen auf die Abbildungstheorie. Comm. Math. Helv. **31**, 152–183 (1956).
[20] Grauert, H., Remmert, R.: Singularitäten komplexer Manngifaltigkeiten und Riemannsche Gebiete. Math. Z. **67**, 103–128 (1957).
[21] Grauert, H., Remmert, R.: Zur Spaltung lokal-freier Garben über Riemannschen Flächen. Math. Z. **144**, 35–43 (1975).
[22] Grothendieck, A.: Sur la classification des fibrés holomorphes sur la sphère de Riemann. Amer. Journ. Math. **79**, 121–138 (1957).
[23] Igusa, J.: On a Property of the Domain of Regularity. Mem. Coll. Sci., Univ. Kyoto, Ser. A, **27**, 95–97 (1952).
[24] Jurchescu, M.: On a theorem of Stoilow. Math. Ann. **138**, 332–334 (1959).
[25] Markoe, A.: Runge Families and Inductive Limits of Stein Spaces. Ann. Inst. Fourier **27**, fasc. 3 (1977).
[26] Matsushima, Y.: Espaces Homogènes de Stein des Groupes de Lie Complexes. Nagoya Math. Journ. **16**, 205–218 (1960).
[27] Matsushima, Y., Morimoto, A.: Sur Certains Espaces Fibrés Holomorphes sur une Variété de Stein. Bull. Soc. Math. France **88**, 137–155 (1960).
[28] Milnor, J.: Morse Theory. Ann. Math. Studies **51**, Princeton Univ. Press 1963.
[29] Narasimhan, R.: On the Homology Groups of Stein Spaces. Inv. Math. **2**, 377–385 (1967).
[30] Oka, K.: Sur les fonctions analytiques de plusieurs variables II. Domaines d'holomorphie. Journ. Sci. Hiroshima Univ., Ser. A, **7**, 115–130 (1937).
[31] Oka, K.: Sur les fonctions analytiques de plusieurs variables III. Deuxième problème de Cousin. Journ. Sci. Hiroshima Univ., Ser. A, **9**, 7–19 (1939).
[32] Oka, K.: Sur les fonctions analytiques de plusieurs variables IX. Domaines fini sans point critique intérieur. Jap. Journ. Math. **23**, 97–155 (1953).
[33] Scheja, G.: Riemannsche Hebbarkeitssätze für Cohomologieklassen. Math. Ann. **144**, 345–360 (1961).
[34] Schuster, H. W.: Infinitesimale Erweiterungen komplexer Räume. Comm. Math. Helv. **45**, 265–286 (1970).
[35] Serre, J.-P.: Quelques problèmes globaux relatifs aux varietés de Stein. Coll. Plus. Var., Bruxelles 1953, 57–68.
[36] Stein, K.: Topologische Bedingungen für die Existenz analytischer Funktionen komplexer Veränderlichen zu vorgegebenen Nullstellenflächen. Math. Ann. **117**, 727–757 (1941).
[37] Stein, K.: Analytische Funktionen mehrerer komplexer Veränderlichen zu vorgegebenen Periodizitätsmoduln und das zweite Cousinsche Problem. Math. Ann. **123**, 201–222 (1951).
[38] Stein, K.: Überlagerungen holomorph-vollständiger komplexer Räume. Arch. Math. **7**, 354–361 (1956).

Subject Index

acyclic cover 42
acyclic resolution 32
additive Cousin problem 137
additive functor 25
algebra, Stein 176
algebraic dimension 20
— reduction 8
algebraized space 15
alternating Čech cohomology module 35, 37
— — complex 35
— cocycle 35
analytic blocks 116
— hypersurface 129
— interior 111
— set 18
— —, irreducible 19
— —, pure dimensional 20
— sheaves 16, 92
— spectrum 180
— stones 111
annihilator sheaves 13
antiholomorphic *p*-forms 70
Approximation Theorem of Runge
90, 122, 170
Attaching Lemma for sheaf epimorphisms 95
— — of Cartan 88
— — — Cousin 84
— sections 92

Banach's Open Mapping Theorem 165
Bergmann Inequality 189
block neighborhood 167
blocks, analytic 116
—, Approximation Theorem for 91
—, exhaustion by 117
branches of an analytic set 19

canonical divisor 214
— — class 214
— flabby resolution 30
— presheaf 2
— resolution relative to a cover 42
— topology 168
Cartan, Theorem of 179
Cartan's Attaching Lemma 88
Cauchy Integral Formula 74
Cauchy–Riemann differential equations 65
Čech cohomology module 35, 37
— — —, alternating 35, 37
— complex 35
— —, alternating 35
character 162, 176
— ideal 163, 176
— Theorem 162
characteristic 209
— class 144
— Theorem 209
chart 194
Chern class 144
— function 216
closed complex subspace 17
Closedness Theorem 169
coboundary 29
— map 28
cochain 28, 34
—, alternating 35
cocycle 29
codimension, complex 19
Coherence Theorem for ideal sheaves 18
— — — finite holomorphic maps 52
— — of Oka 16
coherent sheaf 11, 96
cohomology classes 29
— groups, deRham 63
— —, Dolbeault 79
— modules, alternating Čech 35, 37
— —, Čech 35, 37
— —, (flabby) of an $\mathscr{R}$-sheaf 30
— — of a complex 29
— sequence, long exact 29, 30, 34, 38
completeness relations 188
complex 28
— codimension 19
— dimension 19, 20
— manifold 16

complex (*cont.*)
— space 16
— —, holomorphically complete 118
— —, — convex 109
— —, — separable 117
— —, — spreadable 117
— —, irreducible 21
— —, normal 21
— —, reduced 21
— —, Stein 101
— —, weakly holomorphically convex 113
— subspace, closed 16
— —, open 17
— r-vector 59
— value of a holomorphic function 17
— valued differential form 60
complexes, homomorphisms of 28
conjugation 62
connecting homomorphism 29
constant sheaf 8
Continuation Theorem of Riemann 21
continuity of roots 48
Convergence Theorem 121, 172
Cousin Attaching Lemma 84
— I distribution 137
— I problem (additive) 136
— II distribution 139
— II problem (multiplicative) 138
covering 18
—, acyclic 42
—, Stein 127
cycle 150

$\bar{\partial}$-closed 77
Decomposition Lemma for analytic sets 19
degree equation 210
— of a divisor 206
— — — section 206
deRham cohomology groups 63
—, Formal Lemma 32
—, Theorem of 64
derivation 58
derivative, exterior 62
—, total 62
determinant sheaf 216
$\bar{\partial}$-exact 77
diameter 37
differentiable manifolds 16
— maps 16
— vector field 58
differential forms, complex valued 60
dimension, algebraic 20
—, complex 19, 20
— of a block 96
dimension, topological 19
direct product of sheaves 4
— (Whitney) sum of sheaves 2
distinguished block 173
Division Theorem, Weierstrass 49
divisor 138
—, canonical 214
— class, canonical 214
— classes 146
— group 205
— of a meromorphic function 138
— — section 206
—, positive 139
Dolbeault cohomology groups 79
domain of holomorphy 134
Duality Theorem of Serre 219, 225

embedding dimension 151
—, holomorphic 16
— Theorem 126
equivalent criteria for a Stein space 152
Euler–Poincaré characteristic 210
exact sequence of presheaves 9
— — — sheaves 8
Exactness Lemma 26
— Theorem 26, 47
exhaustion 102
—, Stein 105
— Theorem 108, 127
exponential homomorphism 143
— sequence 142
exterior derivative 62
— product of sheaves 13

finite mapping 45
— sheaves 10
Finiteness Lemma for character ideals 177
— Theorem 186, 202
Five Lemma 11
flabby functor 25
— resolution 30
— sheaf 25
Formal deRham Lemma 32
Fréchet spaces 163
free sheaves 10
function, holomorphic 17
—, meromorphic 21
Fundamental Theorem of Stein Theory 124

Gap Theorem of Weierstrass 229
— value 229
genus 211
glued family of sheaves 5
gluing sheaves by cocycles 147

good semi-norms 118
graph of a holomorphic map 17
Grothendieck, lemma of 75
—, Splitting Theorem of 237
groups, Stein 136
—, toroid 136

Hartogs' Theorem 81
Hodge decomposition 231
holomorphic embedding 16
— function 17
— map 16
— matrix 86
— *p*-form 66
holomorphically complete 118
— convex 109
— — hull 108
— separable 117
— spreadable 117
hull, holomorphically convex 113
hypersurface, analytic 129

Identity Theorem for analytic sets 19
image of a presheaf homomorphism 9
— — — sheaf homomorphism 7
— sheaves 4
inclusion of stones 112
interior, analytic 111
intersection of submodules 7
irreducible analytic sets 19
— complex spaces 21
— component of a divisor 140
Isomorphism Theorem for sheaf cohomology 43

k-algebraized space 15
kernel of a presheaf homomorphism 9
— — — sheaf homomorphism 7
k-homomorphism 8
k-morphism 15
Kronecker symbol 59

Lebesque number 37
Lemma of Grothendieck 75
— — Poincaré 64
Leray Theorem 43
lifting of differential forms 62
— — *r*-vectors 60
linear equivalence 149
locally free sheaf 10
— — subsheaf 213
long exact cohomology sequence 29, 30, 34, 38

manifold, complex 16
—, differentiable 16
mapping, closed 45
—, differentiable 16
—, finite 45
—, holomorphic 16
—, proper 112
matrix, bounded holomorphic 85
maximal subsheaf 233
meromorphic function 21
— section 206
minimum principle 190
monotone 192
multiplicative Cousin problem 138
— set 14

nilradical 8
normal complex space 21
normalization 22
— Theorem 22
nullstellen ideal 18

Oka Principle 145
—, Theorem 145
Oka's Coherence Theorem 16
open complex subspace 17
Open Mapping Theorem of Banach 165
ordering 192, 205
orthogonality relations 188

paracompact space 18
Pfaffian form 61
p-forms, antiholomorphic 70
—, holomorphic 66
Poincaré, Lemma of 64
— problem 139
—, Theorem of 140
point character 176
(*p*, *q*)-form, differentiable 66
positive divisor 139
presheaf, canonical 2
— of abelian groups 9
— homomorphism 9
—, mapping of 2
— — modules 9
— — rings 9
principal divisor 138
— ideal sheaf 129
— part distribution 137
— — — w.r.t. a divisor 219
product of complex spaces 17
— — ideal sheaves 7
— sheaf 146

projection of a sheaf 1
Projection Theorem (local) 52
proper mapping 112
pure dimensional 20

quotient presheaves 9
— sheaves 7
quotients, ring of 14
—, sheaf of 14

radical ideal 19
reduced complex space 21
reducible analytic sets 19
reduction, algebraic 8
— map 20
— of a complex space 20
— Theorem 154
refinement map 18
— of a covering 18
— — — resolution atlas 198
regular point 21
relation sheaves, finite 11
relations, sheaf of 11
representation of 1 161
residue 222
— Theorem 223
restriction map 2
— of a sheaf 2
resolution, acyclic 32
— atlas 194
—, canonical flabby 30
— —, relative to a cover 42
—, flabby 30, 33
—, soft 33
r-form 61
$\mathscr{R}$-homomorphism 7
Riemann Continuation Theorem 21
— domain 134
— —, unramified 134
— inequality 212
Riemann–Roch, Theorem of 211, 226
— —, — —, for locally free sheaves 216
$\mathscr{R}$-module 6
R-resheaf 9
$\mathscr{R}$-resolution 26
$\mathscr{R}$-sequence 7
—, exact 7
$\mathscr{R}$-submodule 7
Runge, Approximation Theorem of 90, 122, 170
r-vector, complex 59

saturated sets 190
Schwartz Lemma 190, 192
section functor 2
sections 2
—, attaching 92
—, meromorphic 206
semi-norm, good 118
Separation Theorem 24
sequence topology 165
Serre Duality Theorem 219, 225
sheaf, analytic 16, 92
— of abelian groups
—, coherent 11, 96
—, finite 10
—, finite relation 11
—, flabby 25
—, free 10
— of germs of antiholomorphic p-forms 70
— — — — complex-valued differentiable functions 16
— — — — — — - — r-forms 61
— — — — — — - — continuous functions 8
— — — — differentiable (p, q)-forms 66
— — — — — vector fields 58
— — — — — — - — r-forms 61
— — — — divisors 138
— — — — holomorphic functions 16
— — — — — p-forms 66
— — — — meromorphic functions 21
— — — — — sections 204
— — — — Pfaffiian forms 61
— — — — principal parts 137
— — — — — part distributions w.r.t. a divisor 219
— — — — real-valued differentiable functions 16
— — — — $\mathscr{R}$-homomorphisms 13
—, gluing 5
— homomorphism 6
—, locally free 10
— mapping 1
— — ideals 7
— — k-algebras 8
— — local k-algebras 8
— — modules 6
— — rings 6
—, reduced 8
—, Stein 229
—, soft 22
Shrinking Theorem 18
singular points 21
Smoothing Lemma 200
soft sheaf 22
spectrum, analytic 180
splitting criterion 235
Splitting Theorem of Grothendieck 237
square integrable 187

stalk 1
Stein algebra 176
— covering 127
— exhaustion 105
— group 136
— set 100
— sheaf 229
— space 100
stone 111
structure sheaf 16
Structure Theorem for locally free sheaves of rank 1 214
subdegree 192
submodule 7
subpresheaf 9
subsheaf 9
—, locally free 213
—, maximal 233
— Theorem 214
subspace, closed complex 16
—, open complex 17
sum of submodules 7
sums of sheaves 1
support of sheaves 1
support of a divisor 6
— — — sheaf 6

tangent space 57
— vector 57
tensor product of sheaves 13
Theorem A for compact blocks 96
— — — locally free sheaves 229
— — — sheaves $\mathcal{O}(D)$ 227
— — — Stein sets 101
— — (Fundamental Theorem) 124
Theorem B for compact blocks 97
— — — locally free sheaves 229
— — — sheaves $\mathcal{O}(D)$ 227
— — (Fundamental Theorem) 124
Theorem of Cartan 179
— — deRham 64
— — Oka 145
— — Poincaré 140
— — Riemann-Roch 211, 226
— — — — — for locally free sheaves 216
Three Lemma 11
topological dimension 19
topology, canonical 168
—, weak 180
toroid groups 136
total derivative 62
trivial extension of a sheaf 12
Tube Theorem 157

unramified Riemann domain 134

value, complex, of a holomorphic function 17
— of a section 8
Vanishing Theorem 210
— — for compact blocks 37
vector field 58
— —, differentiable 58

weak topology 180
weakly holomorphically convex 113
Weierstrass Division Theorem (general) 49
— Gap Theorem 229
— homomorphism 51

zero section 6
— set of an ideal 18

Table of Symbols

$\mathscr{S}_x, \varphi_x$ 1
$\mathscr{S}_1 \oplus \mathscr{S}_2$ 1
$\mathscr{S}_Y, \mathscr{S}\,|\,Y$ 2
s_x 2
$\Gamma(Y, \mathscr{S}), \mathscr{S}(Y)$ 2
r_V^U 2
$\Gamma(\mathscr{S})$ 2
$\Gamma(\varphi)$ 2
$\check{\Gamma}(S)$ 3
$\check{\Gamma}(\Phi)$ 3
$\prod_{i \in I} \mathscr{S}_i$ 4
$f_*(\mathscr{S})$ 5
$\hat{f}_x$ 5
$f_*(\varphi)$ 5
supp $\mathscr{S}$ 6
$\mathscr{R}^p$ 6
$\mathscr{S}' \cap \mathscr{S}''$ 7
$\mathscr{S}' + \mathscr{S}''$ 7
$\mathscr{I} \cdot \mathscr{S}$ 7
$\mathscr{K}er\, \varphi$ 7
$\mathscr{S}/\mathscr{S}'$ 7
$\mathscr{I}m\, \varphi$ 7
$\mathfrak{m}(\mathscr{R}_x)$ 8
$\mathscr{C}$ 8
$s(x)$ 8
$\mathfrak{n}(\mathscr{R})$ 8
red $\mathscr{R}$ 8
s/s' 9
Ker Φ 9
Im Φ 9
$Rel(s_1, \ldots, s_p)$ 11
$\mathscr{C}oker\, \varphi$ 12
$\mathscr{S} \otimes_{\mathscr{R}} \mathscr{S}'$ 12
$\bigotimes^p \mathscr{S}$ 13
$\bigwedge^p \mathscr{S}$ 13
$\mathscr{H}om_{\mathscr{R}}(\mathscr{S}, \mathscr{S}')$ 13
$\mathscr{A}n\, \mathscr{S}$ 13
$\mathscr{R}_M$ 14
$\tilde{\jmath}$ 15
$\mathscr{E}^{\mathbb{R}}, \mathscr{E}_X^{\mathbb{R}}, \mathscr{E}^{\mathbb{C}}, \mathscr{E}_X^{\mathbb{C}}$ 15
$\mathscr{O}$ 16
$\mathscr{O}_X$ 16
Hol(X, Y) 17
$X_1 \times X_2$ 17
Gph f 17
rad $\mathscr{I}$ 19
dim top$_x$ A 19
$\dim_x A$ 19
codim$_x$ A 19
dim A 20
red X 20
red 20
$\mathscr{M}$ 21
$\mathscr{F}(\mathscr{S})$ 25
$Z^q(K^\bullet)$ 28
$B^q(K^\bullet)$ 28
$H^q(K^\bullet)$ 29
$\mathscr{F}^q(\mathscr{S})$ 30
$\mathscr{F}^\bullet(\mathscr{S})$ 30
$H^q(X, \mathscr{S})$ 30
$C^q(\mathfrak{U}, S)$ 34
$H^q(\mathfrak{U}, S), H^q(\mathfrak{U}, \mathscr{S})$ 34
$C^q_a(\mathscr{U}, S)$ 34
$H^q_a(\mathfrak{U}, S)$ 34
$i_q(\mathfrak{U})$ 34
$h^q(\mathfrak{W}, \mathfrak{U}), h^q(\mathfrak{U})$ 36
$\check{H}^q(X, S), \check{H}^q(X, \mathscr{S})$ 36
$\check{H}^q_a(X, S), \check{H}^q_a(X, \mathscr{S})$ 37
i_q 37
$d(M)$ 37
$\mathscr{S}\langle Y \rangle$ 40
$\check{\jmath}$ 46, 48
$\mathring{\pi}$ 51
$T(x)$ 57
$\mathscr{D}(R, M)$ 58
$\mathscr{T}$ 59
$A^r(x), A(x)$ 59
$\varphi \wedge \psi$ 59
$f_*, f^\#$ 60
$\mathscr{A}^r$ 61
$\mathscr{A}$ 61
$\bar{\varphi}$ 62
d 62
$\mathscr{A}^{p,q}$ 66
Ω^p 66
$\partial, \bar{\partial}$ 67, 68
$\bar{\Omega}^p$ 69
Tf 72
supp φ 81
$B(V), B^*(V)$ 86
$d(Q)$ 96
$\hat{M}, \hat{M}_X$ 108
P^0 111
$\mathscr{H}$ 137
$\mathscr{O}^*$ 138
$\mathscr{M}^*$ 138
$\mathscr{D}$ 138
(h) 138
D^+, D^- 140
exp f 143
$c(D)$ 144
$\mathscr{C}^*$ 144
$DC(X)$ 146
$\mathscr{O}(D)$ 146
$G(\mathscr{M})$ 146
$\mathscr{L} \cdot \mathscr{L}'$ 146
$LF(\mathscr{M})$ 147

$LF(X)$ 147
supp o 150
$\mathscr{Z}(o)$ 150
χ_p 176
$\mathscr{X}(T)$ 180
$\hat{T}_\chi$, $\hat{\mathcal{O}}_p$ 182, 183
$\mathscr{T}$ 183
$\|f\|_B$ 187
$\mathcal{O}_h(B)$ 187
$(f, g)_B$ 188
$\mathcal{O}_h^k(B)$ 189
$o_p(f)$ 191
$\omega_p(f)$ 192
$F(\alpha)$, $F(\alpha)^*$ 193
H_j, H_j^* 193
$C^q(\mathfrak{A})$, $C_h^q(\mathfrak{A})$ 196
$C_h^q(\mathfrak{U}, \mathscr{S})$ 196
$\|\zeta\|_{\mathfrak{A}}$ 197
$Z^q(\mathfrak{A})$, $Z_h^q(\mathfrak{A})$, $Z_h^q(\mathfrak{U}, \mathscr{S})$ 197
$\mathfrak{A}' < \mathfrak{A}$ 198
$\mathscr{F}^\infty$ 204
$\mathscr{F}^\infty(X)^*$, $\mathscr{F}(X)^*$ 204
Div X 205
deg D 205
$|D|$ 205
(s) 206
$\mathscr{F}(D)$ 206
$\mathscr{S}(D)$ 207
$\chi_0(\mathscr{S})$ 209
$\chi(\mathscr{S})$ 210
$l(D)$, $i(D)$ 211
g 211
$c(\mathscr{F})$ 215
det $\mathscr{F}$ 216
$\mathscr{H}(D)$ 219
R, $R(D)$ 220
$I(D)$ 220
$J(D)$, J 221
$\text{Res}_x\, \omega_x$ 222
$\langle \omega, F \rangle$ 223
K 225
$\mu(\mathscr{F})$ 233
$\mathcal{O}(n)$ 237

Acknowledgement

The authors are grateful to D. N. Akhiezer who, while translating "Theory of Stein Spaces" into Russian 1989, made the correction to our proof of Theorem B in this Addendum.

Addendum

by D. N. Akhazier[1]

Our goal here is to make more transparent the exposition in §4 of Chapter IV. We retain the notation and conventions introduced in the preamble to the section. In that preamble, as well as in most other places where no changes are required, we use the pieces of the main text. Some formulations of important theorems are changed, but the numbering is the same as in the main text.

1. Good Semi-norms. Topology in $\mathscr{S}(P^0)$. We want to enlarge Section 1 of §4. In particular, we assume that Theorem 1 is proven and the definition of a good semi-norm is given.

This done, we fix an epimorphism of sheaves $\epsilon : \mathscr{O}^l|Q \to \pi_*(\mathscr{S}|P)$ and use good semi-norms to construct a metric on $\mathscr{S}(P^0)$. We take an exhaustion of $\mathring{Q}$,

$$Q_1 \subset Q_2 \subset \ldots \subset Q_\nu \subset Q_{\nu+1} \subset \ldots \subset \mathring{Q}, \quad \bigcup_1^\infty Q_\nu = \mathring{Q}$$

where all Q_i are (compact) euclidean blocks in $\mathbb{C}^m$ with the same center. The map ϵ induces the $\mathscr{O}(Q_\nu)$-epimorphisms

$$\epsilon_{Q_\nu} : \mathscr{O}^l(Q_\nu) \to \pi_*(\mathscr{S}|P)(Q_\nu),$$

which define good semi-norms $|.|_\nu$ in $\mathscr{S}(P^0)$. We put

$$d(s_1, s_2) := \sum_{\nu=1}^{\infty} 2^{-\nu} \cdot \frac{|s_1 - s_2|_\nu}{1 + |s_1 - s_2|_\nu},$$

[1] Prof. Dmitri Akhiezer, Institute for Information Transmission Problems, B. Karetny 19, 101447 Moscow, ussia

where $s_1, s_2 \in \mathscr{S}(P^0)$ (see also Chapter V.6.0), and make the following observation which one should compare with the second lemma in Chapter V.6.3.

d is a metric on $\mathscr{S}(P^0)$. Considered with this metric, $\mathscr{S}(P^0)$ is a Fréchet space.

Proof: The first assertion follows from Theorem 1. To prove the second one, we take any Cauchy sequence $\{s_\nu\}$, $s_\nu \in \mathscr{S}(P^0)$. Then we can find bounded sequences $\{f_{\nu\lambda}\}$, $f_{\nu\lambda} \in \mathscr{O}^l(Q_\nu)$, so that $\epsilon_{Q_\nu}(f_{\nu\lambda}) = s_\lambda|\pi^{-1}(Q_\nu) \cap P$. By Montel's theorem, there is a subsequence $\{f_{\nu\lambda}\}$, which converges uniformly on $Q_{\nu-1}$ to some function $f_\nu \in \mathscr{O}^l(Q_{\nu-1})$. One has

$$\epsilon_{Q_{\nu-1}}(f_{\nu+1})|\mathring{Q}_{\nu-1} = \epsilon_{Q\nu-1}(f_\nu)|\mathring{Q}_{\nu-1} = s|\pi^{-1}(\mathring{Q}_{\nu-1}) \cap P,$$

where $s \in \mathscr{S}(P^0)$. It follows that $\{s_\nu\}$ converges to s in the sense of metric d. □

The topology on $\mathscr{S}(P^0)$, induced by d, does not depend on the exhaustion used in the construction. As we will show, this topology does not depend on the epimorphism ϵ as well. Furthermore, we will see that the topology does not change if (P, π) is replaced by another analytic block $(P, '\pi)$, where $'\pi = (\pi, \varphi)$.

Let $'\pi = (\pi, \varphi)$ be a holomorphic map from X to $\mathbb{C}^{'m} = \mathbb{C}^m \times \mathbb{C}^n$, $n > 0$, and let $Q^* \subset \mathbb{C}^n$ be a euclidean block such that $\varphi(P) \subset Q^*$. Then $P = {'\pi}^{-1}(Q \times Q^*) \cap U$. One can find open neighborhoods $'U$ of P in X and $'V$ of $Q \times Q^*$ in $\mathbb{C}^{'m}$ for which the map $'\pi|_{'U} : {'U} \to {'V}$ is finite. We fix an epimorphism of sheaves $'\epsilon : \mathscr{O}^{'l}|Q \times Q^* \to {'\pi}_*(\mathscr{S}|P)$, construct an exhaustion $\{Q_\nu \times Q^*_\nu\}$ of the above type for the open set $\mathring{Q} \times \mathring{Q}^*$ and denote by $'d$ the associated metric on $\mathscr{S}(P^0)$.

Theorem 1′. *The topologies on $\mathscr{S}(P^0)$ induced by d and $'d$ coincide.*

Proof: Let $e_1, ..., e_l \in \mathscr{O}^l(Q)$ be the standard basis sections. We denote by $'e_1, ..., 'e_l$ the preimages of $\epsilon_Q(e_1), ..., \epsilon_Q(e_l) \in \pi_*(\mathscr{S}|P)(Q) = {'\pi}_*(\mathscr{S}|P)(Q \times Q^*)$ in $\mathscr{O}^{'l}(Q \times Q^*)$ under $'\epsilon_Q$. Further, for $f \in \mathscr{S}(Q_\nu)$, we denote by $'f \in \mathscr{O}(Q_\nu \times Q^*_\nu)$ the holomorphic extension of f to $Q_\nu \times Q^*_\nu$, constant along each fiber $\{q\} \times Q^*_\nu$, $q \in Q_\nu$. The norms of $\mathbb{C}$-linear operators

$$\mathscr{O}^l(Q_\nu) \to \mathscr{O}^{'l}(Q_\nu \times Q^*_\nu), \; \sum_1^l f_\lambda \cdot e_\lambda \mapsto \sum_1^l {'f_\lambda} \cdot {'e_\lambda} \quad (\nu = 1, 2, ...,)$$

are bounded by a constant which does not depend on ν. Therefore the identity map

$$\mathrm{id} : (\mathscr{S}(P^0), d) \to (\mathscr{S}(P^0), 'd)$$

is continuous. By the Open Mapping Theorem of Banach, this map is a homeomorphism. □

2. The Compatibility Theorem. Suppose that along with (P, π) we have another analytic block $('P, '\pi)$ in X which is defined by a holomorphic map $'\pi : X \to \mathbb{C}^{'m}$ and the associated euclidean block $'Q \subset \mathbb{C}^{'m}$.

Theorem 2 (Compatibility Theorem). *If $(P, \pi) \subset ('P, '\pi)$ is an inclusion of analytic blocks in X then the restriction map $\varrho : \mathcal{S}('P) \to \mathcal{S}(P)$ is bounded with respect to good semi-norms.*

For the proof we need the following lemma.

Lemma. *Let $(P, \pi) \subset ('P, '\pi)$ be an inclusion of analytic blocks in X. Then there exists an analytic block (P_1, π) such that*

$$(P, \pi) \subset (P_1, \pi) \subset ('P, '\pi).$$

Proof: By Definition 8 in IV.2.4, we have $\mathbb{C}'^m = \mathbb{C}^m \times \mathbb{C}^n$ and accordingly $'\pi = (\pi, \varphi)$. Furthermore, there is a point $q \in \mathbb{C}^n$ such that $Q \times \{q\} \subset '\mathring{Q}$. We put $Q' := 'Q \cap (\mathbb{C}^m \times \{q\}) \subset \mathbb{C}^m$ and denote by Q^* the image of $'Q$ under the projection map $\mathbb{C}'^m \to \mathbb{C}^n$. We can choose the open neighborhoods $U \subset 'P^0$ of P and $V \subset Q'$ of Q in such a way that the map $\pi|U : U \to V$ is finite and $U \cap \pi^{-1}(Q) = P$. We can also find an euclidean block $Q_1 \subset \mathbb{C}^m$ satisfying $Q \subset \mathring{Q}_1 \subset Q_1 \subset \mathring{Q}'$. Now let $P_1 := \pi^{-1}(Q_1) \cap U$. Then (P_1, π) is the analytic block which we need. □

Proof of Theorem 2: According to the lemma one can decompose the restriction map $\varrho : \mathcal{S}('P) \to \mathcal{S}(P)$ as follows:

$$\mathcal{S}('P) \to \mathcal{S}(P_1^0) \to \mathcal{S}(P), \quad s \mapsto s|P_1^0 \mapsto s|P. \qquad (*)$$

The maps $\mathcal{S}('P) \to \mathcal{S}(P_1^0)$ and $\mathcal{S}(P_1^0) \to \mathcal{S}(P)$ are continuous, if $\mathcal{S}(P_1^0)$ is equipped with the topology defined by $'\pi$ and, respectively, by π. But these topologies coincide by Theorem 1$'$. It follows that ϱ is also continuous and therefore bounded. □

3. The Convergence Theorem. The considerations of the preceding section lead to the following result.

Theorem 3 (Convergence Theorem). *Let $(P, \pi) \subset ('P, '\pi)$ be an inclusion of analytic blocks in X. For every Cauchy sequence $\{s_j\}$ in $\mathcal{S}('P)$ the restricted sequence $\{s_j|P\}$ has a uniquely determined limit in $\mathcal{S}(P)$.*

Proof: We choose an analytic block (P_1, π) as in the above lemma and consider again the decomposition (*). Since the first map in (*) is continuous, $\{s_j|P_1^0\}$ is a Cauchy sequence in $\mathcal{S}(P_1^0)$. This sequence converges and has a uniquely determined limit in $\mathcal{S}(P_1^0)$. But the second map in (*) is also continuous. Therefore, the sequence obtained by restriction to P is also convergent and has a uniquely determined limit in $\mathcal{S}(P)$. □

4. The Approximation Theorem. We start as in the main text and proceed without changes until the decomposition

$$P_1 = P \cup P, \quad P \cap P = \emptyset$$

is proved. This decomposition has a simple (but important) consequence that whenever $(P, \pi) \subset ('P, '\pi)$ the restriction map $\sigma : \mathscr{S}(P_1) \to \mathscr{S}(P)$ is *surjective.*

The rest of the section is as follows.

We extend the euclidean blocks Q, Q_1 and $'Q$ to $\hat{Q}$, $\hat{Q}_1$ and, respectively, $\widehat{'Q}$. Instead of P, P_1 and $'P$ we get $\hat{P}$, $\hat{P}_1$ and, respectively, $\widehat{'P}$. We carry out these modifications in such a way that

$$\hat{\tilde{\tilde{P}}} \cap \hat{P} = \emptyset.$$

By Theorem 1′, the spaces $\mathscr{S}(\hat{P}_1{}^0)$ and $\mathscr{S}(\hat{P}^0)$ carry Fréchet topologies. Furthermore, the restriction map $\mathscr{S}(\hat{P}_1{}^0) \to \mathscr{S}(\hat{P}^0)$ is a continuous epimorphism. We are now in a position to prove the approximation theorem for coherent sheaves on analytic blocks.

Theorem 4 (Runge Approximation Theorem). *If (P, π) and $('P, '\pi)$ are analytic blocks in X with $(P, \pi) \subset ('P, '\pi)$, then for avery coherent sheaf $\mathscr{S}$ on X the space $\mathscr{S}('P)|P$ is dense in $\mathscr{S}(P)$.*

Proof: Let $s \in \mathscr{S}(P)$ be a given section. Since $\sigma : \mathscr{S}(P_1) \to \mathscr{S}(P)$ is surjective, there is a section $s_1 \in \mathscr{S}(P_1)$ with $\sigma(s_1) = s_1|P = s$. Appropriately modifying the blocks as above, we can extend s_1 to some $\hat{s}_1 \in \mathscr{S}(\hat{P}_1)$. Then, as we know, there exists a sequence $\{s^{(n)}\}$ in $\mathscr{S}(\widehat{'P})$ such that $s^{(n)}|\hat{P}_1 \to \hat{s}_1$ in $\mathscr{S}(\hat{P}_1)$. Since the restriction map $\mathscr{S}(\hat{P}_1{}^0) \to \mathscr{S}(\hat{P}^0)$ is continuous in Fréchet topology, one has $s^{(n)}|\hat{P}^0 \to \hat{s}_1|\hat{P}^0$ in $\mathscr{S}(\hat{P}^0)$. By the definition of a good semi-norm, it follows that $s^{(n)}|P \to \hat{s}_1|P = s_1|P = s$. □

5. Exhaustions by Analytic Blocks are Stein Exhaustions. The last section is not changed. Here are the main results again.

Theorem 5. *Every exhaustion $\{P_\nu, \pi_\nu)\}_{\nu \geq 1}$ of a complex space X by analytic blocks is a Stein exhaustion of X.*

Fundamental Theorem. *Every holomorphically complete space $(X, \mathscr{O})$ is a Stein space. For every coherent analytic sheaf $\mathscr{S}$ on X the holomorphic completeness of X implies the following:*

A) *The module of sections $\mathscr{S}(X)$ generates every stalk $\mathscr{S}_x$, $x \in X$, as an $\mathscr{O}_x$-module.*

B) *For all $q \geq 1$, $H^q(X, \mathscr{S}) = \{0\}$.* □

Errors and Misprints

1) p. 3, line 1: ***The Functor*** $\check{\Gamma}$
2) p. 3, line 3: $x \in V$ *(instead of* $x \in X$*)*
3) p. 3, lines 12, 6, and 1 from below: in each mapping, the second S should be $\mathcal{S}$
4) p. 4, lines 21–22: in the sentence on these lines S and $\mathcal{S}$ *should be interchanged*
5) p. 14, line 1: instead of "open in $\mathcal{S}$*" it should be "open in* $\mathcal{R}$*"*
6) p. 38, line 6: the correct formula is this one

$$B(l_1, \ldots, l_m) := \left\{ x \in B \,\middle|\, |x_\mu - x_\mu(l_1, \ldots, l_m)| < \frac{b_\mu - a_\mu}{n} \right\}$$

7) p. 39, lines 1–2 : see Paragraph 1.7 of Chapter A
8) p. 77, line 5 (Statement of Corollary): as in Theorem 6
9) p. 81, line 15 (Statement of Thm. 2): instead of "closed" it should be "$\bar{\partial}$*-closed"*
10) p. 85, line 6: Theorem 2.3.6
11) p. 133, line 2 from below: the correct mapping is this one

$$H^q(X, \mathcal{S}) \to H^q(X \backslash A, \mathcal{S})$$

12) p. 164, line 18 (Statement of Thm. 2): instead of $\overline{U}$ *it should be* $\overline{U_\nu}$
13) p. 168, lines 5–4 from below: the uniqueness theorem (Theorem 4)
14) p. 189, line 8 from below (Statement of Thm. 3): there should be an inequality (not equality)
15) p. 201, line 2: instead of c^{-1} *it should be* $|c|^{-1}$ *(twice)*

Printing: Strauss GmbH, Mörlenbach
Binding: Schäffer, Grünstadt